SCIENCE IN EUROPE, 1500–1800
A SECONDARY SOURCES READER

Science in Europe, 1500–1800: A Secondary Sources Reader

The companion volume in this series is:

Science in Europe 1500–1800: A Primary Sources Reader

Both volumes are part of an Open University course, *The Rise of Scientific Europe 1500–1800* (AS208), a 60 points second level undergraduate course.

Opinions expressed in this Reader are not necessarily those of the Course Team or of The Open University.

Details of this and other Open University courses can be obtained from the Call Centre, PO Box 724, The Open University, Milton Keynes, MK7 6ZS, United Kingdom: tel: +44 (0)1908 653231; e-mail: ces-gen@open.ac.uk.

Alternatively, you may wish to visit the Open University website at http://www.open.ac.uk where you can learn about the wide range of courses and packs offered at all levels by The Open University.

For information about the purchase of Open University course components, contact Open University Worldwide Ltd, The Berrill Building, Walton Hall, Milton Keynes, MK7 6AA, United Kingdom: tel: +44 (0)1908 858785; fax: +44 (0)1908 858787; e-mail: ouwenq@open.ac.uk; website: http://www.ouw.co.uk.

SCIENCE IN EUROPE, 1500–1800

A SECONDARY SOURCES READER

Edited by
Malcolm Oster

in association with

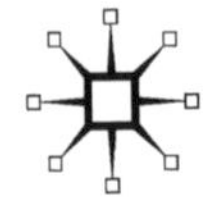

First published 2002 by Palgrave in association with
The Open University

PALGRAVE
Houndmills, Basingstoke, Hampshire RG21 6XS and
175 Fifth Avenue, New York, N.Y. 10010
Companies and representatives throughout the world

PALGRAVE is the new global academic imprint of
St. Martin's Press LLC Scholarly and Reference Division and
Palgrave Publishers Ltd (formerly Macmillan Press Ltd).

ISBN 978-0-333-97005-8 hardback

ISBN 978-0-333-97006-5 ISBN 978-0-230-21461-3 (eBook)
DOI 10.1007/978-0-230-21461-3

This book is printed on paper suitable for recycling and
made from fully managed and sustained forest sources.

A catalogue record for this book is available
from the British Library.

Library of Congress Cataloging-in-Publication Data
Science in Europe, 1500–1800 : a secondary sources reader / edited by Malcolm Oster.
p. cm.
Includes bibliographical references and index.
ISBN 978-0-333-97005-8 (cloth) — ISBN 978-0-333-97006-5 (pbk.)
1. Science—Europe—History. I. Oster, Malcolm, 1952–

Q127.E8 S354 2001
509.4—dc21 2001045149

10 9 8 7 6 5 4 3 2 1
11 10 09 08 07 06 05 04 03 02

Contents

Acknowledgements

The editor and publishers wish to thank the following for permission to use copyright material:

The *American Scholar* for material from Edward Grant, 'When did modern science begin?', *American Scholar*, 66/4 (1997), pp. 105–13; Atlantic Research and Publications and Angel Alcalá, Inc., for material from Virgilio Pinto, 'Censorship: a system of control and an instrument of action', in Angel Alcalá (ed.), *The Spanish Inquisition and the Inquisitorial Mind*, Social Science Monographs (1987), pp. 303–12, 318–19; Boydell & Brewer Ltd for material from William Eamon, 'Court, academy and printing house: patronage and scientific careers in late-Renaissance Italy', in Bruce Moran (ed.), *Patronage and Institutions: Science and Technology and Medicine at the European Court, 1500–1750* (1991), pp. 25–50; Cambridge University Press for material from Toby E. Huff, *The Rise of Early Modern Science: Islam, China and the West* (1993), Introduction and chapters 1–3, 5, 7–8; Robert S. Westman, 'Proof, poetics, and patronage: Copernicus's preface to *De Revolutionibus*', in D. Lindberg and R. Westman (eds), *Reappraisals of the Scientific Revolution* (1990), pp. 175–83, 186–9, 192–4; Charles Webster, *From Paracelsus to Newton* (1982), pp. 48–59; Lawrence Principe, 'Boyle's alchemical pursuits', in M. Hunter (ed.), *Robert Boyle Reconsidered* (1994), pp. 92–7, 100–2; Rob Iliffe, 'Is he like other men? The meaning of the *Principia Mathematica*, and the author as idol', in Gerald Maclean (ed.), *Culture and Society in the Stuart Restoration* (1995), pp. 170–3, 175–6; Betty Jo Dobbs, *The Janus Faces of Genius: The Role of Alchemy in Newton's Thought* (1991), pp. 1, 5–13; M. Biagioli, 'Scientific revolution, social bricolage, and etiquette', in Roy Porter and M. Teich (eds), *The Scientific Revolution in National Context* (1992), pp. 25–32; John H. Brooke, *Science and Religion* (1991), pp. 152–63; Roy S. Porter, 'The scientific revolution: a spoke in the wheel?', in R. Porter and M. Teich (eds), *Revolution in History* (1986), pp. 290–1, 294–303; Margaret Jacob, 'The truth of Newton's science and the truth of science's history: heroic science at its eighteenth-century formulation', in M. Osler (ed.), *Rethinking the Scientific Revolution* (2000), pp. 315, 319, 321–2, 328–32; J. V. Golinski, 'Utility and audience in 18th century chemistry: case studies of William Cullen and

Joseph Priestley', *British Journal for the History of Science*, 21 (1988), pp. 1–32; and P. B. Wood, 'Methodology and apologetics: Thomas Sprat's *History of the Royal Society*', *British Journal for the History of Science*, 13 (1980), pp. 1–26; Cornell University Press for material from Dennis R. Dean, *James Hutton and the History of Geology* (1992), pp. 264–9. Copyright © 1992 by Cornell University; Andrew Cunningham for 'William Harvey: the discovery of the circulation of the blood', in R. Porter (ed.), *Man Masters Nature: 25 Centuries of Science* (BBC Books, 1987), pp. 73–6; Tore Frängsmyr for material from Sten Lindroth, 'Two faces of Linnaeus', in T. Frängsmyr (ed.), *Linnaeus: The Man and his Work*, rev. edn (Watson Publishing International, 1994), pp. 2, 11–29, 31, 33–5, 37; Michael Hunter for material from *Science and Society in Restoration England* (Cambridge University Press, 1981), chapter 4; Istituto e Museo di Storia della Scienza for material from Marco Beretta, 'Humanism and chemistry: the spread of Georgius Agricola's metallurgical writings', *Nuncius*, 12 (1997), pp. 21–7; The Johns Hopkins University Press for material from Bruce Moran, 'German prince-practitioners: aspects in the development of courtly science, technology and procedures in the Renaissance', *Technology and Culture*, 22 (1981), pp. 253–74. Copyright © 1981 Society for the History of Technology; W W Norton & Co. Ltd for material from Bernard Lewis, *The Muslim Discovery of Europe* (Norton & Co., 1982), pp. 222–37, 328–9. Copyright © 1982 by Bernard Lewis; The Orion Publishing Group Ltd for material from Henry Kamen, *The Spanish Inquisition: An Historical Revision* (Weidenfeld and Nicolson, 1997), pp. 103–6, 113–14, 116–20, 131–5; Oxford University Press for material from Paul Wood, 'Science, the universities, and the public sphere in 18th century Scotland', in *History of Universities, Volume XIII* (1994), pp. 101–2, 104, 106–20. Copyright © 1994 Oxford University Press; Princeton University Press for material from Simon Schaffer and Steven Shapin, *Leviathan and the Air-Pump: Hobbes, Boyle and the Experimental Life* (1985), pp. 38–9. Copyright © 1985 by Princeton University Press; and Catherine Wilson, *The Invisible World: Early Modern Philosophy and the Microscope* (1995), pp. 74–5, 81–8. Copyright © 1995 by Princeton University Press; Rodopi for material from J. M. López Piñero, 'Paracelsus and his work in 16th and 17th century Spain', *Clio Medica*, 8 (1973), pp. 119–31; The Society for the History of Alchemy and Chemistry for material from H. G. Schneider, 'Fatherland of chemistry: early nationalistic currents in late 18th century German chemistry', *Ambix*, 36 (1989), pp. 14–19; Maurice Crosland, 'Lavoisier, the two French revolutions and the imperial despotism of oxygen', *Ambix*, 42 (1995), pp. 109–15; and Michael Conlin, 'Joseph Priestley's American defence of phlogiston reconsidered', *Ambix*, 43 (1996), pp. 129–33, 142; Taylor and Francis Ltd for material from Simon Schaffer, 'Newtonianism', in R. C. Olby, G. N. Cantor *et al.* (eds), *Companion to the History of Modern Science* (Routledge, 1990), pp. 613–18; and A. M. Ospovat, 'The importance of regional geology in the geological theories of Abraham Gottlob Werner: a contrary opinion', *Annals of Science*, 37 (1980), pp. 433–40; Thames and Hudson Ltd for material from Paula Findlen, 'Cabinets, collecting and natural philosophy', in Eliška Fučíková *et al.* (eds), *Rudolf II and Prague: The Court and the City* (Prague Castle Administration and Thames and Hudson, 1997), pp. 209, 213–17; Transactions of the American Philosophical Society for material from A. Stroup, 'Utility as a goal', *Transactions of the American Philosophical Society*, 77/4 (1987), chapter 6; University of California Press for material from Gary Deason, 'Reformation theology and the mechanistic conception of nature', pp. 167–79, Robert S. Westman, 'The Copernicans and the churches', pp. 76–85, 89–91, C. Webster, 'Puritanism, separatism and science', pp. 204–13, and William Ashworth, 'Catholicism and early modern science', pp. 136–60, all in D. Lindberg and R. Numbers (eds), *God and Nature: Historical Essays on the Encounter between Christianity and Science* (1986). Copyright © 1986 The Regents of the University of California; and Henry Lowood, 'The calculating forester: quantification, cameral science, and the emergence of scientific forestry management in Germany', in T. Frängsmyr *et al.* (eds), *The Quantifying Spirit in the 18th Century* (1990), pp. 315–32. Copyright © 1990 The Regents of the University of California; The University of Chicago Press for material from Larry Stewart, 'The selling of Newton', *Journal of British Studies*, 25 (1986), pp. 178–92; Mario Biagioli, *Galileo, Courtier* (1993), pp. 1–11, 313; Steven Shapin, *The Scientific Revolution* (1996), pp. 3–10; Steven Shapin, 'The house of experiment', *Isis*, 79 (1988), pp. 379–80, 382–8; A. I. Sabra, 'Situating Arabic science: locality versus essence', *Isis*, 87 (1996), pp. 654–69; Michael D. Gordin, 'The importation of being earnest: the early St. Petersburg Academy of Sciences', *Isis*, 91 (2000), pp. 1–10, 15–17, 19, 21–2, 29–31; and C. Perrin, 'Research traditions, Lavoisier, and the chemical revolution', *Osiris*, 4 (1988), pp. 74–8; Wiley-VCH for material from C. Meinel, '"...to make chemistry more applicable and generally beneficial": the transition in scientific perspective in eighteenth century chemistry', *Angewandte Chemie*, 23, Intnl Edn (English) (1984), pp. 339–47.

Every effort has been made to trace the copyright holders but if any have been inadvertently overlooked the publishers will be pleased to make the necessary arrangement at the first opportunity.

Preface

This Reader provides a historiographical perspective on what is still widely regarded as perhaps the most significant event in human history since the rise of Christianity, namely the Scientific Revolution. The Reader attempts to explore through secondary sources both the conceptual and the institutional foundations of modern science that were themselves built upon the scientific and natural philosophical heritage of classical antiquity. Insofar as an international scientific community was beginning to emerge by 1800, the nature of early modern European societies and states in the preceding three centuries remains a crucial arena for historical investigation. It has been increasingly evident over the last two decades that an ever richer social, cultural and intellectual contextualisation of national and cultural traditions has both sharpened and reshaped our understanding of what was clearly an uneven spread of science across the continent. The distinctive character of national and local contexts has been a signal feature of the Open University undergraduate course and accompanying textbook of the same title, *The Rise of Scientific Europe 1500–1800* ed. David Goodman and Colin A. Russell (Hodder & Stoughton/The Open University, 1991). The emphasis on a cultural topography that explores the interplay of social, political, religious and scientific ideas underpins the notion, reflected here, of distinctive national roles and styles of science. Though this Reader is designed to support students following the course in its newly expanded form, the prominence of key themes embracing ancients and moderns, science and religion, scholars and craftsmen, scientific patronage, the occult sciences, matter and motion, and historiography suggests that this compilation should be of interest to a wider undergraduate and general readership.

The present volume incorporates some materials included in an Anthology compiled and edited for the predecessor course by David Goodman, which was not generally available outside of The Open University. This new Reader covers a considerably more extensive range of readings relating to the key themes already mentioned and signposts both familiar and less familiar dimensions of the Scientific Revolution associated with the rise of new approaches to the physical world across Europe. Virtually all the sciences are touched upon in this Reader at some point, from the mathematically based classical fields of astronomy, optics and mechanics to the more applied sciences, such as chemistry, metallurgy, experimental physics and geology. The rather sparser coverage of medicine, natural history and botany reflects the basic make-up of the course rather than any predisposition to marginalise those very important areas. In practice, the boundaries between these categories were constantly renegotiated, as were the spaces between practical application and theoretical transformation. Similarly, frequent and complex associations with the occult sciences remained fluid and open for much of this period. The Scientific Revolution has been characterised as a process of change and displacement within competing systems of natural philosophy. In general terms, the shape, direction and legitimacy of the sciences were viewed by early modern contemporaries as part of a larger system of one or more of the natural philosophies on offer.

While the Reader can be used on its own, it has been designed to be read alongside a companion volume, *Science in Europe, 1500–1800: A Primary Sources Reader*, ed. Malcolm Oster (Palgrave/The Open University, 2002). The two volumes together provide an illuminating cross-fertilisation between primary and secondary sources. An ever-present danger in readings documenting the period is that they will be judged by a preoccupation as to what constitutes progress and rationality in modern science today. While it would be unrealistic to expect that the historiographical stance of 'whiggism' can ever be fully excised from our current perspectives, the expectation of what early modern thinkers and practitioners themselves believed counted as knowledge should act as the more desirable criterion in understanding supposed gaps between ideal and achievement. This is very much the stance taken by the authors in this Reader, whose work inevitably reflects a multiplicity of perspectives.

The readings are grouped under the chapter headings found in the Goodman and Russell textbook, and an attempt has been made to provide a balance of short, medium and long extracts. Original page numbers or ranges of page numbers used are cited in source references, unless chapters are more appropriate. The decision was taken at the outset, given the inevitable constraints of space, to excise footnotes except on rare occasions and sometimes to edit those notes retained. The 'normal' ellipsis of just three dots . . . indicates a short omission, and a bracketed ellipsis [. . .] shows that larger chunks of text (at least a paragraph) have been left out. Hopefully, the readings will encourage students to delve deeper into the wider literature associated with the history of science represented in this selection.

I owe a considerable debt to Colin Russell, Noel Coley, Michael Bartholomew and Peter Morris, who were involved in the original writing of the course and in compiling documents for the accompanying Reader. I also need to register equal thanks to current departmental colleagues, who were similarly involved in the original writing of the course and compiled items for the accompanying Reader: David Goodman, who was responsible for the final selection in Chapters 1, 4, 5, 7 and 12, Colin Chant for Chapter 13 and Gerrylynn Roberts for Chapter 14, in this new Reader. Similarly, I have to thank Kate Crawley and Michael Honeybone for their patience and hard work in providing invaluable feedback on the selection as experienced tutors of the course, and the external assessor, Stephen Pumfrey, in challenging us to meet our learning objectives. Finally, I have to register my debt to the considerable assistance of Gill Gowans, Alison Kirkbright and Shirley Coulson in preparing the typescript of this volume.

MALCOLM OSTER

Chapter One
Europe's Awakening

1.1 Toby Huff, *The rise of early modern science**

[...] The breakthrough that allowed freedom of scientific inquiry is undoubtedly one of the most powerful intellectual (and social) revolutions in the history of humankind. As the paradigmatic form of free inquiry, science has been given a roving commission to set all the domains of thought aright. Science is thus the natural enemy of all vested interests – social, political, and religious, including those of the scientific establishment itself. For the scientific mind refuses to let things stand as they are....

Given this intellectual commission to investigate all forms and manner of existence, science is especially the natural enemy of authoritarian regimes. Indeed, such regimes can exist only if they repress or otherwise subvert those forms of scientific inquiry that reveal the true nature of the social – economic, political, and medical – consequences of their rule....

Can we...say that men and women in all civilizations have equally shared...the view that science is and ought to be free to state its views on all matters of inquiry? Can we say that the other civilizations of the world equally held a fully rationalist conception of the orderliness of the cosmos and equally valued the rational capacities of man to the extent that they institutionalized the means by which men could fully apply their reason in the interests of advancing the most consistent and theoretically powerful explanatory systems? The fact that modern science arose only in the West – despite the fact that Arabic-Islamic science was more advanced up until the twelfth and thirteenth centuries – suggests a negative answer to those questions. [...]

...Is man,...in a particular time and civilization, thought to be completely rational and hence fully capable of discovering, decoding, and explaining the mysteries of nature? Or is man thought to be too weak in his intellectual

* Toby Huff, *The Rise of Early Modern Science: Islam, China and the West* (Cambridge: Cambridge University Press, 1993), Introduction and chapters 1–3, 5, 7–8.

powers to divine the secret and unknown processes and mechanisms of Nature that the naked eye can rarely see? Is he permitted to speak openly and possibly critically about the wisdom of the ages or about the official and public accounts of nature and its processes? In what forums may these dissenting thoughts be expressed, and can they be freely expressed, discussed, and publicly passed on to wider audiences? These are matters central to an anthropology of man and deserve the highest consideration in the context of the rise of modern science. [...]

Considered altogether, in mathematics, astronomy, optics, physics, and medicine, Arabic science was the most advanced in the world. In different fields it lost the lead at different points in time, but it can be said that up until the Copernican revolution of the sixteenth century, its astronomical models were the most advanced in the world. Consequently, the problem at hand asks why Arabic science, given its technical and scientific superiority built up over five centuries or so, did not give rise to modern science.

An important starting point for such an inquiry is the realization that the sciences we call the natural sciences were called the foreign sciences by the Muslims. In contrast, the so-called Islamic sciences were those devoted to the study of the Quran, the traditions of the Prophet (*hadith*), legal knowledge (*fiqh*), theology (*kalam*), poetry, and the Arabic language.... Moreover, due to ritual and religious prescriptions, time-keepers (*muwaqqits*) found it necessary to use geometry and eventually to invent trigonometry in order to arrive at the requisite calculations to determine the direction to Mecca (the *qibla*) for prayer. In short, driven by both curiosity and religious motives, the Arab-Muslim world from the eighth to the fourteenth centuries achieved significant heights of scientific advance, but thereafter (and perhaps as early as the twelfth century) went into decline and even retrogression. This did not happen uniformly across all fields, but in general scientific advance was on the wane.... [I]n some fields advances continued to be made, yet they did not culminate in a scientific revolution.

This situation is a deep puzzle about which many have speculated for at least the last 150 years. The factors identified as responsible for the failure of Arabic science to give birth to modern science range from racial factors, the dominance of religious orthodoxy and political tyranny, and matters of general psychology to economic factors and the failure of Arab natural philosophers to fully develop and use the experimental method. A common formulation of the negative influence of religious forces on scientific advance suggests that the twelfth and thirteenth centuries witnessed the rise of mysticism as a social movement. This in turn spawned religious intolerance, especially for the natural sciences and the substitution of the pursuit of the occult sciences in place of the study of the Greek and rational sciences. [...]

While many lesser institutions of learning evolved in medieval Islam (including the mosque schools and the elementary schools), the dominating institution of higher learning was the *madrasa*, the prototype of the college (not the university) that developed later in the West. The *madrasas* began to flourish in the eleventh century and, as the premier educational institutions of Islam, came to dominate a significant segment of intellectual life. Two aspects of the organization of the *madrasas* are of particular importance. The first is that the *madrasas* were established as charitable trusts: they were religious endowments that legally had to follow the wishes, religious or otherwise, of their founders. However, the law of trusts (the law of *waqf*) specifically forbade appropriating property and funds through the institution of *waqf* for purposes other than those sanctified by Islam. This stipulation is . . . a major legal impediment to the unfolding of intellectual and organizational evolution in the Islamic world. . . . Secondly, the *madrasas* were schools of law (*fiqh*), and as such centered all instruction around the religious or Islamic sciences, to the exclusion of philosophy, the natural sciences, and theology as well.

In theory the curriculum of the *madrasa* focused on Quranic studies, *hadith* (the traditions of the Prophet), the principles of religion, and the principles and methodology of law. The last of these included disputed questions in law and the principles of argument and disputation, as well as the practical point of view of the school of law to which the *madrasa* or the professor belonged. The deliberate exclusion of philosophy and the ancient sciences obviously stemmed from the suspicion with which such subjects were regarded by the religious scholars. Nevertheless, books on these subjects were often copied and made available in the libraries associated with the schools and mosques, and those law professors who became well versed in the foreign sciences gave private instruction (at home) on these subjects.

When in the eyes of the professor students had mastered the subjects taught in the *madrasa* . . . they were given . . . an authorization to teach these matters to others. . . . And here the stress is on the personal authorization that was involved. . . . It should be stressed that this sort of education was highly personalistic; the authorization or licensing was done by each professor, not by a group or corporate body, much less by a disinterested or impersonal certifying body. Likewise, neither the state, the sultan, nor the caliph had any influence over the recognition of educational competence. . . . To highlight the comparative aspects of this situation, it may be pointed out that educational certification in the Islamic world was the polar opposite of that in China: in the former case it was always (and only) the scholar himself who certified a student's competence, whereas in China it was always (and only) the state, never the corporate scholarly group, which conferred certification of educational competence. [. . .]

In the domain of the natural sciences, since all instruction occurred outside the colleges, specializing in a particular science clearly required that one travel around a good deal, from city to city, in search of scholars versed in the ancient sciences, in order to become a master of the current state of knowledge. Such a system clearly created institutional barriers to specialized scientific training and research. This was less true in the case of medicine where self-instruction as well as tutoring by family members was a possibility. [...]

Insofar as the development of science and scientific thought is concerned, such a system provided neither group support for philosophers cum scientists who held dissident views vis-à-vis the religious and political authorities, nor any mechanisms whereby received wisdom (as understood by the best and most competent experts, or as attested to by experiment) could be separated from the false and disproven. Likewise, the prohibitions against bringing it into the colleges perpetuated the personalized master–student pattern and prevented the efficient cumulation of knowledge by bringing scholars versed in the ancient sciences together in one place. [...]

...Islamic law does not recognize corporate personalities, which is why cities and universities and other legally autonomous entities did not evolve there. [...]

... [T]here was an absence of the rationalist view of man and nature, most thoroughly exemplified in Plato's *Timaeus*, which played such an important role in the philosophical thought of the European Middle Ages. [...]

In contrast to their Muslim counterparts, the European medievals felt especially enabled to study and decipher nature, to systematize, to organize, and to rationally evaluate the merits of the religious and legal texts that lay before them.... This view... did lay the groundwork for intellectual autonomy. Not least of all, it assimilated man's rational capacities to those of God by asserting that man's rational powers were a gift from God given for his glory. Furthermore, there were multiple sources that extolled the rational ordering of nature and the rationality of man. ... [T]he religious scholars, the theologians, and canonists embraced the rationalist images of man and nature implied by Plato and the *Timaeus*. [...]

[A notable] protoscientific institution in Islam is the observatory. For a time, the astronomical school of Marâgha in western Iran actually thrived and made singular contributions to the development of the modern science of astronomy. ...

Founded in 1259 just south of Tabriz, the Marâgha observatory represented new heights in the science of astronomy within Islam and the world. Perhaps because it was founded under the direction of the religious scholar and astronomer Nasir al-Din al-Tusi (d. 1274), Marâgha appears to have been founded under the umbrella of the law of *waqf*, that is, the religious law of endowments. ...

The fact that the observatory was founded under the law of *waqf* would suggest that the institution had suitable legal protection and could be expected to enjoy a long life, indeed, perpetual existence. But such was not the case. By 1304–5 the observatory had ceased to function; it had enjoyed a lifetime of barely forty-five years....

Given the virtues of the religious trust (*waqf*) as an institutional form in Islam, it is not surprising that the Marâgha observatory came to an untimely end, for every step necessary to advance the cause of science was in fact an 'heretical innovation' ... from the point of view of the religious scholars (the 'ulama').... The fact that the Mongol rulers of that period were heterodox believers probably had much to do with why the observatory was constructed in the first place. ... [T]he record of the observatory's founding 'clearly shows that the main purpose for the foundation of the Marâgha observatory was an astrological one'. This suggests that the founders were indeed daring in this undertaking and that it was only a matter of time before a reactionary opposition would be mounted by the religious scholars. [...]

In the case of China,... the disparity between the state of Chinese science and that of the West – but also the disparity with Arabic science – was far greater in regard to the theoretical foundations upon which the scientific revolution was ultimately launched in Europe. The superiority of China to the West... was wholly of a practical and technological nature, not one of theoretical understanding. [...]

If we consider the main fields of scientific inquiry that have traditionally formed the core of modern science – namely, astronomy, physics, optics, and mathematics – it is evident that the Chinese lagged behind not only the West but also the Arabs from about the eleventh century. By the end of the fourteenth century in the areas of mathematics, astronomy, and optics, there was a considerable debit on the Chinese side, despite the fact that there had been many chances for the Chinese to benefit from Arab astronomers and to borrow or assimilate the Greek philosophical heritage through constant interchanges between the Arabs and the Chinese. [...]

The Chinese conception of law contains many nuances that make it different from both Western and Islamic law. It is very unlike Islamic law in that it is not associated with the idea of commandments from God, although it does contain the idea of rites and traditions that are sanctified by virtue of their ancient origins in the practice of sage-kings of the past. ... [O]ne starting point of Chinese law is

> the idea that heaven and earth were governed by one principle – called tao, the way, the creative principle of natural order. Any act contrary to this order in human society resulted in a disruption of the harmony between heaven and earth, and might lead to such calamities as flood, drought, internal disorder. So that the order might be preserved, heaven chose men of outstanding virtue, te, and gave them the mandate, ming, to rule their fellow creatures.

[...] [T]his so-called conception of natural law is different from that of the European medievals. First of all, the Chinese conception is exceedingly concrete and tied to a single group (the ruling elite of China) during a single period of time (the past). It is in this regard ethnocentric despite the fact that China has always known other ethnic groups (usually called barbarians). [...]

[T]here is an absence of legal thought spelling out any theory of autonomous spheres of limited jurisdictions, that is, a theory that treats collective actors as a whole body, or a corporation, with corresponding rights to internal self-regulation [...].

The absence of a legal theory of autonomous jurisdiction – that is, some form of corporate entity...must be seen as one of the most serious weaknesses of Chinese civilization. For without some spheres of autonomy, no groups can emerge as professionals, that is, legitimate specialists who represent the highest levels of thought and action in a particular sphere of human endeavor. [...]

[T]he Chinese educational system was both rigidly controlled and focused on literary and moral learning, while the European universities were both autonomous and self-controlled as well as centered on a core curriculum that was essentially scientific. The implications of these institutional contrasts for the development of science, as in the case of Arabic science, can hardly be overstated. For if science is to flourish over the long run, it must have official approval as well as public support – something that was rarely available in China. ...

The traditional examination system was a unique system.... According to Ho Peng-ti, it 'entailed a wastage of human effort and talent on a scale vaster than can be found in most societies'. This inefficiency is revealed in the fact that it was 'not uncommon for a scholar to have failed a dozen or more times in high-level examinations which were usually held at a three-year interval. The whole life of such luckless scholars was thus wasted in their studies and examination halls.' [...]

... [T]he rigidity of the educational content of the examination system – virtually unchanged from early Ming (ca. 1368) to the twentieth century – and its absolute uniformity, bordering on political indoctrination, make it a colossal failure insofar as science, innovation, and creativity are concerned. ... [I]t did not encourage or tolerate thinkers who were essentially disputatious and critical of the intellectual status quo. ... There was no Chinese equivalent to the scholastic method of disputation, no canons of logic à la Aristotle, and no mathematical methods of proof such as one finds in Euclid's geometry. Derk Bodde points out, 'Throughout its history Confucianism has deprecated the use of debate as a means of advancing knowledge.' ... One might say, therefore,...that the civil examination system also killed off scientific theory (natural philosophy) as a coherent account of the world. It did that simply by standardizing the civil service examination around Confucian literary studies focused on moral and ethical issues of governing and by disallowing any state-sponsored scientific education (other than astronomy and mathematics, both of which were carefully controlled) to be part of the examination system. . . .

To understand the impact of the Chinese educational system, one must remember that in the Sung dynasty the emperor and his advisers sought to establish a universal educational system. To that effect, a program of establishing local schools in all provinces, prefects, and districts was undertaken early in the eleventh century. With the advent of block printing, the government supplied every school with a set of the officially approved versions of the Confucian classics (and commentaries) for the students to study and memorize. [. . .]

In contrast, the Chinese in the seventh century A.D. had a thousand-year-old scribal tradition that considered non-Chinese barbarians. For this reason, the Chinese were exceedingly cautious and selective in their borrowings from other cultures. This was quite noticeable in the domain of science. . . . [E]ven when major scientific improvements and innovations were passed to the Chinese – whether via the Indians or the Arabs – these innovations were either put aside altogether or only adopted after the lapse of many centuries. This was the case with the appearance of the Indian zero in the eighth century and with regard to the Ptolemaic system as used by the Marâgha observatory in the thirteenth century.

The case for the view that the Chinese language . . . has been a great impediment over the centuries . . . has been put most forcefully by Arthur Wright. Reflecting on the experiences of the many foreign individuals who came to China over the centuries and who attempted to bring their ideas forth by translating them into Chinese, Wright sums up as follows:

> Structurally Chinese was a most unsuitable medium for the expression of their ideas, for it was deficient in the notations of number, tense, gender, and relationships, which notations were often necessary for the communication of a foreign idea. . . . Further, the Chinese was relatively poor in resources for expressing abstractions and general classes or qualities. . . .

[. . .] When we approach the question of why modern science did not develop in China from an institutional point of view, we see . . . powerful inhibitions blocking the free and open pursuit of disinterested knowledge. It is these, I believe, that in the long run had the greatest impact on the development of science in China, just as we saw in the case of Arabic science. . . .

The first of these impediments preventing the rise of free and open public discourse . . . is found in the simple lack of spheres of autonomy on any level. . . .

. . . [There was an] absence of cities and towns as autonomous legal units of self-governing citizens. . . .

. . . [I]nstitutions of higher learning, which would be equivalent to either the college in Islam or the university in the West, did not emerge in China. . . .

... [T]here was no conception of a degree-granting institution of higher learning outside the national bureaucracy. Even more remarkable is the fact that there was only one organizational unit of higher education in all of China (with some 120 million people) that had status enough to be (misleadingly) called a university. By way of contrast, Europe from the twelfth to the fourteenth centuries, with half the population of China, had at least eighty-nine universities, not to mention hundreds of colleges with more autonomy than existed anywhere in China. [...]

When we look at official bureaucracy itself, we encounter additional impediments to the free and unfettered pursuit of scientific knowledge. Here I refer to the elements of secrecy and excessive regulation in the study of astronomy and mathematics. Such secrecy obviously worked directly against the scientific norm of communalism, the free and open access to knowledge. Needham's account of the study of astronomy in China is littered with references to the security-minded manner and the semi-secrecy in which these disciplines were kept.

...As a result of Chinese cosmological beliefs, astronomical phenomena were taken to be the most visible and awesome signs of the harmonious state of the heavens. Since there was thought to be a correlation between the heavenly order and the political order, it behooved the emperor and his officials to take special notice of the cosmic realm....

What the Chinese did, however, was to make this study a state secret and thereby drastically reduced the number of scholars who could, legitimately or otherwise, study astronomy. This restriction also greatly reduced the availability of the best and latest astronomical instruments and observational data.... The fact remains that virtually every move made by the astronomical staff had to be approved by the emperor before anything could be done, before modifications in instrumentation or traditional recording procedures could be put into effect. It is not surprising, therefore, that despite the existence of a bureau of astronomers staffed by superior Muslim astronomers (since 1368), Arab astronomy (based as it was on Euclid and Ptolemy) had no major impact on Chinese astronomy, so that three hundred years later when the Jesuits arrived in China, it appeared that Chinese astronomy had never had any contact with Euclid's geometry and Ptolemy's *Almagest*. [...]

Conclusion

...[W]hat happened in Europe...in the twelfth and thirteenth centuries ...did not happen in China.... What happened in Europe was a social and legal revolution that radically transformed the nature of medieval society, in fact laying the foundations of modern society and civilization. Europe experienced a revolution that placed social life on an entirely new footing. From one point of view, it represented the grand fusion for the first time of Greek philosophy and science, Roman law, and Christian theology.

All three structures were unique to the West – as there is nothing in Chinese thought equivalent to Greek philosophy as expressed by Aristotle, to Christian theology, or to the *corpus juris civilis* [Roman law].

At the center of this revolution was a legal transformation that redefined the nature of social organization in all its realms – political, social, economic, religious, and intellectual. In the present context, the development of the law of corporations was foremost among these changes. At the time when the Christian church declared itself to be a corporation, a whole body for all legal purposes, it granted such status to a variety of other collectivities, such as residential communities, cities, towns, universities, and economic interest groups such as guilds. It also laid the legal foundations for professional associations. ...

Focusing on the world of learning, however, it was in the twelfth and thirteenth centuries in Europe that universities arose, establishing neutral zones of intellectual autonomy which allowed philosophers and scientists to pursue their agendas free from the dictates of the central state and the religious authorities. The founders of the universities consolidated the curriculum around a basically scientific core of readings and lectures. This was embodied in the natural books of the new Aristotle that became known during those centuries. ... Anyone who reads these works cannot but be impressed with the extraordinary concentration of energy on the naturalistic understanding of the world in all its dimensions that these works represent. ...

Not only was this new intellectual agenda thoroughly naturalistic, but it was made the core of an evolutionary course of study. Scholars were not only free to raise questions about it, they were taught how to raise questions and were even enjoined to dispute every aspect of it. ... [T]hey asked... whether there were other worlds and, if so, whether the same physical laws would obtain in such worlds. In the domain concerning speculation about time, space, and motion, they asked questions about the existence of a vacuum and its properties. ... It was just this set of naturalistic inquiries ... that set the agenda of scientific inquiry for the next four hundred years in European universities. ...

... [W]hy did the scientific revolution not occur in China? ... [W]e now see a large part of the answer in the deliberate displacement of naturalistic inquiries from the center of higher education in China and the failure of the Chinese legal and political systems to undergo the radical restructuring that took place in Europe in the twelfth and thirteenth centuries. ... More importantly, we see the failure of Chinese civilization to create neutral spheres of intellectual autonomy in which any intellectual agenda could be pursued independent of the interference of state authorities. ...

... [T]he first emperor of the Ming dynasty... T'ai-tsu (r. 1368–98)... 'would put to death any man of talent who refused to serve the government when summoned. As he put it, "To the edges of the land, all are the king's subjects. ... Literati in the realm who do not serve the ruler are estranged from teaching [of Confucius]. To execute them and confiscate the property of their families is not excessive."' The trial and punishment of Galileo (confinement to his villa overlooking Florence) is nothing compared to this.

... Physicians ... were highly esteemed in the Middle East, and their intellectual tradition was one of rich philosophical learning. As a result, they were highly valued as both government officials and community leaders. This appears not to be true of Chinese physicians, and it undoubtedly relates to ... the fact that a system of hospitals equivalent to those developed in the Middle East did not emerge in China. ...

1.2 A. I. Sabra, *Situating Arabic science**

Locality as a focus of historiography

I trust that no one would wish to contest the proposition that all history is local history – whether the locality is that of a short episode or of a long story. All history is local, and the history of science is no exception. There can be no history of science that is not concerned with a localized episode or a sequence of such episodes. ... The thesis outlined [here] is but a generalization of ... the familiar contextualist thesis in scientific historiography; or, to put it another way, contextualism is but an obvious consequence of the simple, undeniable fact of the local character of all events, including historical events. Since a historical event is where and when and how it is, inseparably tied to all the circumstances that combine to define it for us as historians, then, to be genuinely historical, all *history* of science must be contextual, because all historical events are local. [...]

My purpose here is to try to illustrate the advantages of a strict adherence to the axiom of locality in situating the tradition of Arabic science with reference both to the place that this tradition occupies in the general history of science and to its place in the civilization where it emerged and developed. ...

Let us begin with an apparently neutral and innocent definition of Arabic, or what may also be called Islamic, science in terms of location in space and time: the term *Arabic* (or *Islamic*) *science* denotes the scientific activities of individuals who lived in a region that roughly extended chronologically from the eighth century A.D. to the beginning of the modern era, and geographically from the Iberian Peninsula and North Africa to the Indus valley and from southern Arabia to the Caspian Sea – that is, the region covered for most of that period by what we call Islamic civilization, and in which the results of the activities referred to were for the most part expressed in the Arabic language. ... But what about the term *scientific* in

---

* A. I. Sabra, 'Situating Arabic science: locality versus essence', *Isis*, 87 (1996), pp. 654–69.

[this definition]? What does it mean, and can it be regarded as in any way 'innocent'? To me it seems clear that the only correct answer to this last question must be an unequivocal No. *Science* and *scientific* are *our* own terms and they express *our* own concepts... and, therefore, the study of any past intellectual activity can be relevant to what we call 'history of science' only to the extent that such an activity can be shown to help us understand the modes of thought and expression and behavior that *we* have come to associate with the word *science*. ... And so... all history of science is local, and no history of science can ever be neutral.

The character of Arabic science, its strengths and failings, the course of its development, and its ultimate fate have all been variously explained in terms of language as a matrix of thought and expression, of religion as an inexorable shaping force, of natural aptitudes or inclinations of a certain race or inherent mentality, or as one inevitable expression of a world culture of which Islamic civilization was a late embodiment. A perceived emphasis on algebra in the Arabic tradition has been attributed to certain features of Semitic languages that make these languages or their native users prone to 'algebraization', as opposed to Greek 'geometrization'. The persistent attempts of Islamic astronomers to construct kinematic models primarily designed to save the principles and the logical consistency of Ptolemaic astronomy have been seen as a sign of poverty of imagination or of the tendency of the 'Semitic mind' toward things it can easily perceive by the senses. Islamic religion has been cited both as the origin and source of vigor of medieval Islamic science and as the major cause of its final demise. And the 'spirit of culture'... has been invoked to account for every aspect of Islamic civilization, including its scientific products.

It is not difficult to expose the weaknesses from which such explanations suffer. One can refer, for example, to the considerable and highly successful efforts of Islamic mathematicians in the fields of geometry and trigonometry. One can relate the theoretical program of Islamic astronomers to the work of Ptolemy himself and to earlier ideals of Greek astronomy. One can point out the great complexity of the relationship between science and religion throughout Islamic history and in various parts of the Islamic world. And one can easily show the vacuousness of theories born of the spirit-of-culture approach. And, in fairness to those who have advanced explanations of these sorts, it must be said that they tend to be poorly informed (or worse) about Arabic science and, in many cases, about Islamic civilization – a fact that, unfortunately, does not seem to have discouraged their influence on minds that seek ready-made and perhaps comforting explanations.

What is wrong with these explanations, and others like them, is not their consideration of language, religion, and culture as factors in the formation of a scientific enterprise that consciously adopted earlier traditions with markedly different languages and religious and cultural values but, rather, their essentialist character, which has tended to prejudice or obstruct historical research. ... With the sure perceptiveness of a true historian, Richard Southern once described the process of acquisition and adaptation of Greek learning in Islam as 'the most astonishing event in the history of thought'.[1] The event is astonishing because it strikes us as unexpected, and the best way I know to explain the unexpected in history, insofar as it can be explained at all, is to try to understand it, not in terms of essences or spirits or inevitabilities, but as the outcome of choices by individuals and groups responding to their situations as *they* perceived and experienced them. Let me illustrate.

The intersection of Islamism, Arabism, and Hellenism in ninth-century Baghdad

The powerful drive that eventually led to the transfer of the bulk of Greek science and philosophy (as well as elements of the scientific thought of India and Persia) to Islam was launched as a massive translation effort that took place in the context of empire and under the patronage of the confident Abbasid court in Baghdad. Translations into Arabic had been made earlier, and these had been preceded in the Middle East by translations from the Greek into Syriac and Persian, but it was the Abbasids who mounted a concentrated translation effort soon after they came to power in the middle of the eighth century and who further organized and intensified their support during the ninth century. Under their predecessors, the Umayyads, who ruled from Damascus (661–750), the Islamic empire already encompassed large areas – including Egypt, Syria, and Persia – that had come under the influence of Hellenism from the time of Alexander; and before the ninth century was over Islamic rule had reached Kashmir in the east and Khwarazm to the north. In the early Abbasid period the higher administration of the court itself was in the hands of cultivated Persians who had gained much favor and influence with the Abbasid rulers and whose intellectual interests inclined them to various forms of secular learning and to a rationalizing approach for understanding matters of religious belief. Some of these Persian officials acted as translators, especially from Persian, and in general they constituted an important, politically influential part of Baghdad's intellectual elites. Two other groups within the empire (and concentrated mainly in Syria, Iraq, and Persia) had maintained a long-established tradition of Hellenized Syriac learning. One consisted of Christian physicians and Christian theologians, who continued to pursue their interests in Greek logic and philosophy in scattered monastic schools; the other was the pagan Ṣābians of Ḥarrān, in northern Mesopotamia, an ancient Semitic group whose astral religion connected them to Hellenistic

---

1 R. W. Southern, *Western Views of Islam in the Middle Ages* (Cambridge, Mass./London: Harvard Univ. Press, 1962, 1978), pp. 8–9.

astrology and astronomy and to Hermeticism. It was from these two last groups that the Abbasids were able to recruit the scholars who carried out the translations of Greek medical, philosophical, and mathematical works into Arabic, either from preexisting Syriac versions or directly from the Greek.

The survival of these pockets of Hellenic learning during the first centuries of Islamic rule in the Middle East and Asia, although scattered and limited at first in scope and appeal, ensured a certain continuity with the classical tradition – a continuity that was largely lacking, for example, in the case of the 'Renaissance of the Twelfth Century', when European scholars first had to journey to the edges of Western Christendom to acquire Arabic and Greek learning from across the borders with Islam and Byzantium. In the earlier Middle Eastern episode, this continuity meant the immediate availability of texts, in Greek or Syriac versions, and of translators already conversant with these languages and with Greek thought itself in a number of scientific, medical, and philosophical disciplines. And although much additional Greek material was later to be brought over the borders with Byzantium, the continuity with the Greco-Syriac tradition helps to explain the high level of competence, even sophistication, that characterized scientific writings in Arabic from an early period that overlapped the translation movement.

One might then say, and with much justification, that the stage was set, at a certain place and time, for the translation movement that quickly acquired unprecedented proportions – unprecedented not only in the Middle East but in the world at large. But in order to explain the momentum, scope, and multiple dimensions of that movement, it is necessary to go beyond the availability of favorable conditions, and even beyond the important consideration of practical expectations that must have loomed large at least in the minds of the Muslim patrons. Islamic religion had introduced a new ideology with sweeping and universalist claims. Already during the swift expansion of Islamic conquests, that ideology had come into direct contact with a large variety of creeds (Jewish, Christian, Zoroastrian, Mazdian, Manichaean, etc.) with which it inevitably collided and against which it had not merely to defend but – much more importantly – to define itself, often in terms borrowed from its opponents. The result was a huge intellectual ferment, centered especially in multicultural Iraq, to which the movements of Islamic theology, philosophy, and science owed their birth. [...]

Turning now... to leading Muslim intellectuals of the period, one quickly detects an attitude of openness and gratefulness to the recently imported wisdom, mingled with a feeling of high optimism and a certain trust in humanism that was quite pronounced....

... [T]he celebrated Muslim philosopher, scientist, and mathematician al-Kindī (d. ca. 870) was a member of the Arab nobility.... In his work *On First Philosophy*, dedicated to al-Muʿtaṣim, we find strong acknowledgment of the accomplishments of ancient Greece that is combined with the assertion of truth as the universal good that must be sought out wherever it may be found, in addition to a clear concept of the growth of knowledge as a process of accumulation that requires the cooperative effort of different peoples and successive generations. These were the deep convictions of a true devotee, and they carried with them an entire Hellenic or Hellenistic world view, a distinct concept of wisdom in both the theoretical and practical senses (also borrowed from the Greeks), and a sense of mission on the part of the author to do his utmost to disseminate the ancient and especially Greek heritage in his milieu. ... [H]is immediate role, as he also conceived it, was to introduce and, he hoped, to convert his Arabic-reading contemporaries ... to the Greek wisdom that had captivated him – a task that he actually undertook to achieve by producing a huge number of Arabic epitomes and adaptations, with supplementary clarifications and additions when necessary, of a very large number of Greek disciplines of science and philosophy. This was a preposterously optimistic project for anyone to envisage; but not only was al-Kindī able to carry it out, whether single-handedly or with the help of others, it proved to be remarkably successful – so successful, in fact, as to make him truly worthy of the reputation he quickly gained as one of the founders of the Arabic tradition in philosophy and science. [...]

Three loci of scientific activity in Islam: the court, the college, and the mosque

[...] [A]s far as science and philosophy are concerned, the European Renaissance of the sixteenth century was in part a reaction, which became more pronounced in the seventeenth century, against patterns of thought and argument associated with medieval 'scholasticism'. In Islamic history, events followed the reverse order: the 'renaissance' (if that is the right word) came first, in the ninth and tenth centuries, and a form of scholasticism followed, though not immediately and not uniformly in all parts of the Muslim world. My second observation points to another contrast between Islam and medieval Europe that is crucially important.... In Islam, whether in ninth- and tenth-century Baghdad, eleventh-century Egypt and central Asia, twelfth-century Spain, thirteenth-century Marāgha in northwestern Iran, or fifteenth-century Samarkand, the major scientific work associated with the names of those who were active at those times and places was carried out under the patronage of rulers whose primary interests lay in the practical benefits promised by the practitioners of medicine and astronomy and astrology and applied mathematics. Many of these practitioners were also prolific writers on 'philosophy', a mode of thinking known by the Arabicized term *falsafa* and characterized to a large extent by a mixture of Aristotelian and Neoplatonic doctrines and forms of argument – the kind of mixture we find, for example, in the works of al-Kindī, al-Fārābī, and Avicenna. In those circumstances science and 'philosophy', or *falsafa*, were secular activities that were practiced, developed, and propagated as rational inquiries completely independent of any religious authority....

Now, Islamic 'theology', or what has come to be known in Western scholarship by this name, followed a different course – with important, indeed far-reaching consequences for the development of both science and *falsafa*. It began to make a conspicuous appearance in the eighth century (the second Islamic century), well before the patronized translation movement got under way, as the activity of spontaneously sprouting groups of Muslim intellectuals in the urban centers of Baṣra and Baghdad who immersed themselves in probing discussions (*kalām*: speech, discussion, argument), obviously driven by their interest in current religious and political controversies....

There can be no doubt that the early practitioners of *kalām*, the *mutakallimūn*,...represent an important turn in the history of philosophical thought – one that gave rise to new styles of thinking that seriously challenged the Aristotelianism and Neoplatonism of *falsafa* by proposing a thoroughgoing atomism that viewed the world as a creative process. It was this new philosophy...that...later found its way into the colleges of higher education, the so-called *madrasas* that ultimately spread wide and far over the Islamic world as endowed or charitable institutions, having been first introduced on a large scale in the eleventh century by the Sunnite Saljūqs in Iraq and Persia....

The *madrasas*, it should be noted, were first conceived of as primarily schools of law....But, as creations of private endowments, they generally enjoyed a degree of informality that allowed for a variable range of intellectual pursuits that depended on local circumstances and the interests of their professors and their sponsors. Many, perhaps a large number, of the *madrasas* included some teaching in arithmetic, algebra, astronomy, and logic as part of the intellectual equipment of the practicing jurist, along with the indispensable disciplines of language and rhetoric. *Kalām*, as a study of the 'fundamental tenets of religion' (*uṣūl al-dīn*), performed the dual function of supplying a superstructure of theory for the rest of the 'religious sciences' as well as a substitute for Greek metaphysics and natural philosophy.

Combining these two general observations should now help us to appreciate the following result. The sciences of the Greeks, which were first welcomed in Islam along with Greek theories of cosmology and epistemology and metaphysics..., eventually came to be confronted in the *madrasas* by a homegrown religious philosophy that claimed to develop viable alternatives to the Greek paradigms. That much we can say in light of what we already know. But there is no end to the questions that have yet to be examined.... [T]he question of special importance for the historian of science [is] what was the effect of the *kalām* point of view on the dissemination and development of scientific disciplines such as cosmology and astronomy, about which the *mutakallimūn* had a lot to say as an integral part of their own world view?

None of these questions can be answered a priori. They are all empirical questions that require empirical research. Some of my colleagues, I am happy to say, are now beginning to tackle them in earnest. . . .

The *madrasas* were not, therefore, in general a locus where scientific research was promoted for its own sake, but one in which science was interpreted and judged and presented to a large group, indeed the vast majority, of educated Muslims. When we talk of scientific advance in Islam, whether in mathematics, astronomy, or experimental science, we usually have in mind the contributions of men who carried out their work outside of the *madrasas* with the support of kingly patrons. . . .

What we are still far from understanding is how patronage worked in contexts that obviously differed from one center of activity to another. . . .

. . . For example, soon after the Mongol Īlkhāns had captured Baghdad in 1258, thus bringing the Abbasid caliphate to an end, their leader Hūlāgū was persuaded to establish an observatory at Marāgha in northwestern Iran, an event that marked the beginning of one of the longer-lasting and important episodes in the history of Arabic science. Most of the scholars who were soon to be gathered at Marāgha were Muslims (there are reports of one or more Chinese scholars). The man put in charge of organizing the new enterprise was Naṣīr al-Dīn al-Ṭūṣī, a Persian from Ṭūs with serious interests in Shīʿite theology and Avicennan philosophy. . . . At age fifty-five when he surrendered himself to the Mongols . . . he was already famous as a scholar and known to the Mongols as a competent astronomer and astrologer. Two other scholars, both of them Sunnīs, were brought over from Syria. One of them, Mu'ayyad al-Dīn al-ʿUrḍī, had a reputation as a building engineer and instrument-maker. The other, the mathematician Muḥyī al-Dīn al-Maghribī al-Andalusī, was captured by the Mongols during their campaign in Syria in 1259–1260. He managed to save his life only by presenting himself to his captors as an astrologer who could be of use to 'the lord of the earth', the great Mongol Khān. The Mongol patron of the astronomical enterprise at Marāgha was not Muslim, and his interest in the work of the observatory was undoubtedly astrological. The immediate goal was to produce a new set of astronomical tables, based on new observations, of the type that had been used by Arabic astronomers and astrologers for planetary predictions since the time of al-Ma'mūn. The *Zīj-i Īl-Khānī*, as the new Persian handbook came to be known, was not completed until 1272, after Hūlāgū's death. But in the meantime, and for several decades afterward, the scholars at Marāgha and nearby Tabrīz were able to pursue their individual interests in theoretical astronomy and in various branches of mathematics. It was in this singular situation that the . . . research in planetary theory, which was initiated by Ibn al-Haytham before the middle of the eleventh century and had attracted the attention of a few individual scholars in Asia and Syria, first found a sustaining atmosphere; and it was from here that this type of research later spread further east, south, and west, where it was carried on in different terms or with different emphases as scholars with different commitments responded to changing contexts. ...

The phenomenon of the mosque as a significant locus of scientific activity came into being at about the same time as the establishment of the Īlkhānid rule in Iran and Iraq, being associated with the ascension to power of the Mamlūks in Egypt and Syria in 1250. ... It was the longest-lasting episode within the tradition of Arabic/Islamic science, having continuously endured in some of the major mosques all the way up to the nineteenth century, and it possessed interesting features that distinguish it in several ways from court patronage and the *madrasas*, the two other loci of consequence.

Though primarily a place of worship, the mosque, from its inception, and as distinguished from the *madrasa* that was sometimes attached to it, often served as a forum for propagation and discussion of subjects related to Arabic language, grammar, and rhetoric, as well as the vital issues of law, religion, and politics. Through the introduction, apparently for the first time under the Mamlūks, of the office of *muwaqqit*, the timekeeper in charge of regulating the times of the five daily prayers, a place was created for the utilization of one form of scientific knowledge in a permanent religious institution.

Strictly speaking, it would be wrong to consider the *muwaqqit* a 'professional' astronomer. His institutional role in the mosque was not to pursue the goals of astronomy as these had been defined and elaborated by Arabic astronomers since the ninth century but, as is clearly indicated by his status title, to offer reliable guidance to his local Islamic community with regard to definite religious observances (mainly prayer times) as specified by religious law. This function the *muwaqqit* nonetheless performed in his capacity as an expert in what was called 'the science of reckoning time'... by means of exact astronomical computations, and this distinguished him from the traditional *mu'ezzin* (the man who called for prayer), who relied on traditional prescriptions. The main task of the *muwaqqit* was therefore to use the methods of spherical astronomy in order to construct tables, usually computed for a certain locality or latitude, that would enable anyone who could operate a simple observation instrument (such as an astrolabe or a quadrant) to determine the time of day or night from the altitude of the sun or a star. ... And some distinguished *muwaqqits* in the thirteenth and fourteenth centuries accomplished the impressive feat of providing universal solutions of timekeeping problems (indeed, all problems of spherical astronomy) for all latitudes. One *muwaqqit*, the fourteenth-century Ibn al-Shāṭir (d. ca. 1375), who was attached to the Umayyad mosque in Damascus, ventured into the area of theoretical astronomy to produce the most complete solution to the equant problem, which Ibn al-Haytham had forcefully pointed out as a threat to the principles of Ptolemaic astronomy and which was diligently pursued by mathematical astronomers in the thirteenth century. These were all accomplishments that must be regarded as accomplishments in astronomy proper, regardless of their institutional setting. [. . .]

1.3 Edward Grant, *When did modern science begin?**

Although science has a long history with roots in ancient Egypt and Mesopotamia, it is indisputable that modern science emerged in Western Europe and nowhere else. The reasons for this momentous occurrence must, therefore, be sought in some unique set of circumstances that differentiate Western society from other contemporary and earlier civilizations. The establishment of science as a basic enterprise within a society depends on more than expertise in technical scientific subjects, experiments, and disciplined observations. After all, science can be found in many early societies. . . .

Why did science as we know it today materialize only in Western society? What made it possible for science to acquire prestige and influence and to become a powerful force in Western Europe by the seventeenth century? The answer, I believe, lies in certain fundamental events that occurred in Western Europe during the period from approximately 1175 to 1500. Those events, taken together, should be viewed as forming the foundations of modern science, a judgment that runs counter to prevailing scholarly opinion, which holds that modern science emerged in the seventeenth century by repudiating and abandoning medieval science and natural philosophy, the latter based on the works of Aristotle.

. . . What must be emphasized, however, is that the momentous changes in the exact sciences of physics and astronomy that epitomized the scientific revolution did not develop from a vacuum. They could not have occurred without certain foundational events that were unique products of the late Middle Ages. ... Could a scientific revolution have occurred in the seventeenth century if the immense translations of Greco-Arabic (or Greco-Islamic) science and natural philosophy into Latin had never taken place? Obviously not. . . . [S]omething happened between, say, 1175 and 1500 that paved the way for that scientific revolution. What that 'something' was is my subject here.

To describe how the late Middle Ages in Western Europe played a role in producing the scientific revolution in the physical sciences during the seventeenth century, two aspects of science need to be distinguished, the contextual and the substantive. The first – the contextual – involves changes that created an atmosphere conducive to the establishment of science, made it feasible to pursue science and natural philosophy on a permanent basis, and made those pursuits laudable activities within Western society. The second aspect – the substantive – pertains to certain features of

---

* Edward Grant, 'When did modern science begin?', *The American Scholar*, 66 (1997), pp. 105–13.

medieval science and natural philosophy that were instrumental in bringing about the scientific revolution.

The creation of an environment in the Middle Ages that eventually made a scientific revolution possible involved at least three crucial preconditions. The first of these was the translation of Greco-Arabic science and natural philosophy into Latin during the twelfth and thirteenth centuries. Without this initial, indispensable precondition, the other two might not have occurred. With the transfer of this large body of learning to the Western world, the old science of the early Middle Ages was overwhelmed and superseded. . . .

The second precondition was the formation of the medieval university, with its corporate structure and control over its varied activities. The universities that emerged by the thirteenth century in Paris, Oxford, and Bologna were different from anything the world had ever seen. From these beginnings, the medieval university took root and has endured as an institution for some eight hundred years, being transformed in time into a worldwide phenomenon. Nothing in Islam or China, or India, or in the ancient civilizations of South America is comparable to the medieval university. It is in this remarkable institution, and its unusual activities, that the foundations of modern science must be sought.

The university was possible in the Middle Ages because the evolution of medieval Latin society allowed for the separate existence of church and state, each of which, in turn, recognized the independence of corporate entities, the university among them. The first universities, of Paris, Oxford, and Bologna, were in existence by approximately 1200, shortly after most of the translations had been completed. The translations furnished a ready-made curriculum to the emerging universities, a curriculum that was overwhelmingly composed of the exact sciences, logic, and natural philosophy.

The curriculum of science, logic, and natural philosophy established in the medieval universities of Western Europe was a permanent fixture for approximately 450 to 500 years. It was the curriculum of the arts faculty, which was the largest of the traditional four faculties of a typical major university, the others being medicine, theology, and law. Courses in logic, natural philosophy, geometry, and astronomy formed the core curriculum for the baccalaureate and master of arts degrees and were taught on a regular basis for centuries. These two arts degrees were virtual prerequisites for entry into the higher disciplines of law, medicine, and theology.

For the first time in the history of the world, an institution had been created for teaching science, natural philosophy, and logic. An extensive four-to-six-year course in higher education was based on those subjects, with natural philosophy as the most important component. As universities multiplied during the thirteenth to fifteenth centuries, the same science–natural philosophy–logic curriculum was disseminated throughout Europe, extending as far east as Poland.

The science curriculum could not have been implemented without the explicit approval of church and state. To a remarkable extent, both granted to the universities corporate powers to regulate themselves: universities had the legal right to determine their own curricula, to establish criteria for the degrees of their students, and to determine the teaching fitness of their faculty members.

Despite some difficulties and tensions between natural philosophy and theology – between, essentially, reason and revelation – arts masters and theologians at the universities welcomed the arrival of Aristotle's natural philosophy as evidenced by the central role they gave it in higher education. Why did they do this? Why did a Christian society at the height of the Catholic Church's power readily adopt a pagan natural philosophy as the basis of a four-to-six-year education? Why didn't Christians fear and resist such pagan fare rather than embrace it?

Because Christians had long ago come to terms with pagan thought and were agreed, for the most part, that they had little or nothing to fear from it. The rapprochement between Christianity and pagan literature, especially philosophy, may have been made feasible by the slowness with which Christianity was disseminated. ... [I]t was not until 392 that Christianity became the exclusive religion supported by the state. ...

By contrast, Islam, following the death of Mohammad in 632, was carried over an enormous geographical area in a remarkably short time. In less than one hundred years, it was the dominant religion from the Arabian peninsula westward to the Straits of Gibraltar, northward to Spain and eastward to Persia, and beyond. But where Islam was largely spread by conquest during its first hundred years, Christianity spread slowly and, with the exception of certain periods of persecution, relatively peacefully. It was this slow percolation of Christianity that enabled it to come to terms with the pagan world....

...In a momentous move, Clement of Alexandria (ca. 150–ca. 215) and his disciple Origen of Alexandria (ca. 185–ca. 254) laid down the basic approach that others would follow. Greek philosophy, they argued, was not inherently good or bad....They were agreed that philosophy and science could be used as 'handmaidens to theology' – that is, as aids to understanding Holy Scripture....

[...] When Christians in Western Europe became aware of Greco-Arabic scientific literature and were finally prepared to receive it in the twelfth century, they did so eagerly. They did not view it as a body of subversive knowledge. Despite a degree of resistance that was more intense at some times than at others, Aristotle's works were made the basis of the university curriculum by 1255 in Paris, and long before that at Oxford.

The emergence of a class of theologian-natural philosophers was the third essential precondition for the scientific revolution. Their major contribution was to sanction the introduction and use of Aristotelian natural philosophy in the curriculum of the new universities. Without that approval, natural philosophy and science could not have become the curriculum of the medieval universities. The development of a class of theologian-natural philosophers must be regarded as extraordinary. Not only did most theologians approve of an essentially secular arts curriculum, but they were convinced that natural philosophy was essential for the elucidation of theology. Students entering schools of theology were expected to have achieved a high level of competence in natural philosophy. [...]

To appreciate the importance of a class of theologian-natural philosophers for the development of science and natural philosophy in the Latin West, one has only to compare the Western reception of natural philosophy with its treatment in the civilization of Islam, where religious authorities regarded the study of natural philosophy as potentially dangerous to the faith. [...]

Another legacy from the Middle Ages to early modern science was an extensive and sophisticated body of terms that formed the basis of later scientific discourse – such terms as *potential, actual, substance, property, accident, cause, analogy, matter, form, essence, genus, species, relation, quantity, quality, place, vacuum, infinite,* and many others. These Aristotelian terms formed a significant component of scholastic natural philosophy. The language of medieval natural philosophy, however, did not consist solely of translated Aristotelian terms. New concepts, terms, and definitions were added in the fourteenth century, most notably in the domains of change and motion. Definitions of uniform motion, uniformly accelerated motion, and instantaneous motion were added to the lexicon of natural philosophy. By the seventeenth century these terms, concepts, and definitions were embedded in the language and thought of European natural philosophers.

...Medieval natural philosophers produced hundreds of specific questions about nature, the answers to which included a vast amount of scientific information. Most of the questions had multiple answers, with no genuine way of choosing between them. In the sixteenth and seventeenth centuries, new solutions were proposed by scholars who found Aristotelian answers unacceptable, or, at best, inadequate. The changes they made, however, were mostly in the answers, not in the questions. The scientific revolution was not the result of new questions put to nature in place of medieval questions. It was, at least initially, more a matter of finding new answers to old questions, answers that came, more and more, to include experiments, which were exceptional occurrences in the Middle Ages. ...

The Middle Ages did not just transmit a great deal of significantly modified, traditional, natural philosophy, much of it in the form of questions; it also conveyed a remarkable tradition of relatively free, rational inquiry. The medieval philosophical tradition was fashioned in the faculties of arts of medieval universities. Natural philosophy was their domain, and almost from the outset masters of arts struggled to establish as much academic freedom as possible. ...

In the 1330s, William of Ockham expressed the sentiments of most arts masters and many theologians when he declared:

> Assertions . . . concerning natural philosophy, which do not pertain to theology, should not be solemnly condemned or forbidden to anyone, since in such matters everyone should be free to say freely whatever he pleases.

Everyone who did natural philosophy in the sixteenth and seventeenth centuries was the beneficiary of these remarkable developments. The spirit of free inquiry nourished by medieval natural philosophers formed part of the intellectual heritage of all who engaged in scientific investigation. [. . .]

. . . These vital features of medieval science formed a foundation that made possible a continuous, uninterrupted eight hundred years of scientific development, a development that began in Western Europe and spread around the world.

Chapter Two
Copernicus and his Revolution

2.1 Robert S. Westman, *Proof, poetics, and patronage**

[...]

Copernicus the clerical humanist

De revolutionibus is consciously addressed to an ecclesiastical audience. The preface is cast in the idiom of church patronage and reform: hardly surprising when we recall that Copernicus spent his entire career as a church functionary, collecting rents from the peasantry, shoring up military defenses against the Teutonic knights, tending to the health needs of his chapter, and so forth. Of course, he had obtained his own office as canon in the usual manner through which church positions and properties were distributed during the late Renaissance: namely, through family connections. Throughout the sixteenth and seventeenth centuries, nepotism in the church was an important vehicle of upward social mobility. Those sacred offices known as ecclesiastical benefices always had incomes associated with them and were often acquired as a reward for service. Such benefices might or might not involve the care of souls and, in any event, because of the manner in which they were received, tended to be viewed primarily as income and only incidentally as sacred offices. Copernicus did not hold a benefice but received income from church property in absentia after his uncle, the bishop of Varmia, appointed him scholaster (instructor) at the Church of the Holy Cross in Breslau (Wroclaw). Such nepotistic associations certainly suggest that, had he wished to, Copernicus might have moved eventually into the position held by his uncle and that he might

* Robert S. Westman, 'Proof, poetics, and patronage: Copernicus's preface to *De revolutionibus*', in D. Lindberg and R. Westman (eds), *Reappraisals of the Scientific Revolution* (Cambridge: Cambridge University Press, 1990), pp. 175–83, 186–9, 192–4.

have risen from there into the bureaucracy of papal Rome. As we know, however, he did not choose that path, and there is no evidence that he wrote *De revolutionibus* with such an intention in mind.

If the prefatory material to *De revolutionibus* does not reveal a strategy of office seeking, its language nonetheless raises the broader issue of Copernicus's discursive practices as a humanist and the organization of patronage in early sixteenth-century papal Rome. ... After the period of the Great Schism, which left Rome, at the beginning of the fifteenth century, an intellectual backwater, the papacy sought to reestablish its cultural and political leadership throughout Europe. Humanist styles and ideals, which first arose as a new way of interpreting civic experience among the notaries and lawyers in the state chancelleries of Milan and Florence and the princely courts of Mantua and Ferrara, proved to be well suited to papal goals. All papal letters and formal documents were drafted in Latin, and it was the humanists, with their philological emphasis on 'getting the texts right' and their emulation of ancient classical forms of expression who became excellent broadcasters for papal policy. They were also valuable resources in the furtherance of papal foreign policy. Clerical humanists who staffed the papal and cardinalate households made superb ambassadors to hostile, secular courts, where they could present the papal court 'as a cultural force equal to, if not greater than, any secular court'. To be sure, not all the popes were equally supportive of humanism, but under the pontificate of Nicholas V (1447–1455), humanism made major inroads into the Roman bureaucracy. An admirer of Florentine humanism, Pope Nicholas was a great lover of books. He founded the Vatican Library, encouraged translations of Greek classical and patristic masterpieces (among them a substantial quantity of mathematical texts), provided humanists with numerous posts in the Curia, and elevated Nicholas of Cusa to the cardinalate; in short, he established a pattern of reciprocity whereby humanists would spread the glory of their papal patron in return for economic security in the church bureaucracy.

By the end of the fifteenth century, humanists had made Neo-Latin poetry into the favored vocabulary of upper-class Roman society. Such poetry might express praise of the patron during his lifetime or at his funeral; it might satirize his enemies; together with the oration, it was the fashionable genre at formal ceremonies, such as the dedication of a new building; it was a common feature in letters written by humanists or humanist popes; and it was often the language in which hopes for church reform were expressed. Popes and cardinals reveled in verse, and humanists willingly produced reams of it. Exquisite attention to proper Ciceronian style, and translations of Greek philosophical or moral writers, became the cultural badges of ambitious young humanist clerics.

Against this all too brief account, Copernicus's classical learning and humanist church contacts take on new and heightened meaning. Twenty-four different classical authors are cited in Copernicus's writings. Annotated philological works and a well-thumbed Greek dictionary, published in 1499, still survive with the remnants of his library in Uppsala. And it was this somewhat inadequate dictionary that he used when he published his very first book in 1509, a translation of a Greek work – in English, *Ethical, Rustic, and Love Letters* – by an obscure seventh-century Byzantine poet, Theophylactus Simocatta. In the dedication to his uncle, Copernicus reveals an Erasmian appreciation of the use of proverbs: 'Every reader may pluck what pleases him most in these letters, like an assortment of flowers in a garden. Yet Theophylactus disposed so much of value in all of them that they seem to be not letters but rather laws and rules for the conduct of human life.' Although the love letters seemed to Canon Copernicus to 'portend licentiousness', he defended their ethical value: 'Just as physicians usually moderate the bitterness of drugs by sweetening them to make them more palatable to patients, so these love letters have in like manner been rectified, with the result that they ought to receive the label "moral" no less.' ...

Meanwhile, Copernicus's reputation spread to Rome through other humanist channels. It was Pope Clement VII's secretary, a humanist and orientalist named Johann Albrecht Widmanstetter (1506–1577), who first learned of Copernicus's cosmological theory and explained it to the pope, in the Vatican gardens in 1533, before two cardinals, a bishop, and the pope's physician. In return, the pope presented Widmanstetter with a Greek manuscript containing several scientific treatises. Two years later Widmanstetter moved into the service of a recently promoted Dominican cardinal, Nicholas Schönberg (1472–1537), and after the latter's death Widmanstetter became secretary to the succeeding pope. In November 1536, Schönberg wrote to Copernicus, urging him to send a copy of his manuscript to Rome and even offering to provide as amanuensis the representative of the Varmian chapter in Rome, Theodoric of Reden. Copernicus, who well understood the proper epistolary order for seeking approval and protection in pre-Tridentine Rome, could easily have interpreted Schönberg's letter as a sign of eventual papal approbation; at the very least, it indicated support in the highest curial circles. Thus, when Copernicus placed Schönberg's letter in *De revolutionibus* immediately after the title page and just before the preface to the pope, he was allowing the Dominican Cardinal Schönberg to provide the first description of his 'new account of the World': 'In it you teach that the earth moves; that the sun occupies the lowest, and thus the middle, place in the universe.' Hence, Andreas Osiander's famous unsigned letter 'To the Reader Concerning the Hypotheses of this Work' which, without Copernicus's permission, was placed first in the publication, interfered as much with the author's methodological aims as with his intended strategy for seeking the pope's patronage.

Strategies of persuasion

Turn now to the preface itself. Who was the pope to whom the work was dedicated, and what was Copernicus's strategy of presentation? The pope to whom Copernicus's theory had once been described in the Vatican gardens was dead; so was Copernicus's advocate, Cardinal Schönberg. The new pope was now Paul III (1534–1549), the former Cardinal Alessandro Farnese – once a student at the University of Pisa, a poet, well schooled in Greek; as a pope, renowned for his wide learning and love of astrology. Among his institutional accomplishments, he founded the Roman Inquisition and called into session the Council of Trent. Farnese came from a wealthy, noble family and could afford to pay his servants with his own money rather than strictly from ecclesiastical revenues. Indeed, according to a census of cardinals' households prepared in 1526–1527, Farnese's household contained three hundred six persons. Not surprisingly, he was the frequent object of pleas for patronage. Twice, in 1529 and in 1532, Luca Gàurico (1475–1558), a Neapolitan astronomer and author of numerous prognostications, predicted that Alessandro Farnese would become pope. Gàurico soon found himself a regular dinner companion of the cardinal. In 1543, Gàurico presided at an astrological ceremony for the laying of the cornerstone in the Farnese wing of the Vatican Palace. Gàurico calculated the exact hour and zodiacal sign, assisted by Vincenzo Campanacci, a Bolognese astrologer, who 'found the proper time on the astrolabe and announced it in a loud voice'. Three years later Gàurico was rewarded with a bishopric. It is revealing of Copernicus's strategy that he chose not to make predictions about the pope's health, longevity, or political future, nor did he advise him when to make journeys. One may speculate that he kept silent on such astrological matters in order not to create the slightest possibility of offending the pope and, perhaps, to associate himself politically with those humanist elements within the church opposed, on various moral and physical grounds, to divinatory practices.

In appealing for patronal support, Copernicus reveals a rigorous knowledge of common epistolographical and rhetorical resources, such as understatement of his own achievements and exaggerated modesty: all characteristic strategies in a *captatio benevolentiae* designed to capture an audience's attention and good will. The preface utilizes a variety of rhetorical figures – among which irony, confession, and antithesis are most prominent – in order to create a sense of tension and contrast between Copernicus's theory and the cosmologies he hopes to replace. Without breaking his argument, Copernicus displays his rhetorical agility in moving back and forth between opposing themes. One notices the following contraries: coherence and incoherence, order and confusion, praise and ridicule, common sense and absurdity, beauty and monstrosity, novelty and tradition, clarity and obscurity, certainty and uncertainty.

Copernicus begins by presenting himself ironically as someone worthy of laughter and derision, someone who goes against tradition and whose theories will surely be repudiated. . . . Why does Copernicus put himself to such trouble? Well, he says, it is like this: His friends the cardinal of Capua and the bishop of Chelmno had repeatedly urged him to publish. They argued that even if his theory appeared to be crazy, 'so much the more admiration and thanks would it gain after they saw the publication of my writings dispel the fog of absurdity by the most transparent proofs [*liquidissimis demonstrationibus*]'. Finally, he has acceded to their entreaties and will 'permit it to appear after being buried among my papers and lying concealed not merely until the ninth year but by now the fourth period of nine years'. Here is his story: For a long time, he pondered the uncertainties in the traditional astronomies and the disagreements among them. Some use homocentrics; others use eccentrics. The homocentrists cannot get their theories to fit the phenomena absolutely; the eccentrists can deduce the phenomena but violate 'first principles'. Worst of all, neither tradition can deduce what Copernicus calls 'the arrangement of the universe and the sure symmetry of its parts'. In short, we have an ironic reversal: It is tradition itself that is full of monstrous incoherence and absurdity. And here follows the famous metaphor to which Kuhn and others have attached so much significance, which allegedly ties Copernicus to Florentine Neoplatonism through his Bolognese teacher Domenico Maria de Novara:

> With them it is just as though someone were to join together hands, feet, a head, and other members from different places, each part well drawn, but not proportioned to one and the same body, and not in the least matching each other, so that from these [fragments] a monster rather than a man would be put together.

Put logically, the problem is that the old theories have deduced a false universe from false principles. A scientific demonstration of the sort recommended by Aristotle in his *Posterior Analytics*, surely known to Copernicus, stringently required that true conclusions be deduced from true causes. Such *cognitio certa per causas* is certain because it is grounded on causes that are true, proper, and irrevisable.

But, in the following paragraph, Copernicus makes no reference to this strict demonstrative ideal. Instead, he introduces the explicitly humanist theme of reform as a solitary voyage, instigated by the disorder of tradition: 'I reflected on this confusion...', 'I began to be annoyed...', 'I undertook the task of rereading...to learn if anyone had ever proposed...', 'I found in Cicero...', 'I also discovered in Plutarch...', 'I have decided to set his words down here...'. This parable of 'rereading' (*libros relegerem*) the old, pre-Christian philosophers allows Copernicus to valorize the 'absurd' idea of the earth's motion as a thought worth entertaining against 'those who teach mathematics in the schools'. After all, if the old, unreformed astronomies are permitted to make assumptions that are false, 'being granted the freedom to imagine any circles whatever for the purpose of explaining the heavenly phenomena', then perhaps he can be permitted to make a new assumption that, although physically absurd, leads to 'sounder demonstrations (*firmiores demonstrationes*) than those of my predecessors'. Copernicus's search, then, ends in the discovery of order:

> I finally found that if the motions of the other planets are brought into relation with the circular course of the earth, and are reckoned for the revolution of each planet, not only do their phenomena follow therefrom but also the order and size of all the planets and spheres, and heaven itself is so linked together that nothing can be moved from its place without causing confusion in the remaining parts and the universe as a whole.

In logical terms, the deductive outcome of Copernicus's hypothetical premise is a conclusion: that his theory possesses the *symmetria* lacking in rival traditions, and that the consequence of his argument – the fitting together of the universe's parts – is warrant for his putatively absurd conditional premise (that the earth moves).

Now, what kind of audience will be receptive to such a shift in demonstrative standards and to the language in which it is cast? One answer seems to lie in an oft-cited passage: 'Mathemata mathematicis scribuntur': He apparently intends the book for mathematically trained astronomers, 'outside the schools', who alone can follow his theories. Possibly, Copernicus has in mind someone like his fervent disciple G. J. Rheticus, but Rheticus is nowhere mentioned. The audience constructed in the preface provides another explanation. The preface, although dedicated to Paul III, divides the church into two groups: those enlightened by mathematical training, and those not. Copernicus includes popes Leo X and Paul III, Cardinal Schönberg, Bishop Giese, and Paul of Middleburg, bishop of Fossombrone, in the first group. In the second, he places untutored theologians – Copernicus calls them 'idle talkers' – who know nothing about mathematics and whom he imagines will deride him by distorting Scripture for their own purposes. The sole example of the latter is the church father Lactantius, 'otherwise an illustrious writer but hardly a mathematician, [who] speaks quite childishly about the earth's shape, when he mocks those who declared that the earth has the form of a globe'. Copernicus hopes that the present pope, by his authority and learning, will suppress any 'calumnious attacks'. For in fact his book not only promises a reform of the theoretical part of astronomy, namely the principles of planetary motion, but also of the practical part, namely the calendar. In the end, the intended audience is supposed to judge whether Copernicus has succeeded.

...Kepler's old teacher Michael Maestlin (1550–1631) is an ideal example, for by sixteenth-century standards he was a preeminent astronomer; and he was also a prolific annotator of *De revolutionibus* and well steeped in the classics. Concerning the passage where Copernicus mentions that it took him four times nine years to publish his book, Maestlin immediately recognized a line from Horace's famous treatise *Ars poetica*, and in the margin of his own copy Maestlin quoted the passage in full: 'Yet if ever you do write anything, let it enter the ears of some critical Maecius, and your father's, and my own; then put your parchment in the closet and keep it back till the ninth year. What you have not published you can destroy; for the word once sent forth can never come back.' Maestlin's note immediately suggested to me the possibility that other sections of the preface might have Horace as their subtext. The opening verse yields a valuable clue:

> If a painter chose to join a human head to the neck of a horse, – so begins *The Art of Poetry* – and to spread feathers of many a hue over limbs picked up now here now there, so that what at the top is a lovely woman ends below in a black and ugly fish, could you, my friends, if favoured with a private view, refrain from laughing? Believe me, dear Pisos, quite like such pictures would be a book whose idle fancies shall be shaped like a sick man's dreams, so that neither head nor foot can be assigned to a single shape. 'Painters and poets,' you say, 'have always had an equal right in hazarding anything.' We know it: this licence we poets claim and in our turn we grant the like; but not so far that savage should mate with tame, or serpents couple with birds, lambs with tigers.

Copernicus has indeed invoked the aesthetic – but it is a Horatian aesthetic, not a Kuhnian one. And the central theme emphasized by Horace and noticed by his Renaissance commentators was the principle of 'fittingness' or 'belongingness'. Style must fit its subject; diction its characters; characters must preserve decorum, appropriateness; the beginning must fit the end. Significantly, the audience is the custodian of 'appropriateness' and rejects through laughter what it perceives not to agree with nature. What moves or delights or persuades the audience is what makes for good poetry. And it was this rhetorical view of poetry that many Renaissance commentators so appreciated in Horace.

In his preface to the pope, Copernicus is always mindful of the audience whose assent and patronage he seeks. But it is also useful to notice what is not explicitly present in the text. By scholastic standards, Copernicus had made himself vulnerable to a serious objection. He had violated Aristotle's prohibition against *metábasis*, a prohibition that disallowed the transfer of the principles of one discipline into those of another. In this case, Copernicus tacitly transferred the Horatian ideal of good poetry into the domain of astronomical practice: Just as one prefers a coherent to an incoherent literary work, so a theory of the planets possessing mathematical coherence (*symmetria, armoniae nexum*) is to be preferred over one that does not. The implication is that such a world picture is not arbitrary, for art imitates nature; hence, a decorous audience will judge such a theory to be true, while shunning as absurd one lacking in *symmetria*. If such an argument did violence to the *Posterior Analytics* in its rejection of irrevisable knowledge, it was entirely in keeping with humanist commentators on Horace. [...]

The politics of heavenly reform

Judged strictly as an instance of conceptual change in astronomy, Copernicus's work achieved little that was revolutionary in the terms of such writers as Koyré, Butterfield, Kuhn, and Burtt. For them, Copernicus simply discovered how to make a new transformation; others completed the 'revolution'. Our purpose has not been to reappraise the entire epoch denoted by that term. But it is important to point out here that the canonical view of Copernicus as a conceptual conservative – or even Nicholas Jardine's more historically sophisticated and challenging view of him as a kind of 'neo-conservative' natural philosopher – ignores both the subtle politics of Copernicus's reformist strategy within the church and the fact that that strategem initially failed. For *De revolutionibus* was immediately perceived as a resource of disciplinary disruption – a threat not merely to belief about the ordering of the heavens but to the ordering of the disciplines upon which such belief rested. In medieval astronomical textbooks and trees of classification, with their love of divisions and distinctions, astronomy occupied a somewhat disputed zone as a 'middle science' combining mathematics and physics. The parts of this disciplinary couple might coexist in quite different relations to one another. In some works, the mathematical and physical parts have a mutually complementary relation; in others, mathematics is separated from physical considerations and sometimes placed above the latter; but in medieval classification schemes, especially those of Thomas Aquinas and Albertus Magnus, and in many commentaries on Aristotle's *De caelo*, physics is regarded as the superior science from which the mathematical part of astronomy draws its principles – a view still quite prominent in Copernicus's time. Indeed, when Osiander claimed that Copernicus's work would not 'throw the liberal arts into confusion', he was trying to avoid objections from academic philosophers and theologians who associated themselves with this latter tradition, whom he referred to obliquely as 'the peripatetics and theologians'. And Osiander proved to be right: Not long after its publication, a highly placed Dominican in the court of Paul III privately attacked *De revolutionibus* for violating the principles of the 'superior science' of physics.

In his preface to the pope, Copernicus avoids this language of medieval scholastic classification. As we have seen, he draws on a practical knowledge of rhetorical strategies, probably acquired through his association with an artistic culture at Padua blessed by cardinalate patronage. In addition, as Paul Rose has shown, Copernicus was conscious of a discourse of astronomical reform that had already emerged in the work of the Viennese humanist astronomers Georg Peurbach and Regiomontanus and that passed down to him through Domenico Maria de Novara. Evidently Copernicus aimed to solicit reform sentiment from among those in the church who, like the Viennese school, valued the mixed mathematical disciplines but saw them as needing renewal through a return to a purified ancient tradition. Elements of this mathematical humanism appear in the suppressed introduction to Book 1, where he casts his view of the priority of mathematics in the rhetoric of a humanist encomium: Mathematics is the source of astronomy's highest principles by virtue of the dignity of its subject matter, its pedagogical priority, and the kind of knowledge to which it can lay claim. From mathematics one can deduce the true structure of the universe and thus put an end to warring schools of the sort that have kept the universe's true order from being known – an echo, perhaps, of Regiomontanus's sentiment in the *Tabulae primi mobilis* (1467–1468). Mathematics, 'the summit of the liberal arts', prepares us, in short, for a Platonic image of the unity of all knowledge: common principles underlying all disciplines.

Quite apart from the Copernican theory's threatening implications for those interests committed to the Thomist hierarchy of the disciplines, the preface had wider political associations that hitherto have gone unnoticed. Noticing these sorts of political meaning, however, now requires us to make explicit a view of language silently assumed in this paper, a view that rejects as its primary object specific propositions or utterances. As J. G. A. Pocock has observed about the languages of political thought, 'We wish to study the languages in which utterances were performed, rather than the utterances which were performed in them. ... When we speak of "languages", therefore, we mean for the most part sub-languages: idioms, rhetorics, ways of talking about politics, distinguishable language games of which each may have its own vocabulary, rules, preconditions and implications, tone and style.' ... I want to open the possibility that Copernicus's language in the preface shows remarkable convergences with the political vocabularies of humanist curial reformers, as well as with familiar visual images of Reformation popular propaganda. These overlapping arenas of language and image permit another inference: Copernicus's aim was not merely to recommend that the church improve the scandalous state of the calendar but that it reconsider its teachings about the order of the heavens.

Consider how Copernicus writes of the pope. The pope, who, in Copernicus's words, 'holds dominion over the Ecclesiastical Commonwealth', answers to a God of order – 'The best and most orderly Artisan of all.' The pope is also presented to us as a protector – of Copernicus, of truth-seeking philosophers, and of the church's view of the heavens. His authority does not come merely from God but also from his human qualities: 'Even in this very remote corner of the earth where I live you are considered the highest authority by virtue of the dignity of the office and love of the mathematical arts and all learning.' The pope is urged to protect Copernicus against the hostility, uncertainty, and disagreement engendered by certain astronomers and philosophers: 'By your authority and judgement, you can easily suppress calumnious attacks.' ... The pope at the head of his flock is not corrupt; it is he who must expurgate those who are and thereby protect Rome from abuse.

While church reformers used the image of the head to symbolize order and authority, the many-headed monster also functioned as an image of moral disorder at the popular level. In German broadsheets of the 1520s, Lutherans employed the seven-, unequal-headed, biblical beast of the Apocalypse as visual propaganda to attack papal indulgences, and Catholics later portrayed Luther as a many-headed, fanatical wild man. . . .

These 'high' and 'low' reformist points of reference help to situate Copernicus's moral imagery associating head and body, papal authority with celestial reform. Joined in his use of the Horatian aesthetic is a language of natural order and ethical conviction, underwriting a belief that astronomy's first principles are at once true and untainted. By addressing Rome in this idiom of order and reform – resonant with both elite and popular connotations – he also evoked moral and political associations to which fellow clerics from his own local region of Varmia could respond. Indeed, Varmian religious politics was strongly humanist, irenic, and Erasmian in spirit, amidst growing anti-Catholic sentiment and conversions to Luther's doctrines. Copernicus's Varmian patron and closest friend, Bishop Tiedemann Giese, corresponded with Erasmus. Copernicus credited Giese in the preface with encouraging the publication of *De revolutionibus*, and we have already noted Copernicus's debts to Erasmian stylistics. Rheticus portrays Giese, in the *Narratio prima*, as a radical reformer who pushed the cautious Copernicus not to conceal the theoretical principles from which he deduced his new system of planetary motions [.] . . . Giese also wrote a treatise reconciling Copernicus's theory with the Bible (*Hyperaspisticon*, now lost), in which he borrowed the first word (*Hyperaspistes*, or 'shield-bearer') from a polemic written by Erasmus against Luther. And Giese shared the characteristic Erasmian view that gentle persuasion could achieve more than sharp criticism and satire. Differences of opinion could be resolved through love and toleration; Christian unity must come from within the church.

The strategy of persuasion that Copernicus followed in his preface of 1543 undoubtedly reflects the outcome of earlier discussions with Giese. Copernicus attempted to sidestep conservative elements in Rome, and he carefully omitted all references to the Lutheran Rheticus. His reformist rhetoric was not stridently polemical; it was gently Horatian and Erasmian: an end to controversy among astronomers; an internal cadre of humanist mathematicians to reform church teaching on the heavens by providing true principles from which planetary order and calendrical accuracy could be restored, the entire enterprise to be legitimated by papal authority and by appeal to a range of ancient, pagan sources. The approach evokes Erasmus's broad reconciliation of Christian and pagan letters in a *philosophia Christi* – a life of lay piety modeled on the true life of Christ and the earliest sources of Christianity rather than on empty ceremonial practices and overly subtle Scholastic definitions. . . .

In *De revolutionibus*, Copernicus sought to bring the individual parts of the universe into concordance with a sun that he described to his ecclesiastical audience in the most classical, pagan images, not as the generative or emanative force of the Neoplatonists but rather as a properly placed lamp or lantern, an eye, a mind, an enthroned king, a visible god. His choice of language and imagery pointed the church away from mediative, astrological influences and instead returned it *ad fontes*, to an ancient truth: the primitive order of the creation. . . .

Chapter Three
The Spread of Copernicanism in Northern Europe

3.1 Robert S. Westman, *The Copernicans and the churches**

In 1543, on his deathbed, Nicolaus Copernicus received the published results of his life's main work, a book magisterially entitled *De Revolutionibus Orbium Coelestium Libri Sex* (*Six Books on the Revolutions of the Celestial Orbs*), which urged the principal thesis that the earth is a planet revolving about a motionless central sun. In 1616, seventy-three years after its author's death, the book was placed on the Catholic Index of Prohibited Books with instructions that it not be read 'until corrected'. Sixteen years later – and, by then, ninety years after Copernicus first set forth his views – Galileo Galilei (1564–1642) was condemned by a tribunal of the Inquisition for 'teaching, holding, and defending' the Copernican theory. These facts are well known, but the dramatic events that befell Galileo in the period 1616–1632 have tended to overshadow the relations between pre-Galilean Copernicans and the Christian churches and to suggest, sometimes by implication, that the Galileo affair was the consummation of a long-standing conflict between science and Christianity.

...It will be helpful if we can suspend polar categories customarily used to describe the events of this period, such as Copernican versus anti-Copernican, Protestant versus Catholic, the individual versus the church. The central issue is better expressed as a conflict over the standards to be applied to the interpretation of texts, for this was a problem common to astronomers, natural philosophers, and theologians of whatever confessional stripe. In the case of the Bible, should its words and sentences in all instances be taken to *mean* literally what they say and, for that reason, to describe

---

* Robert S. Westman, 'The Copernicans and the churches', in D. Lindberg and R. Numbers (eds), *God and Nature: Historical Essays on the Encounter between Christianity and Science* (Berkeley, CA: University of California Press, 1986), pp. 76–85, 89–91.

actual events and physical truths? Is the subject matter of the biblical text *always* conveyed by the literal or historical meaning of its words? Where does the ultimate authority reside to decide on the mode of interpretation appropriate to a given passage? In the case of an astronomical text, should its diagrams be taken to refer literally to actual paths of bodies in space? Given two different interpretations of the same celestial event, where does the authority reside to decide on the particular mode of interpretation that would render one hypothesis preferable to another? When the subject matters of two different *kinds of text* (e.g., astronomical and biblical or astronomical and physical) coincide, which standards of meaning and truth should govern their assessment? And finally, how did different accounts of the God–Nature relationship affect appraisal of the Copernican theory? Questions of this sort define the issue faced by sixteenth- and early seventeenth-century Copernicans.

Copernicus's achievement

...Nicolaus Copernicus (1473–1543) was a church administrator in the bishopric of Lukas Watzenrode, located in the region of Warmia, now northern Poland but then part of the Prussian Estates. Watzenrode was Copernicus's uncle and guardian, and it was through his patronage that the young man was able to study medicine and canon law in Italy before returning to take up practical duties, including supervision of financial transactions, allocation of grain and livestock in peasant villages, and overseeing the castle and town defenses in Olsztyn. Though a member of the bishop's palace, Copernicus was not a priest but a clerical administrator or canon.

In his spare time Copernicus worried about a problem that had long concerned the church – accurate prediction of the occurrence of holy days such as Easter and Christmas. Now calendar reform was an astronomical problem that demanded not primarily new observations but the assimilation of old ones into a model capable of accurately predicting the equinoxes and solstices, the moments when the sun's shadows produce days of longest, shortest, and equal extent. But predictive accuracy had never been the astronomer's only goal. The mathematical part of astronomy was complemented by a physical part. The object of the latter was to explain why the planets moved, what they were made of, and why they are spaced as they are. According to Aristotle's heavenly physics, the sun, moon, and other planets are embedded in great spheres made of a perfect and invisible substance called aether. The spheres revolve uniformly on axes that all pass through the center of the universe. This model yielded an appealing picture of the universe as a kind of celestial onion with earth at the core; but it failed to explain why the planets vary in brightness. As an alternative, the astronomer Ptolemy (*fl.* A.D. 150) used a mathematical device according to which the planet moves uniformly about a small circle (the epicycle) while the center of the epicycle moves uniformly about a larger circle (the deferent). Such a model could account for variations in both speed and brightness. Ptolemy also invented another device, however, called the 'equant'.... Here the center of an epicycle revolves *nonuniformly* as viewed both from the sphere's center and from the earth but *uniformly* as computed from a noncentral point (situated as far from the center on one side as the earth is on the other). As a predictive mechanism the equant is successful. But now ask how it can be that the planet, like a bird or fish, 'knows' how to navigate uniformly in a circle about an off-center point while, simultaneously, flying variably with respect to the center of the same sphere? In response to objections like this it was quite customary for astronomers in the universities to consider the planetary circles *separately* from the spheres in which they were embedded. This meant that conflict between the mathematical and physical parts of astronomy could be avoided by not mixing the principles of the two disciplines. If, however, an astronomer were determined to reconcile physical and mathematical issues, it would be customary within the Aristotelian tradition (which prevailed within the universities) to defer to the physicist, for in the generally accepted medieval hierarchy of the sciences, physics or natural philosophy was superior to mathematics.

Copernicus, like all great innovators, straddled the old world into which he was born and the new one that he created. On the one hand he was a conservative reformer who sought to reconcile natural philosophy and mathematical astronomy by proclaiming the absolute principle that all motions are uniform and circular, with all spheres turning uniformly about their own centers. But, far more radically, Copernicus argued for the earth's status as a planet by appealing to arguments from the *mathematical part* of astronomy. In so doing he shifted the weight of evidence for the earth's planetary status to the lower discipline of geometry, thereby violating the traditional hierarchy of the disciplines. If anything can be called revolutionary in Copernicus's work, it was this mode of argument – this manner of challenging the central proposition of Aristotelian physics.

We are now prepared to consider the general logical structure of Copernicus's argument. Briefly, it looks like this: *If* we posit that the earth has a rotational motion on its axis and an orbital motion around the sun, then (1) all known celestial phenomena can be accounted for as accurately as on the best Ptolemaic theories; (2) the annual component in the Ptolemaic models, an unexplained mirroring of the sun's motion, is eliminated; (3) the planets can be ordered by their increasing sidereal periods from the sun; and (4) the distances of the planets from the center of the universe can be calculated with respect to a 'common measure', the earth–sun radius (a kind of celestial yardstick), which remains fixed as the absolute unit of reference.... Although they were certainly among the most important consequences, these four were not the only ones to follow from the assumption of terrestrial motion. However, from the viewpoint of the prevailing logic of demonstrative proof, found in Aristotle's *Posterior Analytics*, there was no *necessity* in the connection between the posited cause and the conclusions congruent with that cause. Thus, while Copernicus's premises certainly authorized the conclusions he drew, there was no guarantee that other premises might not be found, equally in accord with the conclusions. In short, Copernicus had provided a systematic, logical explanation of the known celestial phenomena, but in making the conclusions the grounds of his premises, he failed to win for his case the status of a demonstrative proof.

Pre-Galilean Copernicans were thus faced with several serious problems. First, their central premise had the status of an assumed, unproven, and (to most people) absurd proposition. Second, whatever probability it possessed was drawn primarily from consequences in a lower discipline (geometry). Third, even granting the legitimacy of arguing for equivalent predictive accuracy with Ptolemy, the practical derivation of Copernicus's numerical parameters was highly problematic. Fourth, the Copernican system flagrantly contradicted a fundamental dictum of a higher discipline, physics – namely, that a simple body can have only one motion proper to it – for the earth both orbited the sun and rotated on its axis. And finally, it appeared to conflict with the interpretations of another higher discipline, biblical theology – in particular, the literal exegesis of certain passages in the Old Testament.

Under the circumstances Copernicus resorted to a rhetorical strategy of upgrading the certitude available to 'mathematicians' – by which he meant those who practiced the mathematical part of astronomy – while underplaying the authority of natural philosophy and theology to make judgments on the claims of mathematicians. Final authority for interpreting his text, he said, rested with those who best understand its claims. Church fathers such as Lactantius had shown a capacity for error in astronomy and natural philosophy, as when Lactantius declared the earth to be flat. Theologians of this sort should stay away from a subject of which they are ignorant.

Copernicus's strategy of appealing to the autonomy and superiority of mathematical astronomy was undercut by a prefatory 'Letter to the Reader' that appeared immediately after the title page of *De Revolutionibus*. That brief epistle bespeaks the extraordinary circumstances surrounding the publication of the book. It was only at the very end of Copernicus's life that he was finally persuaded to publish his book – not by one of his fellow canons, some of whom were eager to see the manuscript in press, but by a young Protestant mathematics lecturer who had come to visit the old canon from the academic heart of the Lutheran Reformation, the University of Wittenberg. Georg Joachim Rheticus (1514–1574) was permitted by Copernicus to publish a preliminary version of the heliocentric theory (*Narratio Prima*, 1540) and also to attend to the eventual publication of *De Revolutionibus*. But Rheticus lacked the time to oversee the work and so entrusted it to a fellow

Lutheran, Andreas Osiander (1498–1552). Osiander, without permission from either Copernicus or Rheticus, took it upon himself to add an unsigned prefatory 'Letter' written in the third person singular. Upon reading the manuscript, Osiander had become convinced that Copernicus would be attacked by the 'peripatetics and theologians' on the grounds that 'the liberal arts, established long ago on a correct basis, should not be thrown into confusion'. Osiander hoped to save Copernicus from a hostile reception by appealing to the old formula according to which astronomy is distinguished from higher disciplines, like philosophy, by its renunciation of physical truth or even probability. Rather, if it provides 'a calculus consistent with the observations, that alone is enough'. *De Revolutionibus* was thus to be regarded as a strictly mathematical-astronomical text unable to attain even 'the semblance of the truth' available to philosophers; and both mathematicians and philosophers were incapable of stating 'anything certain unless it has been divinely revealed to them'.

Early Protestant reaction: the Melanchthon circle and the 'Wittenberg Interpretation'

When Rheticus returned to his teaching duties at Wittenberg after his long visit to Copernicus, he brought back strongly favorable personal impressions of the Polish canon and his new theory. Rheticus himself was Copernicus's first major disciple, and many of the Wittenberger's students read and studied *De Revolutionibus*. Furthermore, Rheticus composed a treatise, recently rediscovered, in which he sought to establish the compatibility of the Bible and the heliocentric theory.[1] All of this tempts us to ask whether Protestants were particularly well disposed toward the Copernican theory.

To answer this question, we must distinguish between the Protestant Reformers and men who happened to be Protestants and were also well versed in the reading of astronomical texts. The Reformers Luther and Calvin were learned men who knew enough astronomy to understand its basic principles; but neither had ever practiced the subject. It used to be thought that Luther played an important role in condemning Copernicus's theory when, in the course of one of his *Tischreden* or *Table Talks*, he said: 'That fool wants to turn the whole art of astronomy upside down.' But the statement itself is vague on details and, in any event, was uttered in 1539, sometime before the publication of either Rheticus's *Narratio Prima* or Copernicus's *De Revolutionibus*. As for Calvin, there is no positive evidence that he had ever heard of Copernicus or his theory; if he knew of the new

---

1 G. J. Rheticus, *Holy Scripture and the Motion of the Earth* [1540], trans. R. Hooykas, from G. J. Rheticus *Treatise on Holy Scripture and the Motion of the Earth* (Amsterdam: North-Holland Publishing Company, 1984). Cf. Doc 2.2 in *Science in Europe, 1500–1800: A Primary Sources Reader*, the companion Reader to this volume.

doctrine, he did not deem it of sufficient importance for public comment. In short, there are no known opinions by these two leading Protestant Reformers that significantly influenced the reception of the Copernican system.

There was, however, a third Reformer, a close associate of Luther's and the educational arm of the Reformation in Germany, Philipp Melanchthon (1497–1560), known as *Praeceptor Germaniae*. A charismatic man, beloved teacher, and talented humanist, Melanchthon was also a brilliant administrator with a gift for finding compromise positions. In the face of serious disturbances from the Peasants' Revolt of 1524–1525 and plunging enrollments all over Germany, Melanchthon instituted far-reaching reforms that led to the rewriting of the constitutions of the leading German Protestant universities (Wittenberg, Tübingen, Leipzig, Frankfurt, Greifswald, Rostock, and Heidelberg), profoundly influencing the spirit of education at several newly founded institutions (Marburg, Königsberg, Jena, and Helmstedt). Most important of all, Melanchthon believed that mathematics (and thus astronomy) deserved a special place in the curriculum because through study of the heavens we come to appreciate the order and beauty of the divine creation. Furthermore, mathematics was an excellent subject for instilling mental discipline in students. Such views alone would not predispose one toward a particular cosmology, but they did help to give greater respectability to the astronomical enterprise. Thus, a powerful tradition of mathematical astronomy developed at Wittenberg from the late 1530s and spread throughout the German and Scandinavian universities. At Wittenberg itself, three astronomers in the humanistic circle gathered around Melanchthon were preeminent: Erasmus Reinhold (1511–1553), his pupil Rheticus, and their joint pupil and the future son-in-law of Melanchthon, Caspar Peucer (1525–1603). Melanchthon was the *pater* of this small *familia scholarium*. Many of the major elements in the subsequent interpretation of Copernicus's theory in the sixteenth century would be prefigured in this group at Wittenberg.

The 'Wittenberg Interpretation', as we will call it, was a reflection of the views of the Melanchthon circle. Melanchthon himself was initially hostile to the Copernican theory but subsequently shifted his position, perhaps under the influence of Reinhold. Melanchthon rejected the earth's motion because it conflicted with a literal reading of certain biblical passages and with the Aristotelian doctrine of simple motion. But Copernicus's conservative reform – his effort to bring the calculating mechanisms of mathematical astronomy into agreement with the physical assumption of spheres uniformly revolving about their diametral axes – was warmly accepted. Reinhold's personal copy of *De Revolutionibus*, which still survives today, is testimony; it has written carefully across the title page the following formulation: 'The Astronomical Axiom: Celestial motion is both uniform and circular or composed of uniform and circular motions.' As it stands, this proposition simply ignores physical claims for the earth's motion, but commits itself to an equantless astronomy. It is, we might say today, a 'research program', one which Copernicus tried to make compatible with the assumption that the earth is a planet. But the Wittenbergers, with the noticeable exception of Rheticus, refused to follow Copernicus in upsetting the traditional hierarchy of the disciplines. Instead, Reinhold and his extensive group of disciples accepted Melanchthon's physical and scriptural objections to the Copernican theory. In the prevalent mood of reform, Copernicus was perceived not as a revolutionary but as a moderate reformer (like Melanchthon), returning to an ancient, pristine wisdom before Ptolemy.

If Melanchthon and Reinhold saw Copernicus as a temperate reformer, Rheticus saw the radical character of his reform. Rheticus returned to Wittenberg in 1542 as an inflamed convert, writing of Copernicus as of one who has had a Platonic vision of The Good and The Beautiful – though in the harmony of the planetary motions. ... Even more enthusiastically than Copernicus, Rheticus extolled the 'remarkable symmetry and interconnection of the motions and spheres, as maintained by the assumption of the foregoing hypotheses', appealing to analogical concordance with musical harmonies, to the number six as a sacred number in Pythagorean prophecies, to the harmony of the political order in which the emperor, like the sun in the heavens, 'need not hurry from city to city in order to perform the duty imposed on him by God', and to clockmakers who avoid inserting superfluous wheels into their mechanisms. Copernicus's unification of previously separate hypotheses had a liberating, almost intoxicating, effect on Rheticus, which Rheticus expressed almost as a personal revelation fully comprehensible only by visualizing the ideas themselves.

A wide spectrum of early Protestant opinion is defined between Melanchthon's cautious promotion of Copernicus's reform and Rheticus's radical espousal of the core propositions of Copernican cosmology. In general, the Wittenberg Interpretation dominated until the 1580s, while Rheticus's vision was typically ignored in public discussions. By the late 1570s, however, there were signs of the emergence of a cosmological pluralism among Protestant astronomers. A Danish aristocrat named Tycho Brahe (1546–1601) established an extraordinary astronomical castle on the misty island of Hveen, near Copenhagen, where he commenced a major reform of astronomical observations and, by the early 1580s, proposed a new cosmology in which all the planets encircle the sun, while the sun moves around the stationary, central earth. This system – the Tychonic or geoheliocentric – adopted Copernican-heliocentric paths for the planets, causing the orbits of Mars and the sun to intersect, while preserving Aristotelian terrestrial physics...; but in another quite important respect, Tycho departed from Aristotle by abolishing the solid celestial spheres. In 1600 the Englishman William Gilbert (1540–1603) suggested that the earth possesses a magnetic soul that causes it to turn daily on its axis; but he was cryptic about the ordering of Mercury and Venus.

Throughout the second half of the sixteenth century, Copernicus's book was widely read and sometimes studied in both Catholic and Protestant countries. Compared to the fairly large number of people aware of the central claims of *De Revolutionibus*, however, there were relatively few who actively adopted its radical proposals and whom we can justifiably call 'Copernicans' in that sense. To be precise: we can identify only ten Copernicans between 1543 and 1600; of these, seven were Protestants, the others Catholic. Four were German (Rheticus, Michael Maestlin, Christopher Rothmann, and Johannes Kepler); the Italians and English contributed two each (Galileo and Giordano Bruno; Thomas Digges and Thomas Harriot); and the Spaniards and Dutch but one each (Diego de Zuñiga; Simon Stevin). [. . .]

The Copernican theory and biblical hermeneutics

The Protestant Reformers were agreed in emphasizing the plain, grammatical sense as the center of biblical interpretation, thereby making it accessible to anyone who could read. Additional help was sometimes sought from spiritual or allegorical readings, but the literal, realistic meaning always remained central. Now, the literalism of the Reformers was twofold: they believed that the Bible was literal both at the level of direct linguistic reference (nouns referred to actual people and events) and in the sense that the *whole story* was realistic. The Bible's individual stories needed to be woven together into one cumulative 'narrative web'. This required the earlier stories of the Old Testament to be joined interpretatively to those in the New Testament by showing the former to be 'types' or 'figures' of the latter. Luther and Calvin were agreed that there was a single theme, a primary subject matter, which united all the biblical stories: the life and ministry of Christ.

Although Protestants rejected the Catholic appeal to allegorical and anagogical interpretations of Scripture as an illegitimate stretching of the plain meaning, both groups of exegetes had available to them a method of interpretation to which they could appeal: the principle of accommodation. One purpose of this hermeneutic device was to resolve tensions between popular speech, wedded to the experience of immediate perception, and the specialized discourse of elites. The necessity of sacrifices or anthropomorphic references to God as a man with limbs were types of references that could easily evoke appeal to the principle of accommodation. In the seventeenth and eighteenth centuries, Jesuit missionaries in China sparked a controversy over accommodation when they allowed Chinese converts to pray to Confucius, worship ancestors, and address God as *Tien* (sky). Like the Jesuit missionaries, the sixteenth-century followers of Copernicus made use of the option of accommodation. For them, however, the problem was not the alien belief-systems of a foreign society but the disciplinary hierarchy of the universities in which theology occupied the highest rank.

. . . [L]et us look briefly at four specific classes of biblical passages that were relevant to the Copernican issue – references to the stability of the earth, the sun's motion with respect to the terrestrial horizon, the sun at rest, and the motion of the earth. Both Protestant and Catholic geocentrists customarily cited verses from the first two categories and interpreted them to refer literally to the physical world. Consider, for example, Psalm 93:1: 'The world also is stablished, that it cannot be moved'; or Ecclesiastes 1:4: 'One generation passeth away, and another generation cometh: but the earth abideth for ever'; Ecclesiastes 1:5: 'The sun also ariseth, and the sun goeth down and hasteth to his place where he arose'; Psalm 104:19: 'He appointed the moon for seasons: the sun knoweth his going down.' The literal interpretation of these passages springs from different sources for Protestants and Catholics. For Protestants, such as Melanchthon, it came from a steadfast faith in the inerrancy of the grammatically literal text; for Catholics, such as Tolosani, the literal meaning was legitimated by appeal to the (allegedly unanimous) authority of previous interpreters. In both cases the geocentrists ignored verses from categories three and four.

The Copernicans had available to them two hermeneutical strategies. The first, which we may call 'absolute accommodationism', declares that the verses in all four categories are accommodated to human speech. The virtue of this position is that it draws a radical line of demarcation between biblical hermeneutics and natural philosophy, so that the principles and methods of the one cannot be mixed with those of the other. It is also in keeping with the moderate Christocentric reading of Scripture advocated by the Reformers. Far more dangerous was the second strategy, which we may call 'partial accommodationism', according to which the interpreter provides a literal, *heliostatic* or *geomotive*, construal of either Joshua 10:12–13 or Job 9:6 and then accommodates it to verses conventionally read as geostatic. In the Joshua text we read: 'Then spake Joshua to the Lord in the day when the Lord delivered up the Amorites before the children of Israel, and he said in the sight of Israel, Sun, stand thou still upon Gibeon; and thou, Moon, in the valley of Ajalon. And the sun stood still, and the moon stayed, until the people had avenged themselves upon their enemies.' The construction 'stand still' is certainly plain talk to the senses; thus, the heliocentrist, if he wished to pursue a partial-accommodationist line, must point out that we need not intend the horizon as our reference frame and that the sun could be rotating on its own axis, while remaining at rest at the center of the universe. A similar kind of ambiguity of reference frame is present in the Job text: 'Which shaketh the earth out of her place, and the pillars thereof tremble.' The phrases 'out of her place' and 'tremble' could be taken to denote either diurnal or annual motion or simply the earth quaking. The sixteenth-century Copernicans, perhaps taking the lead from Copernicus's brief remarks about Lactantius, tended to adopt the position of absolute accommodation. [. . .]

3.2 Gary Deason, *Reformation theology and the mechanistic conception of nature**

From Aristotelianism to the mechanical philosophy

Historians of the Scientific Revolution have identified the development of mathematical physics as the watershed separating ancient and modern science. Without demeaning the other achievements of the Scientific Revolution, we are compelled to acknowledge the widespread application of mathematical methods to the physical world as the single most significant change made by the seventeenth century in the scientific tradition that it inherited. From Galileo's formulation of the law of falling bodies to Descartes's programmatic reduction of nature to geometry, to Wren's laws of impact and Huygens's law of centrifugal force, to Newton's *Mathematical Principles of Natural Philosophy*, the seventeenth century progressively and successfully described the world using the tools of mathematics.

The mathematization of nature in the Scientific Revolution represented the reassertion of a Platonic view of mathematics over the view of Aristotle, which had dominated natural philosophy since the thirteenth century. For Plato the highest reality was Intellectual Form, or pure Ideas, embodied imperfectly in physical things and more perfectly in mathematics. Because mathematics reflected truth more perfectly than physics, Platonic science exploited it in the analysis of nature, with the ultimate goal of reducing physical reality to numbers and geometrical shapes. Aristotle rejected Plato's mathematicism, believing that mathematics and physics study separate kinds of objects. For Aristotle the quantities and shapes of mathematics were abstractions from physical entities. They captured certain qualities of material things but left unexplained the true natures, which could not be reduced to mathematics.

Among these irreducible qualities was the natural tendency (*nisus*) of objects to change. For Aristotle the world had within it principles and powers of development. Natural things changed as a result of their inherent tendency to embody more perfectly the rational form or essence that defined them. Since rational essences did not change, the end of nature was the perfect embodiment of changelessness, a goal that is never achieved because of the changeability and obstinacy of matter. Aristotle's world of inherent tendencies, continual transformations, and teleological development eluded mathematical description. Consequently, Aristotelian physics employed philosophical tools rather than mathematical ones.

The successful application of mathematics to the physical world in the seventeenth century called in question the Aristotelian conception of the

---

* Gary Deason, 'Reformation theology and the mechanistic conception of nature', in D. Lindberg and R. Numbers (eds), *God and Nature: Historical Essays on the Encounter between Christianity and Science* (Berkeley, CA: University of California Press, 1986), pp. 167–79.

world and necessitated the development of a new conception that allowed the applicability of mathematics to nature. In developing this new conception, seventeenth-century thinkers did not return directly to Plato, whose description of the world in *Timaeus* had fallen into some disrepute as a result of its association with magic and Paracelsian medicine in the sixteenth century. Instead, they constructed a new view of the world by revising and expanding ancient philosophy into a mechanistic conception of nature.

The mechanical worldview rested on a single, fundamental assumption: *matter is passive*. It possesses no active, internal forces. Nothing in matter compels it to develop or to move toward an ultimate goal. The matter of the seventeenth century possessed only the passive qualities of size, shape, and impenetrability. Change did not result from the operation of internal principles and powers, as in the Aristotelian view; instead, motion was explained by the laws of impact and the new principle of inertia. The seventeenth century replaced Aristotle's conception of nature as an organic being achieving maturity through self-development with the view of nature as a machine whose parts undertook various movements in response to other parts doing the same thing.

In the absence of internal principles governing change, material bodies in the mechanical worldview were controlled by external laws. The laws of nature, as understood by the mechanical philosophers, prescribed the movement and interaction of material bodies without themselves being part of the inherent nature of matter. With the exception of the German philosopher Gottfried Wilhelm von Leibniz (1646–1716), who reiterated Aristotle's belief that the laws of nature must be internal to nature itself, virtually every thinker who accepted mechanistic physics claimed that material bodies followed laws imposed on the world much as good citizens followed laws imposed on society. One major difference between the laws of nature and the laws of society, however, was the emerging recognition of the laws of nature as mathematical. The driving force behind the development of mechanism was the belief that recent discoveries by Kepler, Galileo, Descartes, Stevin, and others of mathematical formulae describing physical phenomena could be given conceptual foundation if nature were seen as a collection of inert material particles governed by external mathematical laws. In half a century, nature came to be seen as a machine, and the natural philosophy of Aristotle gave way to the new promise of mathematics.

Until recently, historians have seen the overthrow of the Aristotelian concept of nature as a change of worldviews antagonistic to Christianity. Centuries before the rise of mechanism, Thomas Aquinas (1224–1274) produced a majestic synthesis of Aristotelian natural philosophy and Christian theology that became a prominent form of Christian thought throughout the medieval period. Thomas interpreted Aristotle's principles inherent in nature as powers instilled there by God, which God used in his providential work. God *cooperated* with natural powers in a way that respected their integrity while accomplishing his purposes. When the mechanists rejected

Aristotle's understanding of nature, they simultaneously rejected the theory of God's cooperation with nature. Older histories of the rise of science and some modern textbook accounts interpret the latter rejection as the expurgation of theological dogma from scientific knowledge. Usually influenced by a positivistic conception of science, these studies disallow any contribution from theology to the Scientific Revolution. For example, Andrew Dickson White's *A History of the Warfare of Science with Theology* (1896) claimed that the growth of mechanism 'cleared away one more series of dogmas'. 'As in so many other results of scientific thinking', White added, 'we have a proof of the inspiration of those great words "THE TRUTH SHALL MAKE YOU FREE".'

The antagonism between theology and science portrayed in White's work has disappeared from certain recent studies. These have argued that mechanism did not entail the rejection of theology but represented a furtherance of theological convictions first expressed by late medieval nominalists. Francis Oakley and Eugene Klaaren have placed the seventeenth-century conception of laws of nature in the history of debates about the freedom of God following the Condemnation of 1277. At issue in the debates was the extent to which an Aristotelian conception of nature limited the freedom of God. The Condemnation asserted that certain Aristotelian claims (such as 'God cannot create a vacuum' or 'God cannot move the universe rectilinearly') implied God's inability to choose freely what kind of world to create. Nominalist theologians imbibed the spirit of the Condemnation and scrutinized the cogency of Aristotle's arguments. From their questioning developed the belief that the Creator did not follow Aristotelian principles out of necessity and that he might have chosen to create the world differently. Oakley and Klaaren see the mechanists' insistence that the laws of nature might have been different as a continuation of the nominalist emphasis on divine freedom. Seen in this light, the mechanists' belief that God imposed laws of nature on the world was not simply a timely answer to the need for conceptual grounding of mathematical methods but also a culmination of theological changes begun four centuries earlier.

This essay further qualifies older accounts by tracing ties between the passivity of matter in the mechanical philosophy and the doctrine of the radical sovereignty of God in Reformation thought, especially in the theologies of Martin Luther (1483–1546) and John Calvin (1509–1564). By *radical* sovereignty of God I mean an understanding of sovereignty peculiar to the Reformers and to some of their followers, such as the English Puritans, which held that God's sovereignty excluded the active contribution of lesser beings to his work. Unlike the medieval theory of cooperation, which held that God's cooperation with lesser beings did not compromise his sovereignty, the Reformation believed that an adequate understanding of sovereignty necessitated the exclusion of any contribution to divine providence from human beings or nature. To protect the glory of God and to avoid making God's actions contingent on the actions of created beings, the

Reformers affirmed the concept of radical sovereignty against the medieval view of accommodating sovereignty, or cooperation.

[...] Two themes have been prominent in recent discussions. The first considers the relationship of Protestant biblical interpretation to new scientific hypotheses, especially the Copernican hypothesis. Studies published in the past few years have revised the views of earlier historians who believed that biblical literalism encouraged Protestants to reject Copernicanism more vehemently than did Catholics. John Dillenberger and Brian Gerrish have argued convincingly that Protestant exegetical theory included many elements allowing for rapprochment between biblical claims and astronomical hypotheses. For example, Luther recognized the significance of contexts of discourse in which the same object can be described in either a religious or a scientific way. The light of the moon, he observed, can be seen by the believer as a sign of divine providence, even though the astronomer may understand it as a reflection of the sun. Similarly, Calvin's influential principle of accommodation attributed a degree of poetic licence to the biblical text, by which divine truths were presented in a nontechnical language for the lay reader. ... Recognizing the accommodation of the text to the general reader, the interpreter could avoid conflict with contemporary astronomy by claiming that the biblical author described the heavens as they appear to the unlearned eye, not as they might be understood by the astronomer.

In addition to the subject of Protestantism and Copernicanism, recent discussion has concerned the 'Merton thesis'. In *Science, Technology, and Society in 17th Century England* (1938), Robert K. Merton focused on the ethos of the Puritan branch of English Calvinism and argued that it provided a context of social values promoting the attractiveness of science as a vocation. The Puritan ethos, Merton believed, was epitomized in the often repeated phrase, 'To the glory of God and the good of mankind'. Science seemed predestined to fulfill these goals. It promised to reveal the intricacies of creation and to improve the standard of living. As a result, between 1640 and 1660, when Puritans were the dominant power in English government, natural science became for the first time an attractive vocation. Merton cited figures showing a sharp rise in numbers of students entering scientific fields during and shortly after this period and concluded that Puritan values had contributed significantly to the growth of science in mid-seventeenth-century England.

Since its initial statement in 1938, the Merton thesis has fueled a large controversy among historians of religion and science. Some critics have seen Merton's sociological analysis as unimportant for the history of science, which, they believe, has more to do with scientific ideas than with values that may have influenced the place of science in society. Others have accepted Merton's general approach but have criticized him for restricting his data to the period 1640–1660 and for taking Richard Baxter as a representative Puritan. Both lines of criticism have resulted in heated debates about the definition of *Puritanism* and *science* in the seventeenth century, about the proper focus for writing the history of either, and about the real significance for science of any putative relation between them. Not even Charles Webster's recent exhaustive study has silenced the controversy. Webster's expanded focus and well-substantiated claims leave little doubt about a relation between Puritanism and science, but questions of method, definition, and significance remain and will continue to be debated for years to come.

Although the last word has not been said, the above-mentioned discussions have resulted in the generally accepted view of a *rapport* between certain aspects of Protestantism and the new science. This rapport, however, cannot be taken as grounds for claiming that the Reformation caused the Scientific Revolution, or even that it was a necessary precondition for the rise of science. Numerous factors having no apparent connection with Protestantism influenced the emergence of science, and modern science probably would have developed had there been no Reformation. Recent studies of Protestantism and science, including this essay, have not located the origins of science in Luther's revolt, but they have disputed earlier views that the Reformation and the Scientific Revolution had nothing in common. Whether we focus on Copernican astronomy, on Puritan values implicit in the new science, or on the passivity of matter in the mechanical worldview, Protestantism possessed qualities of thought and practice that had significant affinities with early modern science.

Justification and radical sovereignty in Reformation thought

In a discussion of threads common to the thought of the Reformation and that of mechanical philosophers, it is tempting to turn immediately to the Reformers' view of the natural world. To do so, however, would suggest that the leaders of the Reformation had interests in nature resembling those of seventeenth-century natural philosophers or that their writings gave comparable emphasis to natural philosophy. In fact, the concerns of Luther and Calvin differed widely from those of the mechanists, and these differences must be taken into account. The theology of the Reformation cannot be said to have contained a systematic or detailed view of nature. Very little of the Reformers' work would count as natural philosophy per se, although some of the insights of natural philosophers were employed in their theological reflections, and some of those reflections held implications for natural philosophy.

The Reformers' own emphasis can be maintained if we turn first to the doctrine of justification, for it was the focus of the Reformation and offers the key to its theology. 'The article of justification', Luther wrote, 'is master and head, lord, governor, and judge over all the various branches of doctrines.' According to Luther, human beings are not justified by endeavoring to becoming righteous but by accepting on faith that God through Christ has made them righteous. . . .

Luther's insistence that passive righteousness formed the heart of the gospel departed sharply from medieval teachings on justification, which acknowledged the active participation of human beings in salvation. For Thomas Aquinas, believers initially received a free gift of grace that enabled them to cooperate with God in their salvation. Once in a state of grace, they performed works of charity that engendered righteousness and enabled union with God. For William of Ockham (1285–1349), God provided the initial infusion of grace only when the believer had done everything possible to perform good works without it. Upon receipt of grace, as in the Thomistic scheme, believers actively participated in their salvation by performing charitable deeds, becoming righteous, and accepting eternal life.

To Luther and the Reformation such teaching compromised God's sovereignty. Even though medieval theology recognized that grace was required for salvation, works were required as well. Salvation would not be awarded unless the pilgrim made an active effort. The Reformers argued that this made salvation contingent on human actions and detracted from God's sufficiency. 'For God is He who dispenses His gifts freely to all', Luther wrote, 'and this is the praise of His deity. But He cannot defend this deity of His against the self-righteous people who are unwilling to accept grace and eternal life from Him freely but want to earn it by their own works. They simply want to rob Him of the glory of His deity.' To maintain God's glory the Reformers emphasized his radical sovereignty. Salvation did not depend on human actions but on God alone. Against common sense and church tradition Luther boldly proclaimed this conclusion about salvation: God does everything, human beings do nothing.

[. . .] For the Reformers, human nature could not be improved by virtuous actions, at least not in any way important for salvation. For this reason they resolutely denounced the use of Aristotle's *Ethics* in theology. In his *Disputation against Scholastic Theology* Luther asserted: 'Virtually the entire *Ethics* of Aristotle is the worst enemy of grace. This in opposition to the scholastics.' 'According to him [Aristotle], righteousness follows upon actions and originates in them. But according to God, righteousness precedes works. . . .' This difference was the crux of the Reformation. For the Reformers any effort aiming to perfect human nature for salvation was built on an overassessment of human ability and an atrophied view of divine grace.

The discerning reader may have anticipated that the mechanical philosophers made the same argument against Aristotle's view of the world. Insofar as Aristotle attributed change in the world to the striving of the world for perfection, the mechanists believed that he attributed too much to nature and not enough to God. Against Aristotle they argued that nature contributed nothing to its formation or development. It was not a being capable of any power or purpose apart from the hand of God. To attribute intrinsic powers and purposes to the world, as Aristotle had done, falsely anthropomorphized nature and detracted from the exclusive role of God in forming and sustaining the world....

God and the natural world

Luther and Calvin based their understanding of God's relation to nature on the belief that God created the world *ex nihilo*. In their view, God did not shape preexistent matter; he merely spoke, and the world was. In creating the firmament and giving motion to the celestial bodies, He did not depend on nature, which, Luther says, 'is incapable of this achievement'. By his Word He called the heavens into existence, and by the same Word He created the things of the earth. Calvin saw the Word as the instrument by which God 'preserves all that He created out of nothing'. Were it not for the continuous presence of the Word, the world would slip back into nothing. [...]

In his discussion of Providence in the *Institutes*, Calvin formulated a systematic view of God's relation to the natural world. He made clear that God's activity in nature is ever-present and that nothing in nature can be attributed to natural causes alone. God sustains the existence of creatures, He invigorates created beings with power and movement, and He determines the ends of natural things and of nature as a whole. Under no circumstances can nature be seen as an independent entity running under its own power toward inherent ends. In the *Institutes*, Calvin repudiated all views of nature that made it a complete or even partial cause of events. Natural things are only instruments through which God acts; He could choose to use different instruments, or none at all. For example, in discussing the sun as a cause of propagation of plants, Calvin pointed out that Genesis describes the creation of herbs and fruits *before* the creation of the sun....

As instruments of God's work, natural things do not have an inherent activity or end. Although they may have received a certain nature or property at creation, this constitutes only a 'tendency' that is ineffective apart from the Word of God. For Calvin, as for Luther, the behavior of a thing depends entirely on God. ... Depending on his purpose, God may command natural things such as water, wind, or trees to behave according to their natures or against them. In both cases their action and end depend on Him. For example, Calvin agreed with Aristotle that earth is heavy and has a natural tendency toward the center of the universe. Water, which is lighter, has less tendency toward the center. 'Why then', Calvin asks, 'does not water...cover the surface of the earth?' Because God holds back the waters to make the earth inhabitable: 'In short, although the natural tendency of the waters is to cover the earth, yet this will not happen because God has established, by his Word, a counteracting law, and as his truth is eternal, this law must remain steadfast.' In contrast, at the time of the flood, God removed the counteracting law and allowed water to follow its natural course....

...As a result of their belief in the radical sovereignty of God, the Reformers rejected Aristotle's view of nature as having intrinsic powers. In place of the Aristotelian definition of nature as 'the principle of motion and change', the Reformers conceived of nature as entirely passive. For them the Word or command of God is the only active principle in the world. The Reformers, moreover, rejected Aristotle's belief that the inherent tendencies of things determine their ends. Whereas Aristotle understood teleological change as 'the fulfillment of what exists potentially', Luther and Calvin denied that the potential of a thing determines its end because only God controls its behavior and purpose. In effect, the Reformers' view of God rendered Aristotelian essentialism pointless by denying that essences contribute causality or purpose to nature. Unlike Thomas Aquinas, who asserted that God's respect for created things imparts 'the dignity of causing' to them, the Reformers took away nature's dignity in order to enhance God's.

...The work of the Reformers differed too much from that of the mechanists for us to see in Reformation thought a nascent philosophy of nature that blossomed, as though inevitably, into a mechanical worldview. In fact, many Protestant thinkers after Luther and Calvin departed from their mentors' extreme formulation of divine sovereignty by returning to an Aristotelian view of nature and restoring the balance between primary and secondary causes. Despite this departure, the doctrine of radical sovereignty lived on among some Protestant groups in France, Holland, and England and, under the complex vicissitudes of the next century, became incorporated into the mechanical philosophy.

The basic source of the ideas of mechanical philosophers was the Renaissance revival of ancient atomism. Although the tenets of atomism had been known to the medieval Schoolmen, their source was Aristotle, whose comments about the atomists in *De caelo* and *De generatione et corruptione* were uniformly critical. The rediscovery of Lucretius's *De rerum natura* in 1417 provided a new source for the atomic philosophy, and by the middle of the sixteenth century there was 'fairly widespread interest in the atomic view of matter'.

Yet the revival of atomism was only the first step in the emergence of the mechanical philosophy. Atomism in its ancient form faced serious problems as a philosophy of nature, problems that Plato and Aristotle had been quick to point out. Foremost among these was the difficulty of accepting the claim that the order and regularity evident in nature originated in the chance encounters of atoms moving in the void. A closely related problem of the atomic doctrine was the association of atomism with atheism. For Leucippus, Democritus, Epicurus, and Lucretius, atoms were not created, but eternal. The gods themselves had been created by the concourse of atoms, they were subject to chance, and they were unable to legislate principles that might give nature a rational order or goal. Unlike other ancient philosophies of nature, atomism, because of its emphasis on randomness and materialism, offered no basis for rationality or purpose in this world. For this reason the revival of atomism in Christian Europe was not a serious possibility until significant changes in the ancient doctrine had been made. The move that circumvented these problems and established atomism as a viable worldview was the introduction of God as a cosmic lawgiver, who imposed laws on atoms for the purpose of creating an orderly universe. By giving God this function the mechanists provided cosmic principles and purposes that had been lacking in the ancient doctrine, removed its atheistic associations, and cleared the way for the establishment of a mechanical worldview.

3.3 Charles Webster, *Puritanism, separatism and science**

[...]

Primitive purity of the Scriptures

Insistence on the authority of the Scriptures was fundamental to the Puritans. This position had widespread intellectual and social implications. At the earliest point in their evolution, Puritans became involved in conflict with orthodox churchmen over the relative authority of the Scriptures and natural reason. The Anglican position, as enunciated by Richard Hooker, would seem to allow greater scope for natural reason, with the Scriptures receiving full acknowledgment but in reality playing a background role. The Scriptures were more to the forefront in Puritan thinking. It might appear that this would result in a blind derivation of scientific principles from biblical sources. But in practice this danger was avoided by the operation of the equally strong principle that there exists complete harmony between natural and revealed truth. This latter idea was ultimately developed systematically by Comenius (1592–1670), whose *pansophia*, which had widespread appeal in Puritan circles, was based on the idea of the acceptance of those scientific principles that might be firmly based on reason, experiment, and the Scriptures.

A more immediate point of friction between the Anglican church authorities and the Puritans was the lack of biblical sanction for many of the practices and forms of church organization incorporated in the Elizabethan settlement. To varying degrees Puritans took exception to rules governing ceremony and dress, they opposed much of the content of the prayer book, and they found objectionable the structure of authority within the church. In place of this fabric they wanted a simple order in conformity with their

---

* Charles Webster, 'Puritanism, separatism and science', in D. Lindberg and R. Numbers (eds), *God and Nature: Historical Essays on the Encounter between Christianity and Science* (Berkeley, CA: University of California Press, 1986), pp. 204–13.

ideas about the primitive church, with ministers more truly accountable to their congregations and services more explicitly based on the recital of psalms and exegesis of the Scriptures.

This program struck at the root of the authority of the Church of England. The Puritans repudiated the Anglican view that the church reserved to itself the discretion to legislate on 'inessentials'. As far as Puritans were concerned, these inessentials, having no scriptural basis, fundamentally corrupted the church with heathen magic. In their writings the rambling and degenerate magical edifice of the Egyptians was contrasted with the purity and simplicity of the early church. Tirades by the Puritans and Separatists against relics of pagan magic provoked the grandiloquent assertion of Weber that Puritanism represented the logical conclusion of the 'great historic process in the development of religion, the elimination of magic from the world'.

Weber may have exaggerated this antimagical tendency, but the antimagical polemic of Puritans did carry over into the sphere of science. For instance, the whole of the Puritan apparatus was employed by Richard Bostocke (*fl.* 1580) in his classic defense of Paracelsianism (a new system of medicine self-consciously designed by its author, Paracelsus [1493–1541], to undermine classical medicine on the basis of premises derived from the Bible). Galenism was censured as a corrupt and baseless system of 'heathenish Philosophie' contrasting with Paracelsianism, which he represented as a return to the purity of Adamic knowledge. Bostocke thus inaugurated a line of attack against ancient systems of knowledge that became a commonplace in Puritan circles and was adopted as a standard part of the armory of the Baconians.

Church authorities regarded Puritan scripturalism as particularly dangerous because it contested the standing of the ancient structure of the church and encouraged a process of reappraisal that was seen as the pathway to infinite regress. Every individual would be offered the temptation to challenge all traditional forms of authority. At an early stage the archbishop of York identified the Puritans as those 'who tread all authority under foot'. Puritanism seemed to its critics to involve a reorientation of mental attitudes that would lead inevitably to infinite regress, not only undermining the basic ideal of the historic unity and continuity of the church but also sanctioning all forms of individual assertiveness and anti-authoritarianism. These predictions were to a great degree fulfilled: the church shattered into a multitude of sects, and the spirit of iconoclasm carried over into political and economic life.

One ultimate effect of the Puritan appeal to biblical authority was to increase the intellectual assertiveness of the lower social orders. Increasing numbers of people from humble backgrounds made their voices heard in religious or secular affairs. One of these was William Petty (1623–1687), who offered the opinion that 'many now hold the Plough, which might have been made fit to steere the State'. True to his severely practical turn of mind, Petty elaborated for the benefit of Hartlib a complex form of technical education that might be made available to all children above the age of seven, 'none to be excluded by reason of the poverty and inability of their parents'. The writings of Hartlib, Dury, and Comenius convinced younger associates such as Petty that universal education was the first priority in the social program of the Puritan state, and their elaboration of a comprehensive and detailed educational plan constituted their major corporate endeavor. This venture represents one of the most highly developed practical testimonies to the Puritan faith in the worth of individual experience; it was relevant to science by virtue of the strong 'realistic' or scientific bias of the form of education proposed for all social classes. A further point of emphasis in the educational schemes was increasing the accessibility of the Scriptures, by means of a variety of endeavors for translation and propagation, embracing the Turkish New Testament, the Lithuanian Bible, the Indian Bible, and the Company for the Propagation of the Gospel in New England. Robert Boyle was involved in all of these projects.

William Petty's career presents an extreme example of audacious self-confidence and success. It also betokens the way in which the new climate of opinion offered inducements to the breaking of traditional barriers and to enterprise in both the secular and religious spheres. Petty selected medicine and practical mathematics as his first vocational choices. It is interesting that both areas had experienced rapid expansion by recruitment from below during the first part of the century. And both were, of course, important for expanding the horizons of the English scientific movement. The medical practitioners opposed Galenism, took up Paracelsus, van Helmont, and other innovators, and expanded involvement in practical chemistry. The mathematical practitioners rapidly made England the center of the making of scientific instruments in Europe. Mathematical practitioners were pioneering a new field of artisan activity untrammeled by monopoly, in which the initiative could be taken without impediment. Medicine, in contrast, involved trespassing into an area traditionally controlled by the College of Physicians, where the intruders faced the authority of the entrenched hierarchy. It is notable that here, as in the sphere of monopolies in general, the same arguments were mobilized on the part of the intruders as were used against the authority of the church. It is important to establish whether the known examples of Puritan and Separatist involvement among the new breed of mathematical and medical practitioners are indicative of a pattern. It is at least likely that radical Puritanism contributed to a favorable climate for developments of this kind in the same way as it encouraged the expansion of interloping trade in the face of the trading monopolies.

Edification

With the Puritan mission to purify the church went the parallel mission to purify the personal and social life of the community of believers. This ideal resulted in the idea of the covenanted church, under which the basic unit was a group of members bound together by voluntary contract. Such a congregation was portrayed as a living temple, in contrast with the 'dead' pagan temples of the traditional church. Members of this temple were pledged to cooperate with one another, so as to utilize their gifts to the maximum effect in the cause of perfecting the organic body in which they were participating. Typically, John Dury (1596–1680) described the aims of the collaborative effort of the English followers of Comenius as the introduction of 'all rules, doctrines and inducements which tend unto the apprehension of fundamentall truthes and performances of duties, whereby the consciences of all conscionable men inwardly and their course of life outwardly may be ordered in the feare of God'. This application of the idea of 'edification', deriving from the Pauline epistles, has been designated by one author as the distinctive Puritan contribution to the understanding of Paul's theology.

The ideal of edification, when pursued to an extreme, resulted in an audacious confidence in the attainability of a millennial utopia founded on the Word of God. Edification in the hands of groups inspired by a sense of operating in a covenant relationship with God provides a context for the understanding of motives underlying the institutionalization of science in mid-seventeenth-century England. Pioneers of organized science, such as John Wilkins, John Wallis, Theodore Haak, and Samuel Hartlib, were nurtured in the context of the 'spiritual brotherhood' of Puritan churchmen and their lay patrons. Their scientific outlook was colored by the inspiration of this brotherhood.

Some of the early scientific clubs were analogous to Continental developments, primarily a convenient mechanism for undertaking research requiring collaborative effort, but most betray deeper motivations. Some took on a distinctly utopian flavor, as in the case of Boyle's 'Invisible College' or Samuel Hartlib's 'Office of Address'. These schemes called for the abandonment of self-interest in favor of pooling information, the aim being to make advanced knowledge available to the republic in a form applicable to the solution of pressing economic and social problems. The Office of Address schemes, which attained a degree of official recognition and patronage while attracting a substantial number of committed adherents, presented an ambitious exemplification of the Puritan idea of edification. In this context it is interesting that Robert Boyle, one of the better-known participants in the Office of Address, published as his first work a brief essay calling for the free communication of useful scientific information.

As a final carryover of the utopianism of the Puritan republic, Henry Oldenburg attempted to infuse something of the universalist reformism of the Office of Address into the Royal Society. Realizing the advantages of a transcendental framework for the Royal Society when it came under attack from the High Church, the Society's latitudinarian apologists made good use of millenarian imagery and tried to represent their elitist scientific club as the realization of Solomon's House. Ironically, this propaganda was disseminated partly to protect the reputations of rehabilitated Puritans such as John Wilkins, founder of the Society and brother-in-law of Oliver Cromwell.

Puritans in power

It is now necessary briefly to consider the impact on science of the transmutation of Puritanism from a position of opposition to one of power (1649–1660). In general, the Puritan ascendancy has been regarded as a dismal setback for developments in the cultural sphere, partly by virtue of the inevitable effects of the breakdown of civil order, partly because of the supposed anti-intellectualism of the Puritans. Granted, not all of the changes occurring under the new Puritan regime operated in favor of science. It is easy to enumerate losses to science occasioned by the revolution. William Gascoigne and Thomas Johnson were killed during the hostilities. William Harvey's scientific papers are supposed to have been destroyed by London mobs. The universities were dislocated. Harvey and a host of minor and aspiring natural philosophers were driven out of Oxford and Cambridge. The College of Physicians was not abolished, but its authority was largely destroyed. Numerous scientists went into retirement or exile. It is even arguable that in the two decades before the restoration of the monarchy in 1660, no major scientific work was published. Such adverse trends might be regarded as a characteristic legacy of Puritanism as it is popularly conceived.

More realistically, the above list of casualties is characteristic of any civil war or revolution. And, as for other revolutions, a case can be made on the other side. The loss of life occasioned by the wars was insignificant. Very few intellectuals on either side were killed. Institutions such as Oxford and Cambridge were disrupted much less by the new parliamentarian authorities than by the royalist occupation. The Fellows of the College of Physicians were left to their own devices. Academic traditions were protected, and the enemies of Parliament were treated lightly. Many returned to academic life or suffered no major inconvenience. Because they were excluded from politics and because the door to ecclesiastical preferment was closed to them, many ambitious young men of royalist conviction were diverted into science as a pastime or, alternatively, entered upon medical careers. For the first time the medical faculties at Oxford and Cambridge became viable and modern in outlook, if anything outstripping every other comparable institution in Europe. Royalists remained active in scientific clubs, within or outside the universities.

The civil service under the parliamentarians drew extensively on the abilities of intellectuals, and such policies of the new regime as the suppression of monopolies, the imposition of the Navigation Act, the settlement of Ireland, and the relaxation of censorship provided fresh career opportunities and other positive inducements to effort in science and technology. The Puritan ascendancy marked the beginning of unhindered pursuit of *philosophia libera*, the 'free philosophy', in which adherents of every brand of philosophy, ancient or modern, were allowed totally free expression. This situation contrasts strongly with England before 1640 and with the Continent, where tight restrictions were placed on education, publication, and religious expression by both ecclesiastical and secular authorities.

Perhaps one of the major cultural contributions of Puritanism before 1660 was the constant challenge it offered to authoritarianism within the church and generally in secular life. Before 1640 the Church of England suffered the inconvenience of a crisis of identity; it possessed little internal coherence; and even when assisted by outside authority it failed to control dissent. This open competition was even more pronounced after 1640, and it acted as a spur to initiative for all parties. The support of the people needed to be won by persuasion rather than coercion. The examples cited above suggest that science was one of the beneficiaries of this free trade of ideas brought about by the power of nonconformity and the weakening of the establishment. Detailed examination of the biographies of the leaders of science after the Restoration illustrates the strong formative influence of events associated with the revolution.

Notwithstanding the relevance to science of Puritanism and the Puritan revolution, it is important not to exaggerate the unity of outlook of scientists falling within the orbit of Puritan influence. As in other spheres of religious and secular life, the Puritans and Separatists failed to establish a common platform capable of sustaining the coherence of their movement throughout the revolutionary decades. The greatest unifying factor was provided by a generalized adherence to certain ideas of Bacon, whose anti-Scholasticism, inductive methodology, experimental philosophy, natural histories, and utilitarianism attracted widespread enthusiasm. Otherwise the Puritan advocates of the new science, like their nonscientific brethren, represented an unstable coalition of interests, having very different views on religious, social, and scientific questions.

Thus old associations, formed during the broadly based opposition to the regime of Archbishop Laud, dissolved and ended in indifference or active antagonism. The intellectual alignments of 1660 were thus very different from those evident in 1642. Even at the earliest stage, the embryonic scientific organizations suggest a parting of the ways among the Puritan intellectuals. The '1645 Group', the ultimate ancestor of the Royal Society, possessed, in spite of its instigator, Theodore Haak, a firm 'modern' bias; its membership overlapped very little with the more utilitarian Office of Address of Samuel Hartlib and the 'Invisible College' of Robert Boyle and Benjamin Worsley. The latter two groups coalesced and remained centered in London, while their colleagues from the 1645 Group were conspicuous in securing university preferment. This in turn provided a further point of divergence. Hartlib, Dury, and their colleagues were dissatisfied with the universities, and on matters of policy there was little to separate them from John Webster, the universities' most extreme critic, regarded at the time as a dangerous fanatic. John Wilkins, Seth Ward, Henry More, and their academic colleagues recognized the attack on the universities both as an assault on the idea of a learned ministry and the organized national church and as a threat to social stability. Webster's attack on the Puritan establishment, *Academiarum Examen* (1654), and the reply from Wilkins and Ward, *Vindiciae Academiarum* (1654), are indicators of deep division within the ranks of Puritan natural philosophers.

These differences are reflected in scientific interests and even in philosophical alignments. More and his fellow Platonists moved sharply away from the utilitarianism of Hartlib and Petty; Worsley and the Hartlib group were offended by the 'Oxford professors'' lack of sympathy for astrology. Paracelsianism, Helmontian chemistry, and the pansophism of Comenius were central to the outlook of Hartlib's circle, peripheral to that of their colleagues at Oxford. The ideas of Harvey, Galileo, Descartes, and Gassendi were taken up within the universities but largely bypassed by Hartlib's group. Even approaches to the utilitarian application of knowledge show sharp differences, the effort of Hartlib and his associates being organized with respect to their wider political and social program.

In summary, the more radical Puritans, exemplified by Hartlib and Webster, adopted a more Biblicist metaphysical standpoint, and at the practical level their priorities were dictated by their mission of a godly utopia. Puritans with ties to the universities showed more receptivity to the varieties of mechanical philosophy being developed on the Continent, and their practical work increasingly related to problems in the field of the modern physical sciences. Both sides believed that they represented the central position of Puritanism. In this situation it is not surprising to find that Hartlib and his associates drifted into partnership with Separatists and hence into a minority status, as the semi-official servants of a collapsing republic. In contrast, their colleagues within the universities rapidly made common cause with other occupants of the middle ground within the church, many of whom had remained loyal to the monarchy. Their protection of the fabric of essential institutions earned them rewards at the Restoration. This realignment was symbolized by the association of Wilkins and Ward in replying to critics of the universities. In this process, figures like Petty, Boyle, and Oldenburg were left with divided loyalties, retaining residual links with the more radical Puritans engaged in science for the sake of social engineering, while becoming more deeply involved in problem-solving activities connected with the mechanical philosophy. It is tempting to characterize the divisions that emerged within Puritan science as a dichotomy between occultists and radicals on the one hand and the mechanists and conservatives on the other, the former representing an ideology generated by revolutionary fervor, the latter looking forward to the form of modern science associated with the latitudinarians of the Restoration settlement. The one was eager to generate social change, the other to buttress social stability.

Such an analysis usefully underlines the diversity of Puritan science and emphasizes the degree to which Restoration science was rooted in the experiences of the revolutionary period. It is safe to identify the development of the mechanical philosophy in the 'Anglican' middle ground, as long as it is not claimed that the mechanical philosophy was the prerogative solely of this Anglican center. It would also be wrong to overlook the tendency on the part of more conservative natural philosophers to turn decisively against the mechanical philosophy, as did the Cambridge Platonists; and Anglicans might even occupy a frankly occultist standpoint from the outset, as did Elias Ashmole, John Aubrey, Robert Turner, and Robert Plot.

Regardless of the difficulties pointed out above, the temptation to regard occultism as the property of the more radical Puritans and Separatists, and the mechanical philosophy as the creation of the latitudinarian Anglican middle ground, has proved too difficult for current researchers to resist. Such a thesis transforms the mechanical philosophy into a political weapon, self-consciously forged with a view to sweeping away the republic and restoring a stable monarchy. In this interpretation there is an emphasis on the absorption of latitudinarian proponents of the mechanical philosophy into the Restoration establishment and the consigning of occultists to the radical underground.

As has been indicated, a strong case can be made for the fragmentation of the Puritan movement during the republic. Scientists well illustrate this tendency. But the attempt to view such disparate figures as Robert Boyle and Walter Charleton as joint architects during the 1650s of a new 'Anglican' scientific synthesis is based on supposition rather than direct evidence. It is also doubtful whether before 1660 English natural philosophers had any real sense that the Restoration was about to occur. Even as late as 2 March 1660, Boyle's close friend Oldenburg was confidently predicting that events were moving toward a republican settlement. At this time Oldenburg's more radical Puritan friends were actively planning for a paradise on earth, for the renewal of the church, and for organization of science for an imminent age of 'Universall love, & Universall Commerce, in mutuall peace & noble Communications'. There was no expectation of the restoration of any monarchy other than the monarchy of Christ's personal reign on earth, and for these reformers the restoration of the Church of England and formation of a Royal Society must have seemed contrary to the course of history. History, of course, very soon proved the republicans wrong. The events of 1660 firmly closed the door on further social and political experimentation. Out of the ashes of the scientific movement that evolved under Puritanism may have emerged many of the constituent elements in the science of the new age, but the process of absorption was selective. Many of the dominant personalities and themes of the science of the republicans fell into the background; the creative energies of science were redirected, largely, it seems, to the benefit of the modern worldview.

Conclusion

It would be totally unrealistic to write the history of the emergence of English science in the period 1560–1660 in terms of monocausal derivation with respect to Protestantism, Anglicanism, or Puritanism, however they are defined. But to equate this history with the story of the great discoveries, or to construe science as an entirely autonomous development, unrelated to the Reformation and the Puritan revolution or to the socioeconomic framework of which Puritanism was a constituent element, is to eliminate vital factors in the explanatory mosaic. Any truly historical account of the Scientific Revolution must pay due attention to the deep interpenetration of scientific and religious ideas. It would seem perverse to deny religious motivation in the numerous cases where this was made explicit by the scientists themselves, often with painful emphasis. No direction of energy toward science was undertaken without the assurance of Christian conscience, and no conceptual move was risked without confidence in its consistency with the Protestant idea of providence. Protestantism constituted a vigorous directive force, and within English Protestantism, Puritanism and Separatism generated the challenging new ideas of the day. It is therefore hardly surprising that Puritanism and Separatism should provide a dominant cultural context, reference to which is essential for any balanced characterization of the English scientific movement. Puritanism and Separatism, along with the whole set of conditions associated with the revolutionary decades, assume particular importance in explaining the distinctiveness, diversity, and creativity of English science on the eve of the foundation of the Royal Society.

Chapter Four
Crisis in Italy

4.1 Mario Biagioli, *Galileo as court philosopher**

Court culture and the legitimation of science

That Galileo spent most of his mature life at the Medici court as mathematician and philosopher of the grand duke of Tuscany is certainly no news. However, that Galileo's courtly role was integral to his science is something that has not attracted the attention of historians....

[...] This book [*Galileo, Courtier*] traces Galileo's court-based articulation of the new socioprofessional identity of the 'new philosopher' or 'philosophical astronomer' and analyzes the relationship between this identity and Galileo's work. It does so by reconstructing the culture and codes of courtly behavior that framed Galileo's everyday practices, his texts, his presentation of himself and his discoveries, and his interaction with other courtiers, patrons, mathematicians, and philosophers. *Galileo, Courtier* is neither a biography nor a social history of Galileo's career. Although I follow Galileo through several non-consecutive years and scientific disputes, and analyze several of his texts, my chief aim is to provide a detailed, sometimes microscopic, study of the structures of his daily activities and concerns and to show how these framed his scientific activities. [...]

Also, I do not wish to claim that the point of view adopted here can make sense of Galileo's entire career and texts. Galileo did not begin his professional life as a courtier but became one in 1610 when he was forty-six years old. While I would argue that a move to court was something Galileo may have considered much earlier than 1610, only some of his earlier work was framed by court culture or targeted at a court audience. Similarly, court culture did not equally affect all his many scientific interests. As we will see,

* Mario Biagioli, *Galileo, Courtier: the Practice of Science in the Culture of Absolutism* (Chicago, IL: University of Chicago Press, 1993), Prologue, chapter 6.

while there was a close symbiosis between Galileo's work in astronomy and his court career, his interest in mechanics did not fit the court environment particularly well. ... I do...not...present this interpretive framework as something that can cover the entire scientific revolution, but only to probe the boundaries of its applicability and to identify areas for further research.... I have decided to concentrate on those processes of legitimation that relied on the representation of the new science as something fitting the culture of princes, patrons, and courtiers.

...Galileo is presented not only as a rational manipulator of the patronage machinery but also as somebody whose discourse, motivations, and intellectual choices were informed by the patronage culture in which he operated throughout his life. Not only was Galileo's style embedded in court culture, but, as I hope to make clear by the end of the book, his increasing commitment to Copernicanism and his self-fashioning as a successful court client fed on each other.

...I want to emphasize how he [Galileo] used the resources he perceived in the surrounding environment to *construct* a new socioprofessional identity for himself, to put forward a new natural philosophy, and to develop a courtly audience for it. As indicated by Prince Leopold de' Medici's convening the Accademia del Cimento between 1657 and 1667, Galileo had an impact on Florentine court culture well after his death in 1642.

Obviously, any process of self-fashioning is not without tensions. If the court made possible Galileo's legitimation of his new socioprofessional identity, it also constrained it in ways that, at times, may have collided with Galileo's specific desires. While in certain contexts the fit between Galileo's work and court discourse was remarkable, on other occasions we find irresolvable tensions between patronage strategies and scientific authorship, or between Galileo's attempts to draw the patron on his side to legitimize his scientific claims and the prince's interest in preserving his power and image by not tying them to possibly problematic claims.

These tensions are an ongoing theme of the book and are eventually brought to bear on a reinterpretation of Galileo's trial. Without denying the obvious cosmological and theological dimensions of the events of 1633, I suggest that an understanding of court patronage and culture (and the tensions inherent in them) throws new light on this much-examined event. The very processes that allowed Galileo to fashion himself as a successful philosopher and courtier may have informed the dynamics of his trial.

[...] The first part of the book is about Galileo at the Florentine court, while the second looks at his interaction with the Roman court and with the papal prince.

[...] With Galileo's growing commitment to an increasingly difficult legitimation of Copernican astronomy, the Roman court became his most important theater of operation.

[...] Chapter 6 proposes a reinterpretation of Galileo's trial. ... [T]his chapter looks at a typical courtly event – the fall of the courtier – as described in contemporary court treatises. By bringing some aspects of the fall of the courtier to bear on Galileo's trial, I suggest that the events of 1633 were as much the result of a clash between the dynamics and tensions of baroque court society and culture as they were caused by a clash between Thomistic theology and modern cosmology. [...]

Framing Galileo's trial

[...] Given the major gaps in the available documentary evidence, what I propose here cannot be a comprehensive narrative about 'what really happened' but only a possible alternative interpretive framework based on the analysis of patronage and courtly dynamics.... I will try to show that the dynamics that led to Galileo's troubles were typical of a princely court: they resembled what was known as 'the fall of the favorite'.

...Although we do not know much about what was said in the meetings of the Congregation of the Holy Office in 1632–33, we can reconstruct the processes by which careers were launched and destroyed every day at the Roman court. Although we may never know who 'terminated' Galileo and for what reason, we can understand how patronage dynamics set the stage for such a termination.

[...] Unlike other princely courts, Rome was not the seat of a dynasty. Also, one did not need to be of aristocratic origin to become pope. As shown by the election of Maffeo Barberini as Urban VIII, it could happen that a socially undistinguished subject of the Medici could, for the lifespan that was left to him after the election, rule (with great pleasure) over his former ruler. ... [T]his peculiar demographic pattern turned the papal court into a place of extreme competition and of extremely fragile patronage allegiances. It was a place where 'nobody is well enough supported and connected to be sure of not falling under any circumstances', but one where identities and roles (even ones as unusual [as] Galileo's) could be fashioned (or ruined) much easier than in any other court.

Because of the lack of papal dynasties, the election of a new pope involved a drastic redistribution of power at the Roman court. Together with a new pope, a large group of nephews and close clients...would come to power. Then, because the popes tended to reach the pontificate in their old age and because they did not have biological offspring to benefit from the products of their patronage, they tended to use their power and resources very quickly to achieve a vicarious progeny through patronage and in constructing an image of themselves that could last without the support of a dynastic mythology. ...

Quite literally, the popes' careers (or those of less aristocratic clients who tried to climb with them) were a race against time. Unlike secular princes, who might reach power in their early years, popes usually reached their 'exaltation' late in life. And it was not uncommon for a pope to die soon after having reached the pinnacle of power. ... I think that the patronage frenzy of somebody like Urban VIII, of his secretary, Ciampoli, or of Galileo himself, may be better understood in this context.

Galileo demonstrated he had a good understanding of the periods of the revolutions of Roman patronage. In October 1623 (right after Maffeo Barberini's election to the pontificate) he wrote Prince Cesi that the election of Urban VIII was a ... marvelous conjuncture [that] made him think about the possibility of important changes 'within the republic of letters'. Well aware of his age (he was then about sixty years old) Galileo claimed that if these changes 'do not materialize in this marvelous conjuncture, then they will never come about because – at least as far as I am concerned – there is no point in hoping that a similar situation will come around again'. The changes that Galileo was considering were those connected with the legitimation of Copernican astronomy.

Because of these peculiar power cycles, Rome was a place where exceptional social acceleration, aggressive clients, conspicuous expenditure, and great patronage were the norm. In a sociohistorical period in which birth was the greatest determinant of one's social standing and career, Rome was a 'carnevalesque' place where the legitimation of most unlikely social identities was possible. Private gentlemen like Maffeo Barberini could become popes, and mathematicians like Galileo could try to displace philosophers and theologians.

However, the heavy movement of clients to and away from Rome at the beginning of each pontificate indicated that the frequent cycles that made Roman patronage so powerful were also responsible for the extreme instability of Roman court life. For instance, writing in 1624 to Strozzi, Ciampoli – who was then Urban VIII's [secretary] – described his position at the Roman court as powerful yet precarious: 'Now, it could happen at any moment that I could find myself out of this Grand Hotel – as many here call the Pontifical Palace – and would have to get out of Rome for some time and live in a hut. ... These are events that happen to all sorts of people at every new pontificate.'

[...] Although the patronage crises that followed the election of a new pope often ruined previously established careers, they were eagerly awaited by younger, upward-moving clients. To move from one patron to another was no easy task in early seventeenth-century Italy. Because exclusive patronage relationships were represented as personal bonds, to drop a patron could turn out to be an insulting gesture and one which could prove very dangerous for the client. Given this situation, the frequent patronage quakes that shook the Roman court were looked forward to because they provided relatively safe occasions for changing patronage allegiances. ...

Consequently, courtiers looked forward to new pontificates (unless, of course, their patron was in power). In 1609, the Venetian ambassador,

Francesco Contarini, claimed that the Roman court was not too happy with Paul V 'because they worry he may live long', an outcome contrary to the desires of those who 'would like to see new pontificates frequently'. [...]

...[T]here was an upper limit to the period of the wheel of fortune: the length of the pope's life (and of the clients'). Consequently, clients made bets and developed some medium-range strategies and many more short-term tactics to maximize their chances and reduce their risks. But, most of the time, clients had to adjust or even force their plans into the framework produced by these patronage conjunctures. In some cases... some of these strategies and tactics worked. On other occasions, as when Galileo accelerated his attempts to have Copernicanism legitimized after the election of Urban VIII, the forcing of a program to fit the cycles of patronage may have been partially responsible for its eventual failure.

...Paradoxically, while Rome was the seat of a very conservative institution, it was also a place where change and novelty (within that framework) were the norm. But, for the same reasons, it was also the place in which false steps were most common and dangerous.

The 'fall of the favorite'

I would now like to move from the analysis of the instability typical of baroque courts (and of the pope's in particular) to the analysis of a specific mechanism responsible for the high turnover among top courtiers and so-called favorites.

Although little has been written about Roman court life at the beginning of the seventeenth century, we are lucky enough to have a contemporary analysis of court culture written by a client of the Barberinis and dedicated to the pope's brother, Cardinal Antonio. Published in 1624, *Che al savio è convenevole il corteggiare* was not the work of an outsider. As a member of Cardinal Antonio's entourage, Matteo Pellegrini was well connected and accepted in Roman cultural and political circles. [...]

Several chapters of Pellegrini's handbook for the literati seeking a career at court are dedicated to specific forms of courtly dangers. Among them we find 'The Instability of Favor', 'That the Natural Instability of Favor Is in the Interest of the Powerful', 'The Danger Caused by the Favorite's Arrogance', 'That the Favorite Is at Danger from Envy', 'The Dangers of Being the Favorite', and, finally, 'The Fall of the Favorite'....

Pellegrini's description of court patronage ties is permeated with the discourse of love and flirtation. ... The further one climbed toward the top, the more one's career was represented as a form of kinship with the prince, to the point of being poetically represented as an exclusive love relationship. As Pellegrini put it, 'Two lovers cannot enjoy the pleasures of the loved one at the same time. And there is not space for two on the throne of favor.'

[...] The fall of the favorite was not an accident but a normal process. As indicated by Pellegrini, subordinate courtiers could endure their position and work hard to increase their favor with the prince only if they were allowed to have hope of promotion. Routine ruin of favorites provided that hope. ...

The uncertainty of favor was the most powerful tool the prince had to maintain control of the court. Therefore the fall of the favorite was a mechanism that worked both for the prince and for the upward-moving courtiers. The fall of the courtier did not harm the prince. Paradoxically, the prince's power was increased or maintained both by a courtier's becoming the favorite and by his subsequent falling in disgrace. By being successful, the courtier enhanced the prince's image. If he failed, his fall would also help the preservation of the prince's power by allowing him to display it mercilessly and remind the other courtiers of what could happen to them if they misbehaved. [...]

In short, the fall of the favorite was no accident but rather a routine process of 'seasonal rejuvenation' of the court and of 'cleansing' of the prince's power image. ...

Another crucial feature of the fall of the courtier was its suddenness and inexorability. Pellegrini noticed that 'from the summit of favor one does not descend through the same steps which lead to the top. Often nothing stands between one's highest and lowest status.' [...] Once the prince had decided to drop somebody, nobody could save the victim without joining him in his ruin.

The widespread withdrawal of support from the falling courtier fit the interest of the prince by speeding up the courtier's expulsion. In fact, to be effective, the fall of the favorite had to be quick and inexorable. It was only by being absolute that it would be perceived as a sign of the prince's absolute power in determining the fate of his courtiers. ... Literally, the prince represented himself as 'sacrificing' his favorite courtier to 'justice'. Although he was simply dropping his client, the process was ritualistically represented as involving a mixture of rage and sorrow on the part of the prince. ... The instantaneity of the fall of the courtier and the pretexts the prince might produce in order to represent himself betrayed by the favorite were crucial to represent the sudden extraneity of the prince from the client. Precisely because their relationship had been so close ... the ex-favorite had to disappear as soon as possible and the prince had to be represented as completely apart from him and from his 'misdeeds'. [...]

The fall of Galileo

In February 1632 the *Dialogue* was off the press in Florence. Galileo had initially planned to have the book printed in Rome through the Accademia dei Lincei. In the late spring of 1630, he went there with a complete draft of the manuscript to obtain the necessary permits. Although the Master of the Sacred Palace, Father Niccolò Riccardi, issued a provisional imprimatur for the book (so that Galileo could negotiate with printers), he wanted Galileo to introduce a few changes and to write the preface and conclusion in accordance with the pope's instructions. The revised manuscript was to be sent or taken back to Rome in a few months so that Riccardi could give it a final review. Ciampoli would have taken care of the very last changes and, finally, the Lincei would have printed it. However, in August 1630 Prince Cesi died and the Lincei entered a terminal crisis. Moreover, the outbreak of the plague made it difficult to ship the manuscript safely between Florence and Rome.

Delays intervened and Galileo began to press to have the book printed in Florence.... Galileo used his friendship with Riccardi and Niccolini as well as the power of Medici connections and eventually managed to have the final checking of the manuscript transferred from Rome to the Florentine Inquisitor.... The *Dialogue* was reviewed again by the Florentine Inquisitor, approved, and finally printed in February. In April (before the *Dialogue* had arrived in Rome) Ciampoli fell from Urban's grace. He was to leave the city in the following October.

In the summer of 1632 the pope ordered the book to be taken out of circulation and instituted a special commission to investigate Galileo's possible wrongdoings. In the early fall, having considered the report of the special commission, the pope decided to hand the matter over to the Inquisition, which quickly ordered Galileo summoned to Rome. After much delay, Galileo arrived in Rome in February 1633. The process began in April and concluded in June, with Galileo's condemnation to formal imprisonment and to the recitation of the penitential psalms once a week for three years. He was found

> vehemently suspected of heresy, namely for having held and believed a doctrine which is false and contrary to the divine and Holy Scripture: that the sun is the center of the world and does not move from east to west, and the earth moves and is not the center of the world, and that one may hold and defend as probable an opinion after it has been declared and defined contrary to the Holy Scripture.

Eventually, Galileo was allowed to go back home to Arcetri in the Florentine hills, where he remained under house arrest until his death in 1642.

The interpretation of the sentence is not as straightforward as it may seem. From other available documents we see that the pope and the Holy Office leveled a number of different accusations at Galileo, and it is difficult to assess which could have legitimately led to Galileo's sentencing and which were little more than juridical pretexts. Galileo had been variously accused of having violated the publication agreements, insulting the pope by having the stupid Simplicio express the doctrine of God's omnipotence, violating the injunction not to hold or defend Copernicus given to him by Bellarmine in 1616, and presenting Copernicus's ideas not hypothetically but absolutely. It also has been difficult to evaluate the role of extrajudicial issues such as personal friendship or hostility toward Galileo by the pope, in the Congregation of the Holy Office, and among the theologians and mathematicians of the religious orders, or the effect of the political context on Urban's decisions.

Pellegrini's considerations on the downfall of the great courtier may provide a framework in which to contextualize some of these elements.... I am not claiming that Galileo was Urban's favorite. There were no official favorites in Rome and, although Galileo was well connected at the papal court and visited it every few years, he was not a local courtier. Nevertheless, he did have a special relationship with Urban. Through the years, the two had met and communicated regularly and Maffeo's letters always expressed an unconditional admiration for Galileo. ... When Galileo eventually arrived [in Rome at the beginning of 1624], he was given six audiences with the pope, who presented him with a painting, indulgences, medals, an agnus dei, a promised pension, and a remarkable letter praising Galileo to be shown to the Grand Duke.

The relationship of Galileo and Maffeo Barberini had been one between two independent, non-competing, and very successful individuals, each of whom, by 1623, had reached the top of his career. Urban was Italy's most important prince and Maecenas, and Galileo was its most conspicuous cultural star. Theirs was a relationship that Maffeo seemed to regard almost as a personal kinship, something that bridged gaps in their different statuses and spheres of activity. This was similar to the relationship between princes and favorites. Favorites did not need to be professional courtiers or to emerge through the political ranks. Also, they did not need to have a distinguished background. ... What mattered was their direct and intimate connection with the prince. Although the connection between Maffeo and Galileo cannot be seen in terms of the typical prince–favorite relationship, it shared some of its features. Galileo was not Urban's political favorite, but he certainly was his intellectual one.

...While I am not proposing a strict analogy between the fall of the courtier and Galileo's trial, I want to use their similarity as a heuristic device to uncover patronage-laden aspects of the trial that have been left unnoticed by received interpretations.

The analysis will focus on two dimensions of the fall of the favorite. The first is the patron's use of the trope of betrayal to justify getting rid of a formerly close client. The second is how, in order to preserve the image of the prince's power as absolute, the fall of the favorite too must be presented as 'absolute'....

Galileo's trial is indirectly connected to Ciampoli's falling from grace with Urban in the spring of 1632, an event that still remains largely unexplained. [...]

By the time Ciampoli was dismissed... Urban was facing a serious and delicate political crisis. He was politically weakened and sensitive to accusations of leniency toward heretics (among whom some people could have later put Galileo). He needed to show he was a firm, decisive, and great papal prince. Finally, he seemed to be undergoing a psychologically difficult period in which he perceived himself as surrounded by enemies. . . .

Although there seems to be no direct link between Ciampoli's fall and Galileo's trial (except that they both took place during the same difficult political predicament), Urban later used Ciampoli as a useful scapegoat during the Galileo affair, acting like Pellegrini's enraged prince who drops the 'betraying' favorite while claiming to regret doing so. On many occasions the Florentine ambassador, Niccolini, reported Urban's rage at Galileo and Ciampoli for having 'fraudulently' transgressed their agreements on the publication of the *Dialogue*: 'While we were discussing those delicate subjects of the Holy Office, His Holiness exploded into great anger, and suddenly he told me that even our Galileo had dared enter where he should have not have, into the most serious and dangerous subjects which could be stirred up at this time.' Urban's rages and accusations of betrayal became frequent. . . .

At the same time, in a pattern that accords with Pellegrini's analysis, Urban repeatedly acknowledged the sorrow that all this was causing him because of the closeness and familiarity he had had with Galileo. As he told the Tuscan ambassador, 'Galileo had been his friend, they had conversed and dined several times together familiarly, and he was sorry to have to displease him, but one was dealing with the interest of the faith and religion.' As in Pellegrini, the prince does not represent himself as dropping his clients for his own personal interests but rather as somebody who is forced to give up his close (but betraying) friend in allegiance to some higher ideal (justice, religion, peace, etc.). Self-interest was turned into purity.

[. . .] There is a further point in Pellegrini's analysis that may help us detect an interesting pattern in the later developments of the trial.

. . . Pellegrini . . . stressed the naiveté of a falling courtier who tries to argue for his innocence and of other courtiers who may try to help him. Once the downfall had begun, it could not be stopped. The prince's power was at stake in the absoluteness of the fall. . . . The best strategy was that taken by Ciampoli, that is, to keep quiet and allow himself to be used as a sacrificial lamb to the goddess of Fortune while waiting for the election of a different prince.

Similarly, in October 1632, the Florentine ambassador, Niccolini (somebody who had much more experience than Galileo in Roman courtly matters), wrote to Galileo in Florence that there was no point trying to argue against the claims of the Holy Office. . . . He made the same point a few months later at the height of the trial: 'Galileo tries to defend his opinions very strongly; but I exhorted him, in the interest of a quick resolution, not to bother maintaining them and to submit to what he sees they want him to hold or believe about that detail of the earth's motion. He was extremely distressed by this. . . .'

Because of the papal power invested in the trial, the Holy Office expected Galileo to confess, not to argue. As with Pellegrini's disgraced favorite, it was pointless and even harmful for Galileo to try to stop his fall. The proceedings of the trial support this view. Whenever the Holy Office needed to negotiate with Galileo (rather than confront him with pre-arranged questions) it did so under the table and outside the official trial proceedings. A judicial body representing the power of an absolute prince could not show itself openly negotiating with a criminal. [...] Galileo confessed, and the trial headed for a speedy conclusion....

In a sense, the falling favorite could not be given a fair hearing. Galileo's trial was no trial in the modern sense of the word; like most 'falls of favorites' it was something resembling a ritual sacrifice. Precisely because of his closeness to the pope, Galileo was not just another defendant....

The available evidence indicates that Galileo's earlier attempts to explain his actions to the commission and salvage the salvageable by correcting the book before the matter reached the Inquisition were ignored. When in September 1632 Niccolini mentioned to a representative of the pope that the grand duke would have liked Galileo to have a chance to defend himself *before* any trial was started, he was told that 'the Holy Office is not in the habit of hearing defenses'. While this response may have reflected the Holy Office's procedures, by the time Niccolini made the request the Holy Office had not yet started its proceedings against Galileo but was still gathering preliminary information through the special commission established by the pope. [...]

...The text of Galileo's final condemnation glossed over the alleged frauds that Galileo and Ciampoli had committed in the process of obtaining the imprimatur and focused mostly on Galileo's alleged violation of the 1616 injunction. Quite conveniently, Bellarmine – the person whose orders Galileo was claimed to have transgressed – was no longer there to contradict this representation of the events.

Although the legal status of the injunction of 1616 was quite dubious, the pope and the Holy Office were able to use it to represent Galileo as solely responsible for all that happened. Also, not only did Urban and his collaborators come out clean, but, precisely because of his recovered purity, the pope could represent his condemnation of Galileo as absolutely just. He was not scapegoating Galileo in his own self-interest. Rather, he was sacrificing a former dear friend to prevent the spread of threatening doctrines that might harm the Church. The pope could be represented not as a self-interested human being but as an all-powerful and just papal prince. That was precisely what a good fall of the favorite was supposed to do for the prince. [...]

Although the outcome of the trial of 1633 has embarrassed the Catholic Church for centuries, it certainly did not hurt Urban. Princes did not get rid of clients to do themselves harm. On the contrary, the condemnation of Galileo exonerated Urban from a possible scandal, and might have helped him refute insinuations about his weakness against heretics and reduce the political pressure on him. Cardinal Borgia, the Spanish ambassador who had threatened to impeach the pope because of his behavior during the Thirty Years War, was among the ten cardinals in the Congregation of the Holy Office and his name was on the final sentence. Although it is true that, by dropping Galileo, Urban lost one of his most prestigious clients, the slot was soon filled. Writing to Galileo in the spring of 1634, Raffaello Magiotti told him of the arrival in Rome of an impressive Jesuit polymath, Athanasius Kircher, the new scientific star who – with his numerous 'gems' – was to occupy the Roman stage for many years. In the long run, the Jesuits did manage to become the stars of courtly natural philosophy in Rome.

Although the rest of Urban's long pontificate continued to be politically controversial and financially deadly for the Papal States, the pope held on to his power effectively. He was what we may call a successful tyrant. . . .

Evidently, what Ciampoli had called the 'crazy conjunctures and retrogressions' of Rome's wheel of fortune worked very well for Urban and his family, propelling them from obscurity to exceptional power and wealth in a matter of years. However, those same dynamics did not bring good luck to Galileo or Ciampoli . . . .

The trial and the structural constraints of court patronage

[. . .] While the peculiar features and power of the Roman court made it very attractive to all ambitious clients, Galileo had additional reasons to seek strong patronage ties with Rome. The pope was the prince whose sacred texts Galileo was trying to reinterpret in order to legitimize his new socioprofessional identity, Copernicanism . . . . Consequently, Galileo could gain much more than other clients from Roman patronage, but for the same reasons he also could lose more.

However, he could not seek patronage through a low-key and more discreet approach because that would have not fitted the patronage codes of a princely Maecenas like Urban. As we have seen, great patrons sought and rewarded conspicuous and consequently controversial clients. Urban's appreciation of the *Assayer* indicates that he subscribed to these cultural codes. Paradoxically, although Galileo had to assume conspicuous and controversial positions . . . to succeed in his courtship of the pope, the reinterpretation of the Scriptures entailed by his work was a most delicate matter, one that should have been approached in the most tactful fashion. However, as Galileo told Cesi in 1623, Urban's pontificate was the last conjuncture he could possibly expect. In 1630, his disciple Cavalieri repeated that point, adding that time was running out for Galileo. Consequently, he had to compress a delicate and probably lengthy process of social and cognitive legitimation to fit the cycles of the only process of legitimation he had access to: princely patronage.

Even more paradoxically . . . the type of conspicuous and controversial cultural production Galileo needed in order to court the pope could be developed only at the expense of the very tradition on which the pope and the Church stood. This was a dangerous game and even a breeze could upset it. And, as we have seen, there were all sorts of breezes in Rome. To make things more difficult, Galileo entered the most delicate phase of the game precisely when his patronage connections with Rome had declined in number and importance. Cesarini, Cesi, and Cardinal Del Monte had died and Ciampoli had been expelled. Most of the strong connections with the Roman court that epitomized the 'marvelous conjuncture' of 1623 had disappeared and Galileo was left with the Medici ambassador as his main (if not only) powerful ally. If the publication of the *Assayer* was in perfect tune with a major conjuncture, that of the *Dialogue* was not. [...]

Urban was a sophisticated courtier, humanist, and poet, not a Scholastic theologian. He was a courtier-pope (and that is why he appreciated Galileo so much). The notion of God's omnipotence provided Urban with a perfect trope for the knowledge that best fit his predicament as a papal prince: it brought together both his cultural and theological concerns. I would say that, initially, Urban was less concerned with defending the Scriptures from Copernicus than with setting up a discourse which would have left the Scriptures alone while providing a space in which he and other sophisticated courtiers could enjoy brilliant philosophical 'gems' produced by authors like Galileo. It was not just safety that Urban had in mind when he wanted Galileo to stress the argument of God's omnipotence: it was also an expression of his good courtly taste. He expected the *Dialogue* to be another virtuoso play of hypotheses like the *Assayer*. Consequently, Urban may have perceived Galileo's unbalanced stance in favor of Copernicus not simply as theologically (and politically) dangerous; it was also a sign of bad taste – something that may have helped ruin his intellectual kinship with Galileo. [...]

4.2 William Eamon, *Patronage and scientific identity in late-Renaissance Italy**

One of the central questions raised by a consideration of science in the European courts can be put as follows: How do changes in the institutional locus of science affect scientific culture, personnel, and ideas? To put it in another and simpler way: What bearing does *where* science is done have upon *what kind of* science gets done, and upon *who* does it? Only recently have historians of science begun to concern themselves in any serious way with the places where scientific knowledge is produced, but... without focusing these themes in a direct way upon traditional questions concerning the

* William Eamon, 'Court, academy, and printing house: patronage and scientific careers in late-Renaissance Italy', in Bruce Moran (ed.), *Patronage and Institutions: Science, Technology and Medicine at the European Court, 1500–1750* (Woodbridge: Boydell Press, 1991), pp. 25–50.

Scientific Revolution. In this essay, I want to try to broaden our perspective upon scientific patronage by addressing some of these traditional questions in light of what is now known about science in the courts, and to focus specifically upon the changes in scientific practice and personnel that occurred when science's center of gravity shifted from universities to courts in late-Renaissance Italy.

I begin with an observation first made by Paolo Rossi, that there emerged in the early modern era an entirely new conception of the nature and aims of scientific inquiry. 'In these centuries', wrote Rossi, 'there was a continuous discussion, with an insistence that bordered on monotony, about a logic of [discovery] conceived as a *venatio*, a hunt – as an attempt to penetrate territories never known or explored before.'...Rossi's observation is borne out by numerous sixteenth- and seventeenth-century references to the hunt as a metaphor for science. In the early sixteenth century the Neapolitan astrologer Giovanni Abioso urged scientists to throw out the books of the ancient philosophers and to 'hunt for new secrets of nature'.... Giambattista Della Porta (1535–1615) used the lynx, the legendary keen-sighted predator, as the emblem for his book on natural magic, and thereby inspired the name of the famous Accademia dei Lincei. Explaining the choice of the lynx as the society's emblem, Francesco Stelluti, one of the original Lincei, wrote that the academy's purpose was not to judge nature according to preconceived theories but to 'penetrate into the inside of things in order to know their causes and the operations of nature that work internally, just as it is said of the lynx that it sees not just what is in front of it, but what is hidden inside'. In the seventeenth century, Sir Francis Bacon compared experimental science to 'Pan's hunt', after the ancient god of hunting....

...Bacon's experimental scientist...was not a logician, but a person 'sagacious in hunting out works'. To these images of science – one as logical demonstration and the other as a hunt – corresponded different images of nature. One conceived of nature as a geometrical order, a work of God that composes the world,...a reality written in the language of mathematics. The other viewed nature as a dense forest, an unknown and boundless ocean, an 'America of secrets', a labyrinth in which method offers a thin thread to orient oneself, 'as if the divine nature enjoyed the kindly innocence of such hide-and-seek, hiding only in order to be found'. Since, according to the latter view, nature's secrets are hidden from ordinary sense perception and from the intellect, they have to be sought out by extraordinary means, by using instruments to 'look out at and look into' ...nature, as the motto of the Lincean Academy expressed it, or by using an experimental methodology, 'twisting the lion's tail' to make her cry out her secrets, as Bacon put it in a characteristically colorful metaphor. The logic of discovery implied by the metaphor of the hunt underscored the inadequacy of the medieval encyclopedia of knowledge, which was now to be put to an experimental test, and the results of that trial published to the world at large.

This transformation of the logic of scientific discovery, I shall argue, was in large measure a consequence of the conditions of patronage in Renaissance society. Specifically, it resulted from a shift in the institutional locus of scientific activity from the universities to princely courts, to academies, and to printing houses – that is, to places intertwined with networks of personal patronage. The argument I wish to advance can be stated in the following way: because the conception of science as a *venatio* mirrored the self-image of the court, princes adopted it as a preferred style of research, in contrast to the ponderous and 'ungentlemanly' manner of scholastic disputation. In doing so, they created a potent cultural ideal, *virtuosity*, to which scientists seeking patronage had to conform. Increasingly courts defined the role of the scientist, who was expected to produce works that glorified the prince and legitimized his power. The shift in the sources of institutional support for science away from the universities and toward institutions intertwined with the patronage system resulted in corresponding changes in scientific personnel. New kinds of professionals entered the scientific workforce, individuals whose contributions were formerly considered to be only ancillary to natural philosophy, including engineers, craftsmen, and mathematicians, as well as those whose activities formerly carried the stigma of the forbidden arts, alchemists, magicians, and investigators of the occult sciences.

Although I am deliberately contrasting university and non-university science, I do not claim that the Italian universities of the early modern period were completely retrograde as far as science was concerned, or that talented scientists no longer sought academic careers. Indeed, for many of Italy's universities the sixteenth century was something of a golden age in which students were plentiful, professors relatively well paid, and the quality of instruction such that large numbers of foreign students left home to study in Italy. Princes and city governments, aware of the economic benefits to be expected from an influx of students, hastened to revive and expand their universities. They erected new buildings, bid ever-higher offers for qualified faculty, and promoted the teaching of new subjects more consonant with the demands of the age. However, the sixteenth-century revival was not gained without cost; nor did it result in sustained growth. Indeed, by the end of the century, the decline was already beginning. Gradually the universities lost their former autonomy and became creatures of government, increasingly dependent for their survival upon government favor and support. In the late-sixteenth and seventeenth centuries, when much of this support was withdrawn, the universities found it difficult to survive. The stifling intellectual atmosphere of the Counter-Reformation also had a negative impact, affecting not only the quality of instruction but also the attractiveness of Italy's schools to foreign students. With the loss of government support, the gradual disappearance of foreign students, and the tightening strictures of religious and intellectual orthodoxy, the Italian universities entered into a prolonged period of stagnation, decline, and provincialization. Possibly this state of affairs played some part in Galileo's decision to give up his chair at the University of Padua and seek a place at the Tuscan court. But, as we shall see, there were even more compelling reasons for an ambitious scientist to seek a princely patron. For as the universities declined, the courts emerged as the dominant institutions for the organization of culture, and increasingly science was tailored to match the needs of the courts.

In any case, the conception of science as a *venatio* developed outside of the universities. Indeed, it was conceived *in opposition to* the methodological assumptions of official academic culture. It appears, but faintly, in works on technical subjects such as mechanics, metallurgy, surgery, and artillery, which made up a significant share of sixteenth-century scientific publications. Vannoccio Biringuccio (1480–1539), in his metallurgical treatise, the *Pirotechnia* (1540), ridiculed the philosophers and schoolmen who 'fed on no other food than speculation', and praised the empirics and the 'divine scrutators' of nature who, 'like eager hunting dogs, have always journeyed throughout all the regions and shores of the world, seeking with all possible care to understand the wonders and powers of Nature'. The same emphasis upon empiricism and upon 'peering into' nature in order to glimpse its innermost secrets appears in works on natural magic, in the popular 'books of secrets' of authors like... Girolamo Ruscelli (1500–1566) and Isabella Cortese (*fl.* 1561), and in the writings of many who made a living primarily as scientific authors, churning out various works on 'practical science' for popular readers. The style of science they practiced was scarcely at all concerned with the analysis of the terms of philosophical discourse. It had a tone of naivete when compared to the subtleties of scholastic disputation. And in contrast to scholasticism, it seemed above all to be concerned with the possibility of dominion over nature. For practitioners of this style of science, the universities offered little by way of incentives or opportunities to teach and do research. Nor, given the disciplinary hierarchy that structured the university, could their role be legitimized in the university setting. On the other hand, the style of science they practiced exactly matched the self-image of the Renaissance prince.

Italy in the late Renaissance was a honeycomb of princely courts. From the largest and most powerful, the papal court at Rome, to the dozens of smaller courts of cardinals, dukes, marquises, oligarchs, and feudal magnates, from the great courts of Ferrara, Mantua, and Milan in the north to the twenty-odd feudal baronies in the Kingdom of Naples, courts influenced culture in every region and in every city of Italy, whether princely or republican. In its strictest sense, the court was the space and personnel around a prince, as he made laws, received ambassadors, dispensed justice, made appointments, took his meals, proceeded through the streets, and hunted in the countryside.... In terms of its official functions the court was the microcosm of the state, carrying out the main business of the state. The court was also the ruler's actual, physical home, where he lived with those who served him, and, when he was away from his palace or villa, the space surrounding his person. A centrifugal structure revolving around the prince, the court comprised the prince and those who served him, not as employees but as recipients of his favor, as 'bodies' at the lord's disposal. Everything and everyone served the prince. Men were there solely to carry out his wishes; they had no social identity apart from the one defined by the fundamental relation of service and lordship. 'The courtier's end', wrote Torquato Tasso in a dialogue on courtly virtues, 'is the reputation and honor of the prince, from which his own reputation and honor flow as a stream from a spring.' The vocabulary of this basic relation, expressed on the one side in the lexicon of obedience and adulation, on the other side in that of command and expectation, defined the conditions of courtly patronage.

The key to this relation was power, which radiated like a magnetic force, attracting and repelling, organizing people and spaces into relations of service and overlordship; power, moreover, that rested not just upon military might, but also upon the act itself of exercising it: in other words, upon theater, spectacle, and ritual. The luxurious ostentation at the courts was no mere show. It was a display of power. Without such an exhibition, there was somehow no sufficient claim to the possession of power. All public occasions – marriages, baptisms, funerals, coronations, state visits, religious processions – gave princes the opportunity to self-glorify and to affirm their titles to rule. The cultivation of self-images made courtly art a form of political propaganda, as princes commissioned works of art representing them as Roman emperors, classical heroes, and deities....

These considerations will help to explain the kind of science the courts tended to produce. For scientists, when they served princes, were subjected to the same conditions of servitude as artists and men of letters. Their fortunes rose and fell with the fortunes of princes, as Vannoccio Biringuccio twice found out when his patrons, the Petrucci family of Siena, were thrown out by popular uprisings. Moreover scientists, when they were at court, were there by the prince's will alone. They were there because they had something to offer him.

The princes' most pressing need, in retrospect, was for technical expertise. For the rebuilding of cities and for the construction of palaces, bridges, and fortifications, they needed engineers, architects, and mathematicians. Antonio Lupicini (1530–1598), a Florentine, built flood control devices for the Medici, advised the Venetian government on the drainage of lagoons, was consulted by the Emperor Rudolf II at Prague, and finally served as an engineer in Grand Duke Ferdinand's army in Hungary. Practical concerns also brought mathematicians into the courts. Changes in military technology, above all the introduction of the cannon and the bastion, created the need for experts on ballistics and fortifications....

If, in retrospect, the princes' primary need was for technical assistance, from a contemporary perspective even more important was 'reputation', since what others thought of them was an important ingredient of what they actually were. Machiavelli, discussing the qualities a prince should display, stressed the political use of cultural patronage, advising that 'a prince ought to show himself a lover of ability, giving employment to able men and honoring those who excel in a particular field'. Above all, Machiavelli went on, 'a prince should endeavor to win the reputation of being a great man of outstanding ability'. Princely virtues, described and extolled ad nauseam in the courtly literature and in dedications of books to patrons, gave rise to a potent cultural ideal, the 'learned prince'. The cult of the learned prince, which deployed the Platonic image of the philosopher-ruler who combines power with wisdom, helped to legitimize the political rule of the Medici dukes and the newly-established... rulers at Milan, Mantua, Ferrara, and Urbino. ...

Among the qualities of the idealized 'learned prince', two stand out as particularly important for their influence upon courtly science: curiosity and virtuosity. The positive valuation of intellectual curiosity is one of the most significant contributions of Renaissance culture to the modern outlook. Although it seems 'natural' to think of curiosity as a positive virtue, indeed as a prerequisite for the discovery of knowledge, for centuries in the Christian tradition intellectual curiosity, or *curiositas,* was condemned as a vice. The medieval polemic against curiosity originated with St. Augustine, who in the *Confessions* included *curiositas* in his catalog of vices, identifying it as one of the three forms of concupiscence that are the beginning of all sin, lust of the flesh, lust of the eyes, and ambition of the world. *Curiositas,* a kind of 'lust of the eyes' (*concupiscentia oculorum*) according to Augustine, was an 'empty longing for acquiring experience through the flesh', a 'lust to find out and know' things beyond our ken, just for the sake of knowing them. The 'disease of curiosity' causes people to gawk at monsters, freaks, and mutilated corpses, and to undertake investigations in natural philosophy: 'From the same motive men proceed to investigate the workings of nature, which is beyond our ken – things which it does no good to know and which men only want to know for the sake of knowing.' According to the patristic outlook, the universe is a mystery because God *intended* it to be one. ... Although *curiositas* referred to any kind of intellectual inquisitiveness carried to excess, magic was the medieval world's paradigmatic example of forbidden knowledge. In the thirteenth century William of Auvergne, decrying the widespread curiosity about magic among university students, warned that magic, a 'passion for knowing unnecessary things'..., was a mortal sin. These conventional beliefs were doubtless the source of the various legends concerning medieval men of learning whose insatiable thirst for knowledge led them to sell their souls to the devil in order to know the secrets of nature. The Renaissance legend of Faust drew heavily from medieval tales told of... Roger Bacon, Albertus Magnus, Robert Grosseteste, and Michael Scot. In the Middle Ages, to pry into the secrets of nature was to trespass the boundary of legitimate knowledge, to challenge God's majesty, and to enter into the territory of forbidden knowledge.

This stern attitude toward curiosity was completely absent from the Renaissance courts. Not only was curiosity considered a virtue worthy of a prince, it was an important symbol of his power. A striking visual demonstration of this can be seen in the numerous 'cabinets of curiosities' (*Wunderkammern*) that were collected and put on public display by the Renaissance princes. Through them the court projected an aura of the uncanny, even superhuman. They glorified the prince and celebrated his deeds and his power. Full of strange and exotic *naturalia*, they were meant to delight and to astonish. As if to mock the Augustinian concept of *curiositas*, they invited viewers to gawk at them, to be amused by them, and to be filled with wonder. Indeed, the court itself was a menagerie of bizarre forms: clowns, dwarfs, and exotic animals mingled with princes and courtiers. In a remarkable transvaluation, monsters, freaks, and 'curiosities' became legitimate objects of scientific investigation.

The new attitude toward curiosity, which affirmed the value of inquisitiveness about nature, generated another important cultural ideal: virtuosity .... The virtuoso, one who is curious about all aspects of nature and art, and one to whom learning is an ornament, was the codification of the 'learned gentleman' described by Castiglione at the court of Urbino. Among other things, virtuosity conditioned courtly discourse, with its delight in variety, surprise, and nonchalance, and its distaste for the tiresome logic-chopping characteristic of scholastic disputation.

The valorization of curiosity and of virtuosity gave rise to two characteristic features of courtly science. The first was the fascination with and the display of... marvels, which is best seen in the princely gardens and cabinets of curiosities. A *studiolo* such as that of Francesco I d'Medici appropriated and reassembled all reality in miniature, symbolically demonstrating the prince's dominion over the entire natural and artificial world. Carved gems, watches, antiques, mummies, and mechanical contrivances were displayed side by side with fossils, shells, giants' teeth, unicorns' horns, and exotic specimens from the New World. The juxtaposition of *artificialia* and *naturalia* made the museum a kind of encyclopedia of the exotic, the bizarre, and the marvelous. Similarly the elaborately constructed botanical gardens, with their complex mathematical designs and their fantastic floral displays, impressed viewers with the prince's power to transform nature itself. In its design and arrangement the botanical garden was a microcosm reflecting a measured and orderly universe at the prince's command and disposal. ... Significantly, many of the public museums and gardens built in the late Renaissance had originated as private *studioli*... which were reorganized and opened to the public. Giuseppe Olmi has pointed out that the transition in the status of the princely collections during the second half of the sixteenth century – from private *studioli* to public museums – was part of a strategy by rulers to consolidate their political power. The highlight of the new cultural policy was the opening of the Uffizi Gallery in Florence in 1584 by Francesco I d'Medici, who dismantled his *studiolo* and transfered many of its objects to the new gallery. The need to legitimize the Grand Duke and his dynasty meant that his deeds and his power had to be displayed publicly and impressed upon the minds of subjects.

The second outstanding feature of courtly science was the abiding interest in the 'secrets' of nature, and especially with the subjects of alchemy and magic. Not only in Italy but in courts throughout Europe, 'secrets' were passionately sought after. A Venetian observer at the imperial court at Prague reported that Rudolf II 'delights in hearing secrets about things both natural and artificial, and whoever is able to deal in such matters will always find the ear of the Emperor ready'. The Medici Grand Dukes were avid seekers of alchemical, magical, and technical secrets. Cosimo I and Francesco I both dabbled in alchemical experimentation in a laboratory at the Ducal palace. Arriving there early in the morning and staying until late at night, sometimes carrying on governmental business there, Francesco experimented with preparing porcelain, enamel, and *majolica*, concocting medical remedies, making incendiaries, counterfeiting jewels, and many other secrets. Even on his deathbed, Francesco insisted on treating himself with his own concoctions. . . . These same interests were shared by Cosimo II d'Medici, Galileo's patron, whose court at Florence was a magnet for every sort of 'experimenter' who claimed to have some new 'secret'. Galileo, eager to win Cosimo's patronage, assured the Grand Duke's secretary that he had 'particular secrets, as useful as they are curious and admirable, . . . in great plenty. Their very abundance has worked to my disadvantage, . . . for had I but a single one of these I should esteem it highly, and with that incentive I could have interested some great ruler, which I have not hitherto done or attempted.'

What do all these 'secrets' and experiments signify? On the one hand they attest to an interest in applied science, for many recorded experimental attempts to improve artistic or technological processes. But the preoccupation with secret recipes, magic, and esoterica also had a political purpose, in that it represented the prince as a repository of praeternatural, superhuman secrets, and as the rightful heir to a tradition of esoteric and hidden wisdom. In this regard, the hunt was a particularly suitable metaphor for courtly science. Just as hunting, the best-loved pastime of courtly society, demonstrated in a spectacular fashion that the goods of the earth existed first and foremost for the prince, so science carried out as a hunt, that is, as a capturing of rare secrets, demonstrated that nature's occult qualities existed for the use and delight of the prince. The naked swagger of power that characterized politics in late-Renaissance Italy was acted out wherever the prince showed himself in public, flaunting his eminence and authority. Nowhere was it demonstrated in a more spectacular fashion than in the hunts staged for the court nobility. The expeditions organized by the 'hunting pope', Leo X, were away for months at a time, and included caravans of courtiers, musicians, attendants, and huntsmen, along with their horses, dogs, leopards, and falcons. . . .

The metaphor of the hunt also penetrated courtly scientific discourse, and consequently influenced the fashioning of scientific careers. The hunt displayed nature's variety, pleasure, and surprise. ... So it was with scientific discourse at the court, where the performance counted for more than the content of disputes: unlike the scholastic argument from necessary causes, a 'courtly' argument was characterized by surprise, variety, paradox, spicy witticisms, and 'good sport'. ... In contrast to scholastic argumentation, courtly discourse did not aim to result in necessary logical conclusions, but to present paradoxes and riddles, just as nature continually surprises the observer with its wonders.

The logic of discovery implied by the metaphor of the hunt also influenced the process by which professional identities were defined and fashioned at the courts. As recent research has shown, gift-exchange was the language of patronage, the 'medium through which patronage relationships were articulated and maintained'. When a client approached a patron he was expected to present an appropriate gift. ... The logic of gift-exchange was thus one of the principal drivers of scientific careers: scientists needed to invent or discover things they could use as gifts to present to their patrons. Secrets, paradoxes, allegories, and other forms of privileged information, whose hidden meaning presented a challenge and promised to surprise and delight the discoverer, were among the most appropriate kinds of gifts a client could offer a patron. Rare specimens of plants and animals, 'mysteriously' behaving objects, alchemical recipes, *mirabilia* and exotica of any kind were all more rewarding gifts to give patrons than practical, technological devices. Perhaps the most spectacular scientific gift of the era was that of the Medicean stars, which Galileo 'presented' to his patron Cosimo d'Medici, catapulting him into the Tuscan court.

One man whose scientific career was shaped almost entirely by courtly culture was the Neapolitan philosopher and magus Giambattista Della Porta. Della Porta never held a university post. He did not even attend a university. The son of a minor Neapolitan nobleman,... he built a large and much-admired collection of sculptures, antiquities, and natural curiosities, which he used in his scientific research. He spent several years in the magnificent court of Cardinal Luigi d'Este at Rome and Ferrara. Other princes sought his services as well. He was approached by Cardinal Orsini, who offered him 150 ducats for one of his secrets, by Cosimo II d'Medici, and by Europe's greatest patron of the occult sciences, Emperor Rudolf II, who sent an envoy to Naples to invite Della Porta to Prague. Eventually his research attracted the glowing admiration of the young and impressionable Prince Federico Cesi (1585–1630), the Marquis of Monticello, founder of the Linceian Academy, who in 1611 installed Della Porta as the head of the Neapolitan branch of the society. Della Porta benefitted handsomely from princely largesse: he recorded that his patrons had paid him more than 100,000 ducats to support his research, a sum about ten times the annual salary of the Spanish viceroy of Naples. There can be only one reason for his success, and that is because his research program corresponded exactly with the expectations of his patrons.

Della Porta, by common consent, was Renaissance Italy's foremost hunter and divulger of secrets. His reputation was entirely the product of a research program dedicated to discovering and announcing (to an appropriate audience) the 'secrets of nature', the occult forces that natural magic exploited. If, as I have argued, court culture was a miming of the prince's self-image, natural magic was courtly science *par excellence*. In his famous work on the subject, the *Magia naturalis* (1558 and 1589), which he dedicated to Cardinal Luigi d'Este, Della Porta declared that natural magic enables man (or more specifically, the prince) to better comprehend the cosmos in order to be able to better govern his kingdom. ... Natural magic, an imperious 'survey of the whole course of nature', befitted the image of a prince. Indeed, the magus was in a sense a prince-in-miniature, a prince over the occult forces of nature. Supposedly he could read the secret signs in plants, animals, and the heavens. He understood the 'physiognomy' of things, knew what their uses were and what the heavens portended. He could penetrate into the very secrets of a man's soul, and he knew how to exploit the occult powers of numbers. Della Porta, a prolific author, wrote authoritatively on all aspects of magic. His voluminous works included treatises on alchemy, astrology, human and plant physiognomy, numerology, cryptography, optics, geometry, and the art of memory, in addition to the encyclopedic *Magia naturalis*, which appeared in several dozen sixteenth- and seventeenth-century editions and translations. He had a reputation as wonderworker and seer. Upon being shown a portrait of Henry IV of France, he dazzled the company by predicting the monarch's violent death. Della Porta cultivated the image of the magus, of a courtly Faust who proposed to capture 'demonic' forces naturally and scientifically, without surrendering his soul to the devil. As carefully as he cultivated his reputation as a magus, he nurtured his relations with princes, sending his patrons gifts of his secrets. He dangled his unpublished *Taumatologia*, containing the 'quintessence' of his research, before Cardinal d'Este, Prince Cesi, and Emperor Rudolf II before finally offering it to Cardinal Federigo Borromeo. Always he assured his patrons that he would not reveal nature's deepest arcana to the public, but would give them to princes alone. He proposed sending the *Taumatologia* to Emperor Rudolf II in manuscript 'so that secrets of such great price should not become profaned by the hands of wretched commoners'.

Although the influence of the courts was vast in the late Renaissance, it would be misleading to represent Italian culture as an exclusively 'court culture'. Equally important for the support of the arts and letters were the many public and private patrons who followed – and sometimes anticipated – the example of princes, whether to glorify their own families or to promote culture for its own sake. Wealthy merchants and bankers paid handsome commissions for works of art. ... Pietro Aretino supported poets and humanists. The Senate of Venice endowed public lectureships in mathematics and philosophy.... The printing press also stimulated the production of culture: the center of Venetian intellectual life in the early sixteenth century was the publishing house of Aldo Manuzio. The wide variety of individuals to whom printed books were dedicated – not only to princely patrons, but also to physicians, lawyers, pharmacists, and even surgeons – suggests that courtly patronage was only one form, if indeed the most important form, of cultural organization.

The most widespread alternative form of cultural organization was the academy. Practically every Italian city of at least medium size had one or more of them. ... The academies flourished primarily in the period after 1540. There were two principal reasons for this. The first was a crisis of patronage, which is noticeable from about the middle of the sixteenth century. The extravagant expenditure of Italy's overambitious ruling class eventually took its toll. With alarming frequency princes, pressed for cash and continually in arrears, abandoned their clients, leaving behind a long train of unfinished projects and making the precariousness of the patronage system glaringly obvious. The second and more compelling reason was the stultifying intellectual climate of the Counter-Reformation. The academies responded fairly effectively to this changing cultural climate. The academic organizational model was adaptable to a wide variety of intellectual activities, from the fine arts to literature to science. Moreover, the academies drew men of wide-ranging interests [...].

...[T]he vernacular book market significantly broadened the audience for science. The immense popularity of the 'books of secrets' in sixteenth-century Italy suggests that the conception of science as a *venatio* penetrated popular culture as well as the court and the academies. The printing house may not have been as lucrative or as prestigious as the court, but it offered greater autonomy to writers who could give the public what it wanted, and it fostered a militant sense of independence among those who succeeded at the writing trade. Leonardo Fioravanti (1517–1588), a Bolognese surgeon-turned-*poligrafo*, was one who did succeed, and became through his publications the object virtually of a cult following. Fioravanti skillfully exploited the printing press, advertising with the most extravagant claims his medical secrets, which he dispensed in the local pharmacies. Unlike Galileo, he was never a courtier. It was not for lack of trying, for in 1560, after publishing his first book, he wrote to the Grand Duke of Tuscany announcing his discovery of 'many secrets never before heard of or seen by anyone', which he promised would be useful in the war against 'the great dog of a Turk'. Rebuffed by the Florentine court, Fioravanti devoted himself entirely to writing and to his medical practice, and from his safe haven in Venice lashed out against physicians, professors, Protestants, and courtiers.

Fioravanti succeeded because he knew the rules of the game. In vernacular publishing, the marketplace created certain expectations, which in turn had a direct impact on the production of culture. A general rule of scientific publishing was that the more novel a book claimed itself to be, the greater its chances were of succeeding in the marketplace. . . .

With the establishment in 1657 of the Accademia del Cimento by Prince Leopold of Tuscany, the court, academy, and printing house converged. As already noted, a great deal of scientific activity had been going on in the Florentine court, mostly involving experiments on various 'secrets'. Prince Leopold organized the academy as a way of giving focus to this sporadic activity. Not only did he lavishly fund the academy, he participated actively in its research, and laid the plans for the publication of its experiments. . . .

. . . [W]e seem to be on the verge of something quite new: the conception of science as public knowledge. The methodology by which the Cimento proceeded was to admit only such assertions as were testable by experiment, to 'test and re-test' them (as their motto, 'provando e riprovando', proclaimed), and to communicate the results to the 'republic of letters', inviting other savants to duplicate the experiments. The conception of science as public knowledge has a complex history behind it. Doubtless the invention of printing was momentous, but the court also has an important place in its development. For in no sense was this apparently 'democratic' style of science incompatible with courtly culture. . . . Courtly science may well have been a show, a gesture that served ultimately political ends. However, precisely because of that, because of the need to display, the court promoted its own version of the conception of science as public knowledge. To stretch my analogy of science as a *venatio* a little further, but not I think in an unmeaningful manner, just as the hunt ended with the *massacre*, or public display of the kill, so the secrets of nature were, so to speak, put on display in princely gardens, museums, and in printed books dedicated to princes. Even if it was all meant only to glorify the prince, the secrets of nature became a little less secret thereby, and public scrutiny of discoveries gained the sanction of official culture.

Chapter Five
Iberian Science: Navigation, Empire and Counter-Reformation

5.1 J.-M. López Piñero, *Paracelsian influence in Spain**

At the beginning of the last third of the 16th century the satisfactory panorama of Spanish medicine and science began to change its character. . . .

The most pernicious consequence of the triumph of the Counter-Reformation was . . . the ideological isolation of our country [Spain], imposed with the object of defending it against heterodox ideas. This historical process, which might be symbolised by Philip II's forbidding Spaniards to study abroad, deprived Spanish medicine and science of all the normal means of communication with Europe, precisely at a time when the latter was undergoing a decisive phase of transformation. [. . .]

The incommunication, however, was not complete enough to prevent the works of Paracelsus and those of his more outstanding followers from being read in Spain. In order to gain an idea of the works in circulation, it is useful to consult the records of a Spanish medical library such as that of the Faculty of Medicine of Madrid, which possesses more than a thousand tomes dating from before 1600. Limiting ourselves to the 16th century, we can find in it the Latin edition of Paracelsus' work printed by Perna (Basle, 1575), a Latin translation (Strasbourg, 1573), and a French one (Lyon, 1589) of *Die grosse Wundarznei*, two Latin editions of *Labyrinthus medicorum errantium* (Nuremberg, 1553 and Hannover, 1599), *De Vita Longa libri V* with an introduction by Bodenstein (Basle, 1562) and the *Fasciculus Paracelsicae medicinae* of G. Dorn (Frankfurt, 1581). There are also works by other Paracelsists like the Italian Fioravanti, the Swiss Penotus, and the Englishman

* J.-M. López Piñero, 'Paracelsus and his work in 16th and 17th century Spain', *Clio Medica*, 8 (1973): 119–31.

Muffet, the writings of the Frenchmen Grevin and Launay on antimony.... Very similar to these are the works of Paracelsus in the National Library of Madrid....

Against what one might expect, the Spanish Inquisition treated Paracelsus' works with moderation during the 16th century. The *Index et Catalogus Librorum prohibitorum* (1583) of the Inquisitor Gaspar de Quiroga merely demanded that his surgical writings be expurgated of particular paragraphs with religious allusions. In the *Index Librorum expurgatorum* (1584) of the same author, the expurgation is limited to passages of the *Chirurgia minor*, and the same applied to the *Index* of 1608. The role of the Inquisition in the scientific isolation of Spain was more complex than might be imagined from old romantic historiography. Fables such as the supposed persecution of Vesalius for having carried out autopsies can only be maintained by those who are ignorant of the fact that dissection was a common and perfectly regulated practice in 16th century Spain. In the same way, Paracelsus was not considered a heretic, unlike, for example, Miguel Serveto, whose *opera omnia*, including the strictly medical and scientific works, were absolutely forbidden. The expurgation to which Paracelsus' work was subjected in the 16th century was even milder than that imposed on the celebrated *Examen de Ingenios* of the Spanish physician Juan Huarte de San Juan. All of which does not mean, however, that the Inquisition did not weigh heavily in the process which left our country in a state of cultural incommunication. As the major repressive instrument of the Counter-Reformation, it contributed in a thousand ways, directly and indirectly, towards impeding the free circulation of new ideas and towards making critical and independent thought a calculated risk. It was, without doubt, one of the factors that prevented a Paracelsist movement from developing in Spain and his work from having a more important repercussion among scientists and physicians. [...]

In spite of everything, it is logical that the chief repercussion of Paracelsus' work in Spain should have been in the sphere linking chemistry to medicine. The complex scientific ambience surrounding Philip II is of particular interest in this respect.

Together with numerous scientists and technicians, many alchemists formed part of this circle. There is documental proof that in 1557, 1559 and 1567 several 'maestros' had been engaged to serve Philip II. We also know that in the last years of the century there worked in the Escorial a certain Richard Stanihurst, who in 1593 dedicated to the monarch a work titled *El toque de Alquimia*. It was a hand-written statement destined 'to declare the true and false effects of the art (alchemy) and to aid others to know the false practices of the deceivers and vagabond liars'.

In this ambience we find as well the Bolognese Leonardo Fioravanti, renowned as the leading Italian Paracelsist. Fioravanti's relations with Spain dated from his years in Naples, during which he had become the favourite physician of the Spanish governors of the province. At this period 'alchimisti di diverse nationi' used to gather in his house to carry out their practices. In 1551 the Viceroy of Naples named him private physician to his son, García de Toledo, and in the company of the latter, Fioravanti sailed for Africa in the fleet of Charles V. Years later, he dedicated to Philip II his work titled *Della Fisica* (1592), of which the fourth volume is devoted to alchemy. This book allows us to reconstruct interesting details of his time spent in Spain during the years 1576 and 1577. ... He had dealings with many scientists and physicians and, for example, praises Monardes most highly. But, according to his own statements, in Madrid as well as in Barcelona and Navarra, he was considered at times a 'great doctor' and at others an 'alchemist' and a 'necromancer'. He did in fact maintain close relations with various Spanish alchemists, interchanging a variety of information with them, and reproduces at the end of his *Fisica* the text of the *Coplas sobre la piedra philosophal* of the Valencian Luis de Centelles, which had been supplied to him by an alchemist in Madrid. It is reasonable to assume that Fioravanti contributed to the diffusion among Spanish alchemists of Paracelsus' works, which, as we shall later see, ended by occupying a leading position in their literature.

Within strictly academic medicine, interest in chemistry was still centered on the preparation of medicaments, but with certain significant differences as regards the preceding period. The prevailing traditional 'materia medica' had not prevented a number of new chemical applications to the preparation of medicaments from becoming widely known, especially the production of 'quintessences' by distillation. ... But, above all, the group 'distillers to His Majesty', who turned the pharmacy of El Escorial during the last years of the reign of Philip II into a famous centre for obtaining 'quintessences', stand out from the rest. [...]

In this scientific atmosphere surrounding Philip II, there was a scientist of real importance who had been directly influenced by Paracelsus: the Extremeñan Diego de Santiago, one of the 'distillers to His Majesty'. His book *Arte Separatoria y modo de apartar todos los Licores, que se sacan por vía de Destilación* (1598) is not merely another monograph on distillation, but a work in many ways of interest in the history of European chemistry. Nevertheless, it has been ignored by foreign historians, even in studies of the history of distilling, and Spanish historians themselves are either unaware of its existence or devote it scant attention.

Diego de Santiago, born in the middle of the 16th century in a small village of Extremadura, spent his life working in a laboratory in his place of birth, in Zamora, in Madrid in the service of the king, and in Seville, where he published his book. He includes a detailed study of the instruments, techniques and materials used in chemistry, a summary of the theoretical foundations of chemistry itself and an ample description of its applications of the preparation of medicaments as well as to matters bearing on preserves, wines, the analysis of water, poisons. [...]

The book is, of course, the result of a lifetime's work, 'especially the last twenty years spent with the Distillers to His Majesty, conferring with physicians and carrying out innumerable experiments, on which, with the various instruments of my invention, has been spent what my work has brought me'. If he is posed problems of which he has no practical experience, Santiago is able to state: 'I say nothing of them for I have not experienced them.' Consequently, it is reasonable that his work should be almost totally lacking in quotations from other authors, since 'when something is seen, there is no need of authorities or allegations'. Because of this, the sole reference in his book acquires a greater relevance: 'Those who follow the doctrine of the Ancients, when offered the occasion to treat the effects of spiritous medicines, of which the Ancients had no knowledge (this being the reason for their not having treated of them), they wish not to believe what the Moderns say of them; the latter with just cause having come to a knowledge of them attained by means of our separatory art, this being well understood by Arnaldo de Vilanova and Raymundo Lulio and Theophrastus Paracelsus and Ubequero and Joannes de Rupescissa, and many others who have followed this separatory art, by which means what was occult in Nature has been brought to light; in this way come about the effect which those who follow corporal medicine know not.' The tradition to which Santiago belongs becomes obvious: he is one of the three great names in Low-Medieval alchemy, together with Wecker, a typical representative of the literature 'de secretis', and Paracelsus. The influence of the last, evident throughout his work, is assimilated by a scientist of critical bent and astonishingly modern mentality....

Another interesting mark left by Paracelsus' influence in Spain at the end of the 16th century was the creation of a Chair of Chemical Medicaments at the University of Valencia, whose brief duration does not diminish its historical significance.

During the 16th century the Faculty of Medicine of the University of Valencia was the most important medical centre in Spain. Richly endowed by the prosperous urban bourgeoisie on whom it depended economically, it came to possess nine Chairs plus two 'regencias' or minor Chairs. Open to the new movements of the Renaissance, it had the first Chair of Anatomy in Spain, which – as we have previously mentioned, acted as a vigorous focus for the diffusion of the Vesalian reform – and also the first Chairs of Surgery and 'Medical Botany'. ...

Lorenzo Cózar, then the most outstanding personality in Valencian medicine, was appointed titular professor of this Chair. He had been professor of surgery since 1585 and, in 1589, was named 'Protophysician' of the Kingdom of Valencia by Philip II. ... The content of his teaching seems to be beyond doubt, since 'remedios secretos' was a term habitually used to designate chemical medicaments. One need only recall the title of Gessner's influential work *De remediis secretis liber*. What is not so easy to gauge is the degree of Cózar's dependence on Paracelsus. In any case, the Chair only lasted for the academic year 1591–92 and Cózar never managed to publish any of the works he had in mind. Most probably, in both cases, this was due to the death of the Valencian doctor, a fact which appears implicit in the documentation. After the death of the famous 'Protophysician', the Chair was presumably left without a suitable occupant and interest in it must have declined.

In the Spain of the first half of the 17th century, the negative factors which had hindered the development of Spanish science and medicine during the last third of the preceding century reached their culmination. The isolation from the rest of Europe became more impenetrable and the scholasticism more sterile and bookish while the situation of the dissident minorities and the political and economic crisis worsened. [...]

Spanish alchemy at this period continued to be a marginal activity emphasizing, as in the rest of Europe, occultism and allegory. In the literature of this subculture – a literature rarely appearing in print – Paracelsus became a figure of central importance. Academic scientific culture, on the other hand, maintained a highly critical attitude towards his work. [...]

We do not know how Paracelsus' work was incorporated into the Spanish alchemists' milieu, although everything leads us to believe that the process of assimilation took place during the last third of the 16th century through Fioravanti and other foreign Paracelsists. What is undeniable is the fact that Paracelsus' influence attained such importance that some of his works, authentic or apocryphal, were translated into Castilian....

The first is a manuscript in octavos of nearly two hundred folios, originally from Barcelona and, at present, to be found in the Wellcome Historical Medical Library in London. It contains the Spanish translation of two of Paracelsus' works titled 'De natura rerum' and 'De la naturaleza del hombre', a series of fragments from other writings of his.... [...]

Nothing can be said of the anonymous translator of these Paracelsist texts although it appears to have been a Spaniard living in the first half of the 17th century, who possessed an overriding interest in alchemy. [...]

The second manuscript is partially different. It is kept in the National Library of Madrid.... It is a Castilian translation of the treatise entitled *De lapide philosophico* attributed to Paracelsus, each one of whose fragments is amply commented upon by the translator. ... [O]n the cover there is also the signature of Luis Amigó y Beltrán, lawyer of the Royal Councils. In fact, we are dealing with one of the most outstanding figures in Spanish alchemy at the end of the 17th century. The presence of this manuscript containing pronouncedly allegorical and occultist elements in his library is an example of the historical continuity of Spanish alchemy throughout the century.

5.2 Virgilio Pinto, *Censorship and the inquisitorial mind**

Censorship is in a sense peculiar to the modern age. Examining the intellectual history of the Western world, from classical antiquity onward, we come across references to the repression of ideas as much by the civil powers as by the church. As opposed to these instances, beginning in the sixteenth century censorship became an act of systematic control. Various circumstances contributed to this important transformation: the restructuring and centralization of power in the state as well as in the church, the religious conflicts generated by the Protestant rebellion, and the great development experienced with the invention of the printing press. In a way, censorship expresses the conflict between established power and its dissidents over this basic instrument for the diffusion of ideas.

...Heresy, a transgression against faith, went counter to the church (unity) and to Catholic truth (orthodoxy) in that it represented the ultimate separation from the church. Thus it confronted an ecclesiastical reality in which the process of centralization urged ever greater ideological control and struggle against heresy, strictly speaking an intellectual crime in that it represented the point where freedom of thought became punishable.

...[C]ensorship was but one of several instruments with which an attempt was made to counter the expansion of heresy, and in Spain the Inquisition soon monopolized jurisdiction over this activity.

The juridical support of this effort proceeded from the creation by law, during the sixteenth century, of a particular crime, that of distributing, owning, or reading prohibited books. The authority to proceed against the violators of the prohibition was granted by successive popes to the inquisitor general of Spain. Paul III in 1539 empowered Inquisitor General Juan de Tavera to proceed against those who had books by heretics in their possession. Julius III in 1550 empowered the inquisitors to punish those who owned or read forbidden books, thereby considerably broadening their jurisdiction. Finally, Paul IV in 1559 empowered Inquisitor General Fernando de Valdés to absolve those who might for this reason have been punished with excommunication. This authority was extended to the subsequent inquisitors general and endowed the Inquisition with enough of a juridical foundation practically to monopolize censorship.

The book as an object of inquiries

In the middle of the sixteenth century, the Spanish Inquisition succeeded in erecting an efficient censorship apparatus....

---

* Virgilio Pinto, 'Censorship: a system of control and an instrument of action', in *The Spanish Inquisition and the Inquisitorial Mind*, ed. Angel Alcalá, Social Science Monographs (Boulder, CO, and Atlantic Research and Publications, Highland Lakes, NJ, 1987), pp. 303–12, 318–19.

The objective of this apparatus was twofold, to identify heterodoxy and to prevent its diffusion. The apparatus centered primarily on one of the most important means of the diffusion of heterodoxy, the book or pamphlet in either printed or manuscript form. In accordance with the thinking of the Inquisition's theoreticians, the heterodox book was considered by the censors a mute heretic. Around it they created a dense network of surveillance embedded in the natural medium in which the book might evolve. The Inquisition's watchful presence imbued the atmosphere surrounding the book with the threat of punishment. What is more, the Inquisition developed several mechanisms for identifying the heterodox book that converted this process into a prosecution similar to the judicial proceedings to which presumed heretics were subjected. We may therefore speak of the 'trial' of a book as one aspect of the apparatus of censorship. In this process a series of practices whose objective was to have the social body share in them were condensed. On the one hand, the judgment of the orthodoxy or heterodoxy of a book was similar to the ritual of academic disputes, consisting of isolating propositions and submitting them to verification, which at that time meant the logically rigorous determination of their truth or falsity in terms of the official orthodoxy, thus giving truth a dogmatic meaning. On the other hand, the issuing of the verdict conformed to the ritual that the Inquisition used to create its own public image. It aimed at treating the heterodox book as a totem, the bearer of punishment, that is, creating a distance between the popular masses and the written culture, reserving it for those very academic circles that monopolized the instrument by which orthodoxy was determined.

The 'trial' of a book might be initiated by denunciation or by interception by one of the controls established by the Inquisition. ... Denunciations poured in from all the areas where books were present – academic, ecclesiastical, and bureaucratic. The rigor and nature of the denunciations varied – for example, those who had mastered the academic technicalities tended to denounce concrete propositions, phrases, lines, or paragraphs of works – and the final verdict tended to be conditioned by the type of denunciation. With the receipt of a denunciation of the confiscation of a work, the inquisitorial apparatus was set in motion. The work was delivered to the censors, who in their capacity as experts pronounced a judgment (*calificación*) that tended to be accepted by the Inquisition. If the book had been judged by one of the local tribunals, its documentation was forwarded to the Council. Once the Council had this information, it either tried to complete it or pronounced its verdict, which might allow or forbid the circulation of a book either totally or in part. In any case, the Inquisition monopolized the function of censure, superseding the universities that had until then performed this task. Within the Inquisition, the Council monopolized the making of decisions related to censorship. If it decided to forbid the circulation of a work, it notified the local tribunals by means of a resolution. The local inquisitors in turn publicized the decision by issuing edicts that were read during the principal Sunday mass at the most important churches of the towns and cities and then affixed to the portals of the churches. Prohibiting books by edict was the norm at least from 1532 on. Such edicts announced the pertinent facts about the affected books – author, title, place, and date of publication – and the reason for their prohibition. They abounded in the sixteenth and seventeenth centuries....

[...] Inquisitorial standards concerning censors were not very demanding. Two conditions were fundamental, purity of blood and theological training. [...]

The identification of heterodoxy was only one of the censorship apparatus's functions; the other was its interception. In the course of the sixteenth century, a series of controls was established whose objective was to cover the various stages of the normal cycle of the book from the time it left the author's hands until it reached the reader's....

The controls centered on the surveillance of printing presses and printed matter, of the movement of books, especially in ports and at borders, and of bookstores and libraries, both institutional and private. Even readers were the object of controls through the canon or moral penalties that they incurred by owning or reading forbidden books. The Inquisition tried not to leave a single crack through which prohibited or heterodox books might crawl.

[...] The bookstores were systematically examined...when indexes were published and whenever the inquisitors, always without previous notification, decided to do so. At the beginning of the seventeenth century it was decreed that the bookstore had to place a list of the books it had at the disposal of the inquisitors, and in 1605 it was ordered that a list of their buyers also be compiled.

Finally, beginning in the 1530s libraries were also subject to control. Quite frequently the Inquisition empowered an individual to examine the libraries of a city. These examinations were more frequent in cities with universities, such as Alcalá or Salamanca. Aside from these sporadic controls, the libraries had to be examined whenever a new index was published. In such a case the examination of the convent libraries was entrusted to their own superiors.

All these controls functioned with a certain efficiency until well into the seventeenth century, in accordance with the Inquisition's restrictive attitude toward the concessions of licenses for reading prohibited books. In the eighteenth century prohibited books abounded in the libraries, as can be deduced from several inquiries made by the Inquisition itself, but this was due, I believe, more to the inquisitorial prodigality in conceding licenses for reading probibited books than to the inefficiency of the controls.

. . . [P]rinters and booksellers at times . . . expressed their dissatisfaction to the Inquisition. The publication of the indexes, the visits to bookstores, the suspension of editions, the examination of packages of books that arrived at the ports translated into losses that they were powerless to remedy. This was a side effect, but nevertheless an important one, of the controls, since they disrupted the functioning of the structures of the publishing industry, a lucrative but always uncertain business. [. . .]

The Indexes of forbidden books

The indexes were an important instrument of censorship, though not the only one, and one that took shape gradually over time. They had various purposes: to compile prohibitions first announced by edict, to redefine the scope of the prohibition, and to correct the discrepancy between the rhythm of publication and the rhythm of the prohibition of books. The Inquisition initiated the publication of the indexes under outside influence, to the point that, despite its jurisdictional monopoly of these matters and its presumed autonomy of action, the majority of the indexes' prohibitions were in principle a simple acceptance of prohibitions decided abroad. The indexes, on the other hand, arose from the need to counter the Protestant literature, which was growing vertiginously. Consequently, the lists of prohibited books, initially rather simple and relatively short, became heavy volumes providing a special cosmic view of heterodoxy in that they delineated its limits – the sphere of what was prohibited – and fastidiously pointed out its manifestations. The expurgatory indexes now forbade not works but particular propositions. [. . .]

. . . In 1584 a bold astrologer, Mendoza de Porres, dared to defend the scientific and autonomous nature of the astrological 'science', arguing that it had its own principles and rules whose veracity and rationality should be judged by astrologers and not theologians. An anonymous censor responded that the judgment of truth was the theologians' task. The incident was a consequence of opposing conceptions with regard to the value of the science and moreover reflected another behind-the-scenes drama, the manipulation of the censorship apparatus by those who opposed the intellectual regeneration. We find it on other occasions as well, for example, in the conflict that broke out in 1735 at the University of Valencia over the spread of rationalistic philosophy and in the attention paid by the Inquisition in 1786 to the denunciation by Fray Diego José de Cádiz of the chair of political economics established by Normante in Zaragoza, a denunciation which delayed the introduction of economics in Spain. In these . . . different areas, we see the same dynamic: the fear of change. . . . [O]utside the Spanish ecclesiastical and academic world, a new language was shaping itself, a sign of the emancipation and progress of thought.

5.3 Henry Kamen, *The Inquisition and Spanish intellectual life**

The impact on literature and science

The times are such that one should think carefully before writing books.
Antonio de Araoz, SJ, September 1559

Almost from the beginning the Inquisition took an interest in aspects of literature. When Hebrew books and the Talmud were found in the possession of conversos, they were seized and destroyed. The inquisitors also seem from an early period to have frowned on magic and astrology. There is a reference, probably from the late 1480s, to the burning of a large quantity of such books found in the university of Salamanca. The diffusion of the printing-press in Europe at the end of the fifteenth century made authorities in both Church and state aware of the need to oversee book output.

... But printing was still a novelty, printed books were few, and controls in the early years of the century were lax. The coming of the Reformation by contrast unleashed a flood of controversial literature, which authorities everywhere attempted to curb. In England the government produced licensing laws in 1538, and in the 1540s various Italian authorities passed similar edicts. Spain came late into the field of censorship. [...]

In the 1530s and 1540s the Inquisition attempted to stop the entry of heretical literature into the peninsula. As the only Spanish tribunal with authority over all Spain, it was able to act in areas (such as seaports) where state officials could not. The government took no direct initiative over controlling literature until the shock discovery of Protestants in 1558. That event stung the regent Juana into action. On 7 September 1558 she issued a radical decree of control. The law banned the introduction into Castile of all books printed in other realms in Spanish, obliged printers to seek licences from the council of Castile (which in 1554 had been granted control over such permits), and laid down a strict procedure for the operation of censorship. Contravention of any of these points would be punished by death and confiscation. At the same time the Inquisition was allowed to issue licences when printing for its own purpose. According to the new rules, manuscripts were to be checked and censored both before and after publication, and all booksellers were to keep by them a copy of the Index of prohibited books. ...

... Heresy was spreading through European universities. ... When he [the king] arrived in the peninsula in 1559, he issued an order ... to all subjects of the crown of Castile studying or teaching abroad to return within four months. An exception was made for those studying at particular

* Henry Kamen, *The Spanish Inquisition: An Historical Revision* (London: Weidenfeld and Nicolson, 1997), pp. 103–6, 113–14, 116–20, 131–5.

named colleges at Bologna, Rome, Naples and Coimbra. No Castilians were in future to be allowed abroad to study except at these.

The censorship law of 1558 and the ban on studying abroad were harsh measures because they had no precedent. . . .

. . . [T]he biggest loophole in both measures was that they only affected Castilians. Philip was able to issue his decrees through the council of Castile; in the other realms by contrast he would have had to summon the Cortes, which he did not do. The entire eastern half of the peninsula, and the whole length of the Pyrenees, were consequently exempt from control. Any author who had difficulties getting a licence to publish in Madrid could therefore simply go to a printer in one of the other non-Castilian peninsular realms and publish there. In Catalonia, the king complained in 1568, 'the printers publish many new books without having our licence'. Not until the later years of the reign did the crown manage to claim some degree of control over licensing in the crown of Aragon: in Catalonia from 1573, in Valencia from the 1580s, in Aragon from as late as 1592. Even in Castile the 1558 law exempted most ecclesiastical books (which constituted the most important part of regular book production), and Inquisition publications, from the need to obtain state control. Over a large part of Spain, consequently, the 1558 law had no force. In these areas, printing normally had to be licensed by the local bishop.

Second, the control of imports was operative in Castile alone. The 7 September law regulated the import of books *only* 'into these realms' (Castile and León). The other realms, namely 'Aragon, Valencia, Catalonia and Navarre', were excluded. Books coming *from* them to Castile were subject to control, but there was no legislation to restrain books coming to them from outside. The bookshops of Barcelona imported freely books published abroad in Spanish and other languages. Outside Castile, as a consequence, the government was obliged to rely on the Inquisition in its attempts to oversee the book import trade.

Third, the printing controls had to contend with the reality that Spain relied heavily on foreign imports for its access to literature. When the tutor of the future Philip II went book-shopping for the prince in Salamanca in the 1540s, most of the books he bought were printed abroad. Imported volumes on humanities (including the complete works of Erasmus), literature, science and art featured in the list. Effective application of the law governing imports was not possible, because bookshops in Spain depended for their living on supplies from abroad. Foreign presses continued to dominate the printing of religious works not only in Bibles but also in mass-books and works of devotion. The flow of literature never stopped, as the case of Barcelona shows. No attempt was ever made by the Inquisition to interfere with trade in this city. Ten years after the decrees of 1558–9, booksellers continued to rely for their income on the uninterrupted import of foreign books, many of which went on to Castile. 'The books that enter through this frontier are very numerous', the inquisitors reported from Catalonia in

1569, 'and even if there were many inquisitors we would not be enough to deal with so many volumes.'

Fourth, the biggest problem with the legislation of 1558–9 was that, true to form, many Spaniards simply ignored it. The printer of a new book normally preferred to apply for the licence issued by the council of Castile, because it carried with it a 'privilege' or exclusive right to publish and sell. Reprints on the other hand did not require a new licence. Printers and authors therefore felt free to bring out 'reprints', even if important changes had been introduced into the text. But many authors tried successfully to avoid the censorship process, which could involve interminable delays. They published without permission, or (more frequently) published abroad, in Italy or in France. In the 1540s, most books by Spaniards were published outside Spain, notably in Antwerp, Paris, Lyon and Venice. Despite the apparently restrictive nature of the 1558 law, Spanish writers continued throughout the latter half of the century to publish as much abroad as they did at home. It was a freedom enjoyed, ironically, by no other European country. The works published abroad were, of course, imported into Spain. No intention of heresy arose. In the late sixteenth century at least sixty leading Spanish writers published their works abroad, in Lyon in France, rather than in Spain. The fact was that the quality of presses outside Spain was much better, and controls less onerous. As a consequence, the penalties laid down by the 1558 law often remained a virtual dead letter. Enjoying the ability to publish with impunity in the realms of Aragon, Italy, France or the Netherlands, Spaniards could boast that they had more freedom of literature than their neighbours. In Valencia at a later period, some forty per cent of publications had no licence or permission of any sort. Despite all the unlicensed publishing, not a single author or printer in Spain – other than those condemned as Protestants – is known to have suffered the death penalty. By contrast, in England and France the risk of punishment was real. [. . .]

The scope of the 1583 [Quiroga] Index was, in appearance, staggering. By its sheer size it drew into its ambit the whole of the European intellectual world, both past and present. ... At first glance it would appear that the Inquisition was declaring war against the whole of European culture.

In reality [this] Index was much less aggressive than appears. For the most part it simply took over existing condemnations in the Catholic world. It . . . did virtually nothing to affect the literary or reading habits of Spaniards. The overwhelming bulk of books it prohibited was unknown to Spaniards, had never entered Spain, and was in languages that Spaniards could not read. ... More directly relevant may have been the expurgations listed in the 1584 Index. Authors and printers may have been irritated by these, but they were hardly a blow to creativity. [. . .]

The Index of 1632 was issued by Inquisitor General Zapata, and that of 1640 by Inquisitor General Antonio de Sotomayor. Similar to the 1612 compilation in scope and content, Sotomayor's Index offered a general survey of the intellectual advances of the seventeenth century.... It is not surprising to find Francis Bacon and other major writers condemned in the first class as heretics.... Despite its coincidence with the early period of the Scientific Revolution, moreover, the 1640 Index was tolerant towards some aspects of science. Johannes Kepler and Tycho Brahe, as heretics, were classified as *auctores damnati* and therefore appeared in the first class; but virtually all their works were permitted in Spain after very minor expurgations. Some were allowed without any expurgation, but with the proviso that a note on the book should state that it was by a condemned author. Into this category fell Kepler's *Astronomia nova* of 1609 [and] his *Epitome Astronomiae Copernicanae* of 1618....

With these Indices ended the first great period in the censorship of the Inquisition. The great compilations of 1583 and 1640 were not by their nature repressive weapons, and served more to dissuade Spaniards from reading foreign authors whom none but a few could have read anyway....

Any evaluation of the role of the Indices also needs to take into account the practical question of whether they were put into effect. We may consider the situation in Catalonia. Twelve copies of the Quiroga Index arrived in Barcelona in October 1584. They were at once redistributed, a copy being sent to each bishop in Catalonia. The bishop was asked to collaborate with the *comisarios* (local clergy who helped the Inquisition) of that area. The comisario in his turn had to communicate the contents of his single copy to the leading persons of his district, and to the main booksellers. The booksellers for their part refused to buy copies of the Index 'because they say they are very expensive'. In this instance, it appears that a single copy of the Index had to serve for an entire bishopric. If the example is typical, it would appear that in many parts of Spain the Index remained unknown. 'There must be a great many books that are not corrected', the inquisitors of Barcelona mused in 1586, when they commented on the lack of available copies of the Index. Certainly, its availability was an excuse given by some booksellers in the city in 1593, when they were accused of selling prohibited books. [...]

In general, the operation to control book imports was riddled with inefficiency. The inquisitors of Barcelona in 1569, unable to process the great number of books entering, reported that 'to entrust the work to friars and experts is not enough to keep people happy, and annoys the booksellers'. They therefore proposed 'a commission of two persons to look at the books, paid by the booksellers, whose suggestion it is'. Orders were sent out periodically by the Suprema for books to be seized; but, the council complained in 1606, 'it is reported that many of the books ordered to be picked up are not being collected'. Inevitably, some condemned books filtered into the country. In Barcelona in 1569 the bookshops were still selling 'many forbidden books'. The continuing entry of books is demonstrated by the case in Madrid of Joseph Antonio de Salas, knight of the Order of Calatrava, whose library was offered for sale to the public on his death in 1651.

It was then found that among the 2,424 volumes in the collection, to quote the censor, 'there were many books prohibited or unexpurgated or worthy of examination, either because they were by heretical authors or were newly published abroad by unknown writers'. There were 250 prohibited works in the library – a proportion of one in ten – confirming that foreign books were smuggled regularly and often successfully into Spain, despite the death penalty attached to the offence. The second major control was at the point of contact between a book and its potential reader. Libraries and bookshops were at intervals visited and checked. Bishops were encouraged to inspect all libraries in their dioceses, and at Salamanca University a score of the staff went carefully through the library to weed out any dangerous books. As early as 1536 Thomas de Villanueva was employed by the Inquisitor General to visit bookshops in Valencia. A lightning check in 1566 in Seville is described thus by an inquisitor: 'at a fixed hour, nine in the morning, all the bookshops of Seville were occupied by familiars of the Holy Office, so that they could not warn each other nor hide nor take out any books, and later we came and made all the shops close and are visiting them one by one'. In reality, such visits were few and far between. They also took place only in big towns where there was an inquisitorial presence. And even there, as the inquisitors of Barcelona admitted in 1569, the bookshops 'have not been visited for many years'. Bookshops, moreover, pleaded ignorance if found with books that needed censoring. In Barcelona in 1593, as we have seen, they said that no copies of the Index were available, and they were consequently unable to monitor forbidden items. On this occasion, some booksellers were fined. It is the only recorded case of any action being taken against bookshops in that city. [. . .]

There are two distinct opinions about the impact of the Inquisition on literature. One, strongly supported by traditionalists, denies any negative influence. Menéndez y Pelayo asserted that 'never was there more written in Spain, or better written, than in the two golden centuries of the Inquisition'. The other, reflected in many modern studies of literature, claims that Spaniards virtually ceased to write and think. 'It would seem superfluous to insist', argued Lea, 'that a system of severe repression of thought by all the instrumentalities of Inquisition and state, is an ample explanation of the decadence of Spanish learning and literature.' For the English Catholic historian Lord Acton, the injury inflicted on literature by the Inquisition was 'the most obvious and conspicuous fact of modern history'. Américo Castro put the argument succinctly. For him, 'not to think or learn or read' became habitual for Spaniards faced by 'the sadism and lust for plunder of those of the Holy Office'.

The available evidence does not support either of these extreme views. Both assume that the censorship system functioned effectively in Spain; one view claims it worked for the better (purging heresy), the other that it worked for the worse (suppressing creativity). In reality, neither the

Index nor the censorship system produced an adequate machinery of control.

...[T]he bulk of creative and scientific literature available to Spaniards never appeared in the Index. ... The vast riches of scholarship opened up by the imperial experience during the Inquisition's great period were never affected:...the treatises on mathematics, botany, metallurgy and shipbuilding that flourished under Philip II never came within the ambit of the inquisitors. Long after the measures of 1558–9 Spain continued to profit from a world experience vaster than that of any other European nation. Its contribution to navigation, geography, natural history and aspects of medicine was highly valued in Europe, leading to some 1,226 editions of Spanish works of the period 1475–1600 being published abroad prior to 1800.

...Those who really wished to obtain banned books of special interest – in astrology, medicine, scholarship – faced few obstacles. They brought books in personally, or through commercial channels, or asked friends abroad to send them. Total freedom of movement between the peninsula and France and Italy guaranteed an unimpeded circulation of people, books and – at one remove – ideas. Finally, no evidence has ever emerged that the book controls eliminated promising new life among intellectuals, or prejudiced existing schools of thought. [...]

...There are no convincing grounds for believing that the Spanish Inquisition was unique in Europe in its efficiency at imposing control,[1] or that Spaniards were subjected to a regime of 'thought control' which 'fossilized academic culture' for three hundred years.[2]...

The impact on science was largely indirect. Spaniards in the early modern period had possibly the least dedication to science, measured by the university affiliation of scientists, of any nation in western Europe. Those who took learning seriously went to Italy. Thanks to access to Italian and foreign scholarship, scientific enquiry did not collapse. Technology filtered into the country: foreign treatises were translated, engineers were imported by the state. Foreign technicians – all of them Catholic – came to the peninsula with their expertise. The Inquisition, for its part, did not normally interfere.

Scientific books written by Catholics tended to circulate freely, though the works of Paracelsus and a few others were disapproved of. The 1583 Quiroga Index had a negligible impact on the accessibility of scientific works, and Galileo was never put on the list of forbidden books. The most

1 Pardo Tomás, *Ciencia y Censura* (1991), however, feels (p. 269) that 'the efficiency of control was considerable' up to the seventeenth century. His view is based exclusively on the Inquisition's own papers, which are obviously optimistic about the success achieved.

2 Virgilio Pinto Crespo, 'Thought control in Spain', in *Inquisition and Society in Early Modern Europe*, ed. Stephen Haliczer (Totowa, NJ: Barnes and Noble, 1987), p. 185.

direct attacks mounted by the Inquisition were against selected works in the area of astrology and alchemy, sciences that were deemed to carry overtones of superstition.

If there was, then, an imbalance between scientific progress in the peninsula and in the rest of Europe during and after the Renaissance, the Inquisition was not perceptibly responsible. The range of books it prohibited, of course, may well have dissuaded a few readers. Arguably, this would not have had serious consequences for learning during the sixteenth century. By the late seventeenth century, on the other hand, it was clear that English and Dutch intellectuals had become the pioneers in science and medicine. They were Protestants, and their books automatically fell within the scope of inquisitorial bans. Logically, Spanish intellectuals from the mid-seventeenth century began to look on the Holy Office as the great obstacle to learning. The complaint of the young physician Juan de Cabriada in 1687 echoed the outlook of his generation: 'how sad and shameful it is that, like savages, we have to be the last to receive the innovations and knowledge that the rest of Europe already has'. Those who could read French managed to import scientific and philosophical works privately. Descartes was being read in Oviedo, Hobbes in Seville. During the eighteenth century, however, intellectuals in the peninsula faced an uphill struggle against the attempts of the Inquisition to block the spread of the new learning.

At no time was the peninsula cut off from the outside world by the decrees of 1558–9, or by any subsequent legislation. Under Habsburg rule the armies of Spain dominated Europe, its ships traversed the Atlantic and Pacific, and its language was the master tongue from central Europe to the Philippines. Tens of thousands of Spaniards went abroad every year, mainly to serve in the armed forces. Cultural and commercial contact with all parts of western Europe, especially the Netherlands and Italy, continued absolutely without interruption. It is consequently both implausible and untrue to suggest that Spain (and with it Portugal) was denied contact with the outside world.

The image of a nation sunk in inertia and superstition because of the Inquisition was part of the mythology created around the tribunal. Scholars continue still to suggest that Spaniards 'had to guard their speech carefully'. These views must be set against the reality that, like other European states, Spain had active political institutions at all levels. Free discussion of political affairs was tolerated, and public controversy occurred on a scale paralleled in few other countries. . . .

The freedom in Spain was a positive side of the picture. The negative side was the unquestionably isolated state of peninsular culture. Spain remained during the early modern period on the fringe of the major currents in western European philosophy, science and creative art. In the great age of empire, Philip II had to rely for technological expertise on Italians, Belgians and Germans. . . .

Chapter Six
Science from the Earth in Central Europe

6.1 Bruce Moran, *Science in the courts of German princes**

The expansion of mathematical and technical literature during the 15th and 16th centuries is only one aspect of a many-sided relationship between European courts, technology, and science. A deeper understanding of aristocratic involvement in technology and science may be attained by developing a picture of a specific princely type – one who not only patronizes technical and mathematical projects, but who is himself a technical and mathematical practitioner. The study of prince-practitioners in Germany will serve to illustrate the function of several European courts as institutional nodes of technical activity. ...

... Certainly the most influential courts in northern Europe in the development of aristocratic attention to technology and science were the imperial Hapsburg courts in Vienna, and, in particular, the fabulous court of Rudolf II in Prague. For a brief period, the Prague court took shape as a center of scientific activity and mechanical artisanry. Astronomers, alchemists, naturalists, and magicians resided at Prague for various lengths of time, enjoying the patronage and protection of the emperor. Rudolf secured as well the talents of numerous clock and instrument makers. ... The design of machines, both fanciful and practical, became an overall preoccupation at court and involved Rudolf himself in the construction of a self-orienting travelers' chart which was operated by a concealed compass. In Hesse, the Palatinate, Württemberg, Braunschweig, and Saxony, Protestant princes adopted the values of Hapsburg patronage and extended them to accommodate princely involvement in practical mathematics, instrument making, and observation.

---

* Bruce Moran, 'German prince-practitioners: aspects in the development of courtly science, technology and procedures in the Renaissance', *Technology and Culture*, 22 (1981), pp. 253–74.

From his court in Kassel, the *Landgraf* of Hesse, Wilhelm IV (1532–92), organized projects of exact stellar measurement which have become well known to historians of science. Wilhelm's participation began with observations of the fixed stars in 1566–67 and involved him in the observation of solar meridians at Kassel even after his mathematician, Christoph Rothmann (ca. 1550–ca. 1605), received overall charge of the court's measurements. Above all, Wilhelm's observations of the new star of 1572 were frequently compared by 16th-century astronomers and later gained the attention of Galileo, who viewed them as among the most accurate measurements of the star. [...]

Comparison of observations of the comet of 1585 for instance, measured both at Kassel and by Tycho Brahe at Uraniburg, indicates a level of observational discrepancy of diurnal motions in longitude and latitude of little more than one-half to one minute of arc with measurements frequently coinciding exactly. Much of the reason for the court's observational success derives from procedural innovations and technical improvements in the accuracy of the *Landgraf*'s observational instruments. Through the temporary residence at the Kassel court in 1584 of Paul Wittich (1550–87), a mathematician-astronomer who had earlier visited the observatory of Tycho Brahe, a cross-fertilization of technical ideas made possible the refinement of the Kassel instruments. [...]

Princely interest in fine mechanical engineering nurtured a variety of projects at Kassel leading to a close association of technical and scientific pursuits. The invention of a new, more accurate means of time keeping by the *Landgraf*'s court clockmaker Jost Bürgi was thus adapted to the court's astronomical projects and provided the Kassel observers with additional flexibility in the measurement of stellar right ascensions. Aside from the construction of instruments useful for observation, however (the reputation of which occasioned the Parisian philosopher Peter Ramus to liken Kassel to Alexandria and the *Landgraf* to Ptolemy), Wilhelm's court won far-reaching renown as a center for the production of clockwork-driven celestial spheres and astronomical clocks. The most splendid of these is the famous *Wilhelmsuhr*, manufactured by the Marburg mechanician Eberhart Baldewein (1525–92) for Wilhelm in 1561. Numerous artisans were employed in the construction of machines like the *Wilhelmsuhr*. Yet Wilhelm himself was never far removed from the activities of the workshop. [...]

Among prince-practitioners, however, interest in precision and in the production of machines and measuring instruments has a substantive basis in practical problems arising from efforts toward political consolidation and exploration as well as territorial and commercial expansion. More than artistic style, these concerns emphasized skills pertaining to surveying, cartography, mining, and fortification and attached important political and economic functions to the projects of mathematicians and artisans. At such courts, similarities of procedure, reaffirming precise description and the organized collection of information, characterized both scientific and essentially administrative programs.

Economic as well as scientific projects coincide at Hesse-Kassel with the values of systematic collection and observational accuracy inherent in both. While the *Landgraf's* attempt to construct a literal catalog of reliable stellar positions depended upon precise astronomical measurement, Wilhelm's desire to produce a sound basis for determining economic policy in Hesse resulted in the organization of detailed statistical surveys structured personally by the prince. Through individual inquiry, officials of the Kassel court determined the population, resources, and possessions, as well as property and forest rights of each village, town, and estate throughout the principality, compiling this information also into a form of catalog referred to by Wilhelm as his *Ökonomische Staat*. [. . .]

At court, pragmatically motivated interests in cartography increased princely involvement in new technology and directed their participation in the construction of mathematical instruments useful in surveying operations. From workshops in Nürnberg, Augsburg, and Dresden, Elector August of Saxony commissioned various odometers and pedometers. With other Protestant princes, including Christian of Denmark (1503–59) and Wilhelm IV of Hesse, August exchanged surveyors and surveying instruments and involved himself with the construction and improvement of measuring devices. To the odometers manufactured by his own mechanicians, Valentine Thau and Johannes Homelius (1518–62), August attached his own innovations. Measurements which he obtained with these instruments, and descriptions of their use, were recorded by August in manuscripts preserved at the elector's Dresden library.

Of the various practitioners of surveying, the mining surveyor held a preeminent position at princely courts. Since mining often contributed significantly to the wealth of emerging centralized principalities, it does not seem surprising to find that economic interests in mining helped to shape the technical projects of prince-practitioners. At least four compass dials were designed by Elector August for use in the Saxon mines. In Braunschweig, the interests of Duke Julius in mining and metallurgy inspired plans for the construction of several machines which the prince intended to be employed in the mining operations of the Harz. [. . .]

The responsibilities of mathematicians and technicians varied immensely at princely courts in the 16th century. Rarely were roles specifically defined, and they often crossed institutional (court–university) as well as professional (mathematician–physician–astronomer–geographer) lines. Type and duration of employment at court depended entirely upon the personal designs of individual sovereigns. Princely projects were therefore role directive and could lead the mathematician or technician to confront problem situations which may not have arisen necessarily outside the context of the court. In this way the general skills of the Kassel court mathematician, Christoph Rothmann, came to be shaped and fitted to Wilhelm IV's project of astronomical observation.

The need for more accurate measurement of the stars and planets was noticed by many observers in the 16th century, and Wilhelm's correspondence makes frequent reference to the lack of predictive certainty within both classical and contemporary catalogs. In 1585 therefore, eight years after Rothmann's appearance at Kassel, Wilhelm revealed his plan for a renewed effort at observing stellar positions utilizing the talents of his mathematician. [...]

In the 16th century, the political dependency of many German universities upon the secular authorities of increasingly centralized territorial states contributed to the use of university teachers by princes for purposes of consultation or for services related to special projects at court. Mathematicians proved especially versatile since, besides developing mathematical skills, many also earned university degrees in the medical arts. The combination of mathematical and medical backgrounds among university professors probably arose from economic necessity. ... Professors in the arts, at least in Protestant universities, knew to expect less income than their colleagues in the faculties of theology and law, and many chose to expand their financial base by combining the roles of mathematician and physician at the university and, if the opportunity arrived, at court as well. [...]

Princely projects demanded role versatility from both academician and artisan. Whatever educational background, whether university or workshop, mathematicians and technicians were often joined in new pursuits at court, combining and sharing skills, experience, and ideas; linking practice and theory, science and technology. [...]

As patrons and practitioners, princes, both within and outside Germany, fused together courtly and technical roles which both bridged the gulf between scholar and craftsman and helped to narrow the distance between the activities of the social elite and scientific and technical operations traditionally reserved for the artisan. That the privileged aristocracy would willingly collaborate with academic and artisan practitioners became a basic feature of utilitarian science. [...] In the courts of Renaissance prince-practitioners, aristocratic participation in technical and scientific projects accentuated the development of many of the procedural values characteristic of 17th-century experimental science.

6.2 Marco Beretta, *Humanism and chemistry in Agricola's metallurgical writings**

[...]Although he was not strictly speaking a humanist, Georgius Agricola followed in his youth the educational program imparted by the *studia humanitatis*, he studied Greek at the University of Leipzig and, between

* Marco Beretta, 'Humanism and chemistry: the spread of Georgius Agricola's metallurgical writings', *Nuncius*, 12 (1997), pp. 21–7.

1524 and 1525, he worked at Aldus' atelier in Venice on the edition of Galen's collected works under the supervision of Gianbattista Oppizzoni, a famous Italian humanist. He also collaborated to the Aldine editions of the works by Oribasius, Hippocrates and Aetius. Aldus' printing atelier was the cradle of Renaissance humanism and between 1507 and 1508 it hosted Erasmus, one of the first non-Italian scholars to be accepted in the circle of humanists who collaborated on the editions of the sumptuous manuscript codex owned by Aldus. The contribution by Agricola to the medical editions mentioned above should not be overestimated but his active presence in Venice at a mature stage of his scientific education is significant in order to understand his later approach to metallurgy.

The first edition of Agricola's *Bermannus, sive de re metallica* appeared in Basel in 1530, little more than three years later than his return from Italy to the mining district of Chemnitz. The publisher of this short dialogue on mining was Froben, the printing house of Erasmus, and it was under the warm recommendation of the Dutch humanist that this booklet came to be published by one of the most prestigious European presses. *Bermannus* even contained a letter written by Erasmus to the brothers Andreas and Christopher Cornitz in which he explained the importance of Agricola's work by emphazising the precision and simplicity of its terminology. Instead of reading about the valleys, the hills, the shafts and the machines used for the extraction of metals, Erasmus could see them through the clarity of Agricola's descriptions. Such a positive judgement from a scholar like Erasmus who was normally indifferent to scientific topics cannot be understood if we do not regard *Bermannus* as product of a humanistic approach to science. Furthermore, it should be emphasized that Erasmus had been critical towards the aims and methods used by the alchemists and that he depicted the supreme art as a 'ridiculous charlatanry'. In one of his *Colloquia*, first published in 1519, the Dutch humanist derided the pretentious ambitions with a satirical narrative which left no doubt about Erasmus' negative opinion. The authoritative judgment by Erasmus was substantiated by the fact that he had been personally acquainted with Paracelsus but was not fully persuaded of the validity of the approach to medicine and alchemy supported by the Swiss physician. To a consequent humanist the resort to medieval sources, which was a typical feature of Renaissance alchemists, was an unacceptable principle. Furthermore the use of the vernacular and of a terminology filled with an inconsistent mixture of classical words and German neologisms was a linguistic approach which betrayed the true aims of philosophy, i.e. the simplicity and clarity of the language. [...]

The attempts to understand the meaning of the mineralogical terms employed by the ancients could not be a mere philological exercise but it necessarily implied a constant comparison between the names and the objects they referred to. This comparison confronted a purely humanistic discipline such as philology with a technological and practical expertise which could not be achieved anywhere else than in the mining district. But this peculiar investigative combination of erudition and observative skills was not the only requirement of a good metallurgist. Agricola was in fact well aware that the ancients did not discover everything in the mineral world and that many new substances as well as compounds were unknown to them and that they therefore needed to be defined and named *ex novo*. To this aim Agricola established five rules for naming the new minerals: 1. the use of classical Latin terms with a new meaning, usually aimed at a more restricted and specialized definition; 2. the specialization of general terms which acquired a meaning connecting them to technical concepts not known to the classical authors who had introduced these terms; 3. the use of circumlocutions to define in a more accurate way those compounds which could not be named by using one term; 4. the Latin translation of German mining expressions and finally 5. the use of German mineralogical terms adapted to the Latin language through the introduction of Latin terminations and suffixes.

Interestingly these same rules were shared by those Aristotelian naturalists of the sixteenth century who were engaged in classifying botanical and zoological species. The major concern of naturalists like Andreas Caesalpino and Conrad Gesner was also to find a continuity with the approach and language they inherited from the classical sources. Agricola was therefore approaching mineralogy from an established scientific perspective which offered the epistemological coordinates of the official natural philosophy. It is not surprising that his views on the scientific importance of naming in mineralogy were in total opposition to the Paracelsian approach. Paracelsus in fact vehemently criticized the use of the Aristotelian logic in science and believed that the disputes of the definition of words were a major hindrance to its progress. Agricola by contrast dismissed the ambitions of the alchemists and the Paracelsians and regarded their nomenclature as an intolerable mixture of *inaudita vocabula* coined without any given rule or order. Furthermore, by resorting to the works written by the Arabs and deviating from the classical sources, the *indocti et inepti Chymistae* of the sixteenth century corrupted the true art of investigating mineralogy. In the *Bermannus* Agricola, like Erasmus before him, derided the scientific validity of alchemy and dismissed with absolute conviction and commitment its metallurgical and mineralogical sources.

At the end of the *Bermannus* Agricola's friend, Plateanus, collected the 127 lexical innovations which were introduced in the mineralogical nomenclature in conformity with the linguistic rules set out by Agricola. If we consider the modest dimension of Agricola's first mineralogical work and its explicit introductory character, the presentation of such a number of linguistic innovations was highly significant. Many of the new terms not only referred to newly discovered minerals and compounds but also to technical operations, to machines and to the differentiated legislative duties which governed the exploitation of mines. This specialization of the technical nomenclature was the real innovative contribution which brought mineralogy onto a new path. In the treatment of mining and mining techniques, the *Bermannus* did not go far beyond the experimental results published at the beginning of the sixteenth century in the German mining booklets *Eyn Nützlich Bergbüchlein* (ca. 1500) and *Probierbüchlein* (Augsburg, 1510), but, by classifying and ordering this same knowledge with a renewed approach, it marked a most significant difference. Agricola's background as a refined and educated humanist and physician opened a completely new scientific horizon to the practitioners and engineers who so far had studied metallurgy and mineralogy only in the field.

In 1544 Agricola wrote a geological treatise entitled *De ortu et causis subterraneorum* in which he violently and systematically attacked and refuted the works of alchemists and astrologers. Two years later he published a letter, addressed to Wolfgang Meurer, professor of Greek at the University of Leipzig, in which he presented a Latin–German mineralogical dictionary. This dictionary is of great interest because if we compare it with the list of 127 terms presented as an appendix to *Bermannus* in 1530, we see that Agricola now lists 480 new terms. This impressive quantitative increase of names did not of course always correspond [to] the discovery of new substances or compounds. In most cases the new names defined in a more descriptive and appropriate manner compounds that until then had been confused under generic names. This fact should not lead us to diminish the crucial importance of Agricola's linguistic innovations. In *De natura fossilium*, also published in 1546, Agricola demonstrated the utility of his linguistic and philological approach to nomenclature. Silver ores and minerals, for instance, were designated by the term *argentum rude* (crude) followed by a third name which should have characterized the differences between species of silver. The first name designated the substance, the second its chemical characteristics and the third name was derived from the external physical qualities of the minerals, especially from their colours. Accordingly, Agricola mentioned the *argentum rude rubrum, argentum rude album, argentum rude nigrum, argentum rude cineraceum, argentum rude purpureum, argentum rude iecoris colore* etc. (By *rude* is meant an ore rich in any designated metal.) Each of these names corresponded in effect to a mineral compound with different chemical properties.

Agricola also provided a new classification of what he defined as the mixed minerals: 'The "mixed" minerals,' he wrote, 'which are composed of those same simple minerals, differ from the compounds in that the simple minerals each preserves its own form so that they can be separated one from the other not only by fire but sometimes by water and sometimes by hand. As these two classes differ so greatly from one another I usually use two different words in order to distinguish one from the other. I am well aware that Galen calls the metallic earth a compound which is really a mixture, *but he who wishes to instruct others should bestow upon each separate thing a definite name*' (my italics).

An interesting feature of Agricola's classification and naming of minerals is his constant emphasis on the operational and technological aspects of metallurgy. The German physician distinguished different classes of minerals and metals on the basis of their different reactions when submitted to the action of chemical agents like fire or by other laboratory techniques. This kind of differentiation, which was not totally unknown to the ancients, was nevertheless neglected by the classical authors who defined and classified minerals on the sole basis of their external features.

Agricola's classification and nomenclature of minerals served as a basis for all the approaches which eventually were followed by eighteenth-century mineralogists, and it is striking, in certain respects, that his emphasis on the importance of the comparative study between the chemical composition and the external physical qualities of the minerals was revitalized during the eighteenth century by Torbern Bergman and other prominent metallurgists and chemists.

6.3 Charles Webster, *Paracelsian prophecy and natural magic**

[. . .] In the course of the Scientific Revolution the idea of recovery from the adverse effects of the fall and restitution of man's dominion over nature contained in Christian eschatology was reinforced by an analogous set of ideas inherited from ancient theology. The more magical Neoplatonic writings and the Corpus Hermeticum made current by the Florentine Neoplatonists contained more than a veiled expectation that initiates of hermeticism would ascend to a level of mystical illumination offering a variety of benefits, including the possibility of being transformed into powerful Magi. ... Both experimental science and natural magic were involved in the understanding and conquest of the forces associated with spiritual magic. The latter is construed as the non-demonic form of magic utilizing the powers of the *spiritus mundi*, and reaching no higher in its mediations than the human spirit. Natural magic or science could thus be viewed not only as a manifestation of skill or knowledge, but also as a sign of election and as a reward for the elect.

At very few points in his writings did Paracelsus fail to relate his subject matter to the origin and destiny of things. His ideas on life and matter are rooted in his interpretation of the opening verses of Genesis. Paracelsus expressed himself variously on eschatology, reflecting on occasions most of the points of view known within the Christian church. Sometimes he placed emphasis on the prospects for Christ's imminent return. The frequent appearance of comets in particular, and the increasing instability of the times, whether climatically or politically, seemed to point towards this end.

---

* Charles Webster, *From Paracelsus to Newton: Magic and the Making of Modern Science* (Cambridge: Cambridge University Press, 1982), pp. 48–59.

The world seemed to be running down, to have reached its last stage of life. The logical outcome was thus a cosmic catastrophe and the initiation of a new world of the spirit after the shattering of the old material world.

However, for the most part he placed an emphasis on the enduring nature of the trial of mankind mounted by God during the last Monarchy on earth, giving man the fullest opportunity to accept salvation, to be raised to the highest degree above the level of animals and beyond control of the stars. It was therefore more important to seek the rule of God than to engage in vain prophecies. It was granted to man to know what would occur, but not when. Paracelsus preached against those desperate folk who were predicting the end of the world and warned that we had no idea how many generations would elapse before the final outcome. His remarks carry the obvious implication that the Christian was not justified in abandoning his longer-term responsibilities on the assumption of an immediate break-up of the present order.

Paracelsus equalled any renaissance optimist in his views on the dignity and capacities of man. Man was cast in the image of God; he was the centre part of creation, the summation of all the elements in the macrocosm. His status was so high and so great that he is provided with *arcana*, *mysteria* and *magnalia* without number from the heavens. As God's legitimate heir, the effects of the Fall notwithstanding, man was destined to inherit the kingdom on earth. The Devil would stand in the way of this inheritance, and attempt to prevent man from transcending his earthbound nature. But man would attain this definite goal or 'number' and achieve perfection, so signifying the return of paradise and final victory over the Devil. The circle of the world would then be completed, with man established at a position of supremacy in its centre.

It might be anticipated that at this point Paracelsus would declare for the breaking up of the present order of things. However he allowed for an indefinite boundary and even for an extensive interval between the last phase of the world and the future state of the resurrection. Thus paradise might spill over even into the present monarchy, so creating a period during which the elect would enjoy the benefits of both forms of existence. A golden age or New Hebron (Eden) would be established on earth. His writings strongly implied that the transition towards the earthly paradise had already begun.

Paracelsus believed that from a practical point of view man would only learn to exploit nature to its fullest extent if he were to learn by the mistakes of the past. God had granted all men since the Fall the capacity to overcome their disadvantages and perfect their arts and sciences. By following the light of nature, those capacities, implanted in man like a seed, could be developed to perfection. Nature was so organized that all things were arranged in a 'concordance', shared out on earth according to human needs, and awaiting initiative for the development of all necessary crafts and industries. [. . .]

Apart from regarding the moral example of the sages, practitioners of the sciences and arts could learn nothing from the past. They should look forward to a new age. Besides, their own times or monarchy presented hitherto unparalleled problems, owing to the pressure of population and the scientific and medical needs of an advanced society. Books written two thousand years ago for an entirely different culture could scarcely be relevant to this situation.

But there was no reason for lack of optimism. The latest ages were predestined to see the greatest achievement, by virtue of the predestined intellectual rebirth. The nearer man's approach to Judgment Day the greater would be the development of his learning, acuteness, wisdom and reason. For these reasons Paracelsus felt justified in breaking with tradition in order to develop a form of science and medicine consistent with the needs of the last monarchy, and durable until the dawning of the final age. Although little as yet seemed to have been achieved it was God's will respecting the immediate future that we should have experience of all of his works, and come to possess knowledge of all the secrets of nature, with nothing being withheld. There would be revealed to mankind great and wonderful arts and improvements, and civilisation would achieve an unprecedented understanding of the firmament, sea and earth. Gradually the quality of life would improve; the seasons and weather would be favourable; the land would be fruitful and the harvests rich; animals and man would prosper; disease and misfortune would vanish.

None of this was very different from the imagery of the New Jerusalem prevalent in the early church, and cultivated in prophecies from Papias onwards, and which surfaced periodically during phases of millenarianism. Each revival placed the emphasis slightly differently. For obvious reasons of advantage the millenarianism of the generation of Paracelsus, exploited extensively by the radical sects, placed its emphasis on the overthrow of the civil and ecclesiastical establishment, with the substitution of a fellowship of the poor, bound together by ties of mutual service and sharing of property. Paracelsus keenly advocated this platform both in its negative and positive aspects. His assault on the values of the medical establishment merely betokened an application to a special case of his feelings concerning the learned professions and the princely courts. On the other hand his respect for the usefulness of folk medicine reflects his acknowledgment of the basic virtue of simple people. Exaltation of the poor to the detriment of the rich is a theme introduced by Paracelsus at many points in his scientific and religious writings. The missing ingredient compared with other preachers developing the same theme is the sense of any desire to affiliate with any sect or specific religious movement. Like certain other spiritual reformers such as Sebastian Franck and Caspar Schwenckfeld, Paracelsus steered away from groups such as the Anabaptists. Indeed he regarded all sectarian groups, whether supporters of Luther,

Huss, Calvin, Zwingli, Bucer, Schwenckfeld, or the Anabaptists, as detracting from the ideal of a unified spiritual church.

The social ethic adopted by Paracelsus strongly influenced his ideas for the mapping out of the future course for intellectual progress. Nothing would be achieved in the degenerate world of the scholar. His own restless wanderings had taught him that the model for productive knowledge was provided by the best practices of the manual arts. It was the humble practitioners in these areas who were exposing the full extent of the treasures of nature and exploiting them to the best advantage.

Only by travelling, and systematically and minutely studying the diversity of things would the true extent of the variety of diseases, living organisms, or minerals be revealed. The chemist could not expect the mountains to come to him; he must visit them and there study the metallurgical and chemical methods of the workmen. It would then immediately be realized that species within any group, whether diseases or minerals, were infinitely more numerous than had been anticipated by the learned commentators. The traveller would also learn what seemed to be overlooked by the medieval compilers, that species and diseases were localized phenomena, Providence having provided in any one locality and for any one nation, everything required for material needs. This adaptation of the design argument was used to cut the ground from under classical sources, which even at their best, Paracelsus urged, were remote from the experience of the northern peoples.

Paracelsus was giving forceful expression to an instinct which was rapidly spreading among his contemporaries and particularly among his compatriots. There was growing impatience and doubt concerning the scientific usefulness of current fashionable compilations such as *Gart der Gesundheit*, *Hortus Sanitatus*, and Rösslin's *Kreuterbuch*, all of which claimed distinction on grounds of derivation from classical and Arabic sources. The investigative instincts and intense naturalistic detail awoken in Dürer at the turn of the century were passed on to his disciples and imitators. They in turn worked with such authors as Brunfels, Fuchs, Bock, Gesner, and Agricola, whose works transformed the level of commentary on the plants, animals, minerals and technology of Europe, and paved the way for the various types of investigative natural history produced during the next century.

Paracelsus himself made no direct contribution to this systematic effort. Even those of his writings which might have lent themselves to the listing of types tend to drift off in other directions. Nevertheless a vast amount of authentic data concerning chemistry, pharmacology and other aspects of medicine is buried in his writings, much of it judged by lexicographers and historians of the specialized sciences to constitute pioneering recording or observation.

Perhaps the exhaustive analysis of the older literature and organization of data which appealed to such encyclopaedic mentalities as Conrad Gesner seemed to Paracelsus like a secondary priority. He was more concerned with the effective exploitation of the properties of commonly available materials and with the understanding of the dynamics of natural processes, normal and pathological. By analogy with the sophisticated routines of the mineral workers, and in expectation of regaining the arcana of the ancient *magi*, Paracelsus placed his emphasis on argument by analogy, acute observation and the revelation of new properties by experimental manipulation.

Paracelsus argued that the light which provided the path to grace was also the light of nature given to guide man to the *secreta, mysteria* and *magnalia*, which were offered as rewards to those accepting the rule or monarchy of God. Throughout Paracelsus contrasted his active empirical approach to knowledge with the sterile authoritarianism of his opponents. His discourse is peppered with terminology signifying his faith in proof by 'experiment', for which he favoured analogy with the proving work undertaken by metalworkers. His terminology – Experienz, Experiment, Erfahrung etc. – was to dominate the experimental philosophy of the seventeenth century, and there is a substantial degree of overlap between the epistemology of Paracelsus and the Baconians. However the underlying science to which experiment contributed was viewed very differently. 'Science' was not merely the explanatory model built up by the observer on the basis of his observations, it was also in the opinion of Paracelsus a possession of the system under investigation. Science for instance was the property which enabled all things to retain their type. And since the terrestrial being was in concordance with the heavens and also subject to emanations from celestial bodies, its 'science' could be regarded as an emanation from the stars.

Despite his use of much of the vocabulary of modern experimental science, Paracelsus's view of science was firmly rooted in the tradition of spiritual magic. For Paracelsus the operative side of science involved bringing into action forces derived from the heavens. Mankind was in a unique position to exploit these forces by being situated at the 'boundary' or at the 'centre' between the heavens and the rest of creation; man was the 'medium', and operative science could be regarded as a form of magic. Only mankind was granted the capacity to unleash the virtues hidden in stones, plants, words and characters. The astronomer could be said to draw powers from the firmament into the individual. At one level scientists would be involved in chemical manipulations, separating out the active principles of drugs, thought to be the equivalent of the pure essence derived from the stars; at another they would be counteracting with magical formulae, damage done by evil incantations. The model for this impressive command over nature was the *magus* of the scriptures, especially as exemplified by Moses, Solomon or the three Kings of the East, an image which the renaissance magicians were accustomed to conflate with its kinsman, the *magus* figure of hermeticism.

Although hermetic elements inevitably impinged on Paracelsus's rendering of magic, this aspect of his worldview retained its fundamentally scriptural centre of gravity, and it was firmly related to his particular Christian social ethic. The magus only expressed natural powers by virtue of his status as a believer; the saint and the magus were really two sides of the same coin. Paracelsus firmly believed that the essential capacities of the biblical magus were within reach of the renewed Christian of the scientific age. Not through books or secular learning, but only through the true faith could the powers of the magus be derived from the stars. A more hermetic view of the magus ideal of Paracelsus derives from the corpus of doubtful or spurious writings attributed to Paracelsus, and to the works of the early Paracelsians, sources which have deeply influenced historical accounts of Paracelsus himself and which prove that the Paracelsian movement was swept up in the hermetic tide which engulfed Europe in the late sixteenth century.

The Paracelsus idea of magic extends across a broad spectrum lying between traditional magic and experimental science. For the most part, at least in the genuine writings, the emphasis was firmly towards the latter. The intervention of the magus in nature was seen to be successful by virtue of his knowledge of natural processes, skill in manipulation, and direction of the forces inherent in nature, rather than by the employment of intelligences, or miraculous powers. Thus, for the most part Paracelsus was an advocate and practitioner of natural magic, and much of what constituted natural magic for Paracelsus represented a fund of sound observations of a kind prerequisite for work in the experimental sciences developed in the course of the ensuing century. Much of what was rejected later was also rejected by Paracelsus as credulous vanity. He fully appreciated that the repute of legitimate magic was placed at risk by foolish or unscrupulous exploitation of ceremonies, conjurations, blessings, and curses. There remained a penumbra having less certain status: the doctrine of signatures, physiognomy, and chiromancy persisted and seemed to be eminently reasonable areas for scientific speculation; hieroglyphs, characters, and emblems lost none of their effectiveness as a means of conveying and organizing information, and they continued to be regarded as a vehicle for the transmission of arcane truths from the ancient to the modern world.

Although expressed in his own characteristic language the idea of natural magic developed by Paracelsus coincided in major respects with the natural magic of the Neoplatonists. A sequence of authors from Ficino onwards, and including Pico della Mirandola, Trithemius, and Agrippa, placed magic at the heart of their system of ideas. Like Paracelsus they distinguished natural from demonic magic, emphasized the dependence of physical change on the earth, on forces derived from the heavens. They also believed that the understanding of these natural forces could be turned to operative effect, opening up for man the possibility of achieving by natural means what had hitherto been regarded as miraculous, that is occasioned by good or evil intelligences. All of this was to be attained by the skilful assistance, imitation, or direction of nature, an approach echoed by Bacon in the opening words of *Novum Organum* where mankind was heralded as 'the servant and interpreter of Nature'.

Natural magic became the vehicle for the projection of the term magic into the normal vocabulary of the sciences, so bringing about connotations of the transcendental potentialities of science in both its pure and applied forms. Ficino had firmly distinguished between the magus and the sorcerer, the magus being 'a contemplator of heavenly and divine science, a studious observer and expositor of divine things, a figure respected in the gospels, not signifying a witch or conjurer, but a wise man and priest'. Similarly Trithemius emphasized that his own practice of magic had nothing to do with the popular tradition, being based on the sophisticated knowledge of mathematics and concerned with the analysis of mathematical harmonies within nature. This 'natural magic' was pure, solid, and permissible, in contradistinction to the worthless magical cults of the illiterate. As Pico declared: 'Magic is the sum of natural wisdom, and the practical part of natural science, based on exact and absolute understanding of all natural things.' According to Pico, by the employment of acute observation rather than demons, the forces of the celestial world might be brought to bear on the terrestrial world in order to perform natural works rather than to seek miracles. Agrippa affirmed that

> Magic comprises the most profound contemplation of the most secret things, their nature, power, quality, substance, and virtues, as well as the knowledge of their whole nature. It instructs us concerning the differences and similarities among things, from whence it generates its marvellous effects, by uniting the virtues of things by the application of one to another, joining and knitting together appropriate inferior subjects to the powers and virtues of superior bodies.

The idea that the essence of magic was the application of 'agents to patients' or 'actives to passives' became the hallmark of definitions of natural magic from Roger Bacon to Francis Bacon. At one extreme this idea involved an experimental approach to nature; at the other it implied that the full operational utility of natural magic entailed the magi immersing themselves in the harmonies of nature.

Agrippa complained that because the natural magicians were capable of performing operations that were 'above human reason', or 'before the time ordained by Nature', they were thought by the uninformed to aspire to perform miracles, and hence to be in league with the Devil. This much was suspected of Agrippa, and the supposition that mathematicians and natural magicians were merely learned conjurers persisted and was used to stir up enmity against them. Conrad Gesner acknowledged that Paracelsus was a clever physician, but warned that he was also a magician who was guilty of consort with demons. This charge is more understandable in the case of those natural magicians hypnotized by the possibility of individually attaining universal wisdom, a direction taken by Guillaume Postel, Giordano Bruno or, in his later career, by John Dee. This mentality was responsible for generating such esoteric fantasies as the Rosicrucian brotherhood. By contrast Paracelsus had placed emphasis on the operative aspect of natural magic, and he anticipated that the growth of knowledge and amelioration of social conditions would be consequent upon a massive collaborative effort of the kind exemplified by the division of labour among craftsmen. This tendency towards the democratisation of magic found its classic expression in such figures as della Porta, Bacon, and Samuel Hartlib. At the social level this attitude towards natural magic was averse to the exclusivism of secret brotherhoods of illuminati, pointing instead in the direction of coordinated effort involving all classes of operator, and the free exchange of information. [...]

6.4 Paula Findlen, *Rudolf II and the Prague court**

Shortly after Rudolf II moved from Vienna to Prague in 1583, court architects and engineers began to construct long galleries in Hradčany Castle to house several thousand paintings and equally vast numbers of sculptures, coins, vases, gems, natural rarities, precious medicines, scientific instruments, clocks, books and other curiosities that became the imperial *Kunstkammer*. Slowly the spaces known as the Spanish Room and the New Room emerged on the first floor, above the stalls on the ground level. In this northern part of the palace, Rudolf placed the majority of his artistic possessions – Titians, Correggios, Brueghels, Dürers, the best that Renaissance Europe had to offer – and an antiquarium. Court artists and craftsmen decorated the galleries, creating elegant chests to display the Habsburg treasures and embellishing the rooms with appropriate motifs that portrayed Rudolf as a new Maecenas who commanded the world from the vaults of his museum.

Construction of the rooms which housed Rudolf II's marvels took somewhat longer; these artefacts remained in the southern wing of the palace until approximately 1605–6. By then, the imperial artisans had completed the three vaulted rooms on the first floor known as the 'anterior *Kunstkammer*'. Just on the other side of the Mathematics Tower lay the *Kunstkammer* itself, containing an extensive display of some of Europe's finest scientific instruments, a handful of books and manuscripts and a desk where the Holy Roman Emperor sat amidst his treasures. Nearby was the library; below lay the workshops in which the imperial artisans laboured and possibly an alchemical laboratory. Visitors usually entered the *Kunstkammer* through an antechamber decorated with images of the four elements and twelve months, a microcosm of nature supervised by Jupiter, Rudolf's mythological

---

* Paula Findlen, 'Cabinets, collecting and natural philosophy', in Eliška Fučíková *et al.* (eds), *Rudolf II and Prague: The Court and the City* (Prague Castle Administration and Thames and Hudson, 1997), pp. 209, 213–17.

alter-ego. Thus in the last decade of Rudolf II's reign, he epitomized the image of the learned monarch who created and consumed all forms of knowledge, inhabiting a setting that the English statesman and philosopher Francis Bacon might have described as the realization of his image of a palace fit for a 'philosopher's stone' had he ever made the voyage from London to Prague. In fact, this was an apt image for the castle of the 'German Hermes' whose library contained many alchemical and cabbalistic treatises and whose rooms were the site of many attempts to engender the philosopher's stone.

Few visitors enjoyed the privilege of seeing these rooms of wonders, yet word of their contents spread throughout Europe, so that many travelled to Prague with the aspiration of gaining an audience with the Holy Roman Emperor by presenting him with a gift worthy of his collection. By the time of Rudolf's death in 1612, it was hard to distinguish his own reputation from that of his artefacts. They had come to represent the Emperor perhaps better than the Emperor himself: 'Rudolf of few words', as the Venetian ambassador described him, spoke best through the objects in the imperial *Kunstkammer*. Even after the dispersion of the collection in the seventeenth century, it continued to serve as a talisman of Rudolf's reputation as a learned and cultured ruler who saw the world as a precious cabinet of curiosities to be commanded and possessed at will. [. . .]

Even the most casual visitor to Prague could observe Rudolf's wide-ranging curiosity about nature. Around 1603, the physician Johan Eck, a member of the Roman Accademia de' Lincei, which ultimately included the magus Giambattista Della Porta and the astronomer and mathematician Galileo Galiliei among its luminaries, reported to his friends that 'His Majesty [is] inclined towards Lincean things'. A few years later, the Venetian ambassador confirmed Rudolf's interest in 'the secrets of natural matters, as of artificial', observing that 'he who has the chance to treat of these things will always find the ears of the emperor ready'.

In pursuit of these goals, Rudolf assembled a rich array of flora and fauna and an extensive collection of specialized instruments that might help him pursue knowledge of nature. At the height of his reign, a deer park surrounded Hradčany Castle, complemented by an aviary (in which one might glimpse the Emperor's birds of paradise and, after 1598, a dodo) and a botanical garden where distinguished naturalists such as Rembert Dodoens and Carolus Clusius tended exotic plants. From the 1580s onwards, a steady stream of visitors such as the English alchemists John Dee and Edward Kelley, the Italian mystic and Neoplatonist Giordano Bruno, and the physicians and occultists Oswald Croll and Michael Maier enjoyed audiences with Rudolf, bearing gifts of magic talismans, their own writings and promises to unlock the secrets of nature. At the same time, Rudolf looked to the heavens. By 1600, his interest in astronomy had become evident, as he willingly financed the expansion of various palaces to accommodate the astronomical instruments of the Danish noble Brahe and such assistants as astronomer Kepler. Prague had become a microcosm of the scientific activities which characterized late Renaissance intellectual life and, accordingly, housed the artefacts necessary for such endeavours.

The 1607–11 inventory of the *Kunstkammer* reflects the fusion of these interests. Organizing the objects in the imperial collection, the antiquarian Fröschl began with *naturalia*, turned to *artificialia*, and ended with *scientifica*. The entries in the first category reflected the penchant for exotic creatures that characterized a great deal of Renaissance natural history.... Armadillos, chameleons and pelicans crowded between crocodiles, iguanas, tortoises, starfish, blowfish, seahorses and the ubiquitous birds of paradise. Not content to display the living examples in his aviary, Rudolf proudly collected as many dead birds as possible, often announcing the state of the specimen: 'One bird of paradise with its natural wings and feet'. When specimens were inadequate or lacking, he called on painters such as his own artist Joris Hoefnagel or the Medici court artist Jacopo Ligozzi to produce images of natural objects. [...]

The condition of other objects further suggested the idea of a working collection that court physicians and naturalists regularly consulted. It is unlikely, for instance, that Rudolf created displays of 'two dissected frogs', 'seven teeth and one small [tooth] from a hippopotamus' or 'seventeen tigers' and leopards' claws' without the assistance of his naturalists. Nor can we imagine Rudolf alone determining which *terrae sigillatae* (an ancient medicinal clay imprinted with seals), bezoar stones and amulets might be the most efficacious; we know that he lent samples of his medicinal earths to the Paracelsian physician Oswald Croll while Croll was completing his study of the hidden virtues of nature. During his tenure as a court physician, Croll perfected plague remedies such as the Paracelsian *zenexton*, an amulet often presented in a bejewelled gold case containing a cake made of toads, virginal menstrual blood, white arsenic, orpiment, dittany, roots, pearls, coral and Eastern emeralds; the recipe appeared under imperial privilege in his *Basilica chymica* (1609). Most of the ingredients and similar antidotes could be found in Rudolf's *Kunstkammer*....

Natural magic provided an important meeting ground for many of Rudolf's interests. It embodied the image of productive, encyclopaedic knowledge that emphasized the manipulation of nature and the creation of wondrous artefacts – both central images in the Rudolfine *Kunstkammer*. Natural magic occupied an important role in Rudolf's intellectual development. He was an avid reader of della Porta's *Natural Magic* (1558) and, in 1604, sent a courier to Naples to entreat the ageing magus to send an assistant to Prague to teach the Emperor natural magic. '[W]hen our arduous tasks of government permit,' Rudolf confessed, 'we enjoy the subtle knowledge of natural and artificial things in which you excel.' [...]

Just as objects played an important role in della Porta's *Natural Magic*, they also gave testimony to the Emperor's command of natural and artificial secrets. The 'perspective instrument' created by the imperial clockmaker Burgi accompanied an entire category of crystal and glass mirrors as well as a square crystal and ebony table with cabbalistic symbols inscribed on parchment beneath it. Elsewhere, the Venetian lenses for imperial telescopes lay safely tucked away, waiting to be put to use. And possibly an object described as a 'handsome silver cylinder covered with crystalline glass' may have been one of the telescopes built from the instructions in the *Natural Magic* or sent by Galileo to Prague. Given the timing of the inventory, which Fröschl completed one year after Galileo announced his discovery of the moons of Jupiter with his new and improved telescope, it is hard to say which version would have prevailed. Since we know that Rudolf questioned Kepler extensively about his opinions of della Porta's *Natural Magic* in 1610, the sciences of natural magic – in which optics and ciphers figured prominently – were very much on his mind.

As Rudolf's persistent questioning of Kepler suggests, he took an active interest in the production of artefacts that demonstrated power over nature. Natural magic made no clear division between the works of nature and those of humankind, suggesting that both participated in the act of creation. So great was Rudolf's engagement with these pursuits that he often spent afternoons perfecting his knowledge of various arts and crafts by participating in the activities of the imperial workshops. In 1609, the Tuscan ambassador commented unfavourably on this practice: 'For he himself tries alchemical experiments, and he himself is busily engaged in making clocks, which is against the decorum of a prince. He has transferred his seat from the imperial throne to the workshop stool'. Yet Rudolf's decision to practise the sciences he patronized was not uncommon. Ironically, our best image of the 'prince-practitioner' comes from the Medici court that this particular ambassador represented. The image of the prince in his laboratory, surrounded by artisans and advised by an alchemist, evokes an important part of Rudolf's world. Certainly, his *Kunstkammer* inventory echoed the same desire to perfect all sciences in the service of a common goal. [...]

Chapter Seven
French Science in the Seventeenth Century

7.1 William Ashworth, *Catholicism and early modern science**

Whenever the issue of Catholicism and early modern science is raised, it is usually quickly narrowed to the subject of Galileo and the church, which is still such a lively and unsettled question that we have rarely had the time or the stamina to look at the rest of the story. For this one occasion, the great Florentine has been removed from the scene. The background pattern that remains is the subject of this essay.

First a comment about scope: It is clear that my task, if conceived as a comprehensive analysis of the interaction of science and Catholicism in the seventeenth century, is hopeless; such a project, done properly, would take volumes. However, it does seem possible within a limited space to suggest the kinds of questions we ought to be asking and, in a tentative way, the probable nature of the answers. ... It seems to me that the essential questions are these: (1) In the case of the individual Catholic scientist, is there evidence that his religious views affected his scientific work, in either his preconceptions, his motivation, his discoveries, or his conclusions? (2) In the case of Catholic scientists viewed collectively, do any tendencies or interpretations emerge that are distinctive enough to be called a Catholic pattern? (3) Were Catholic scientists helped or hindered by the church in its institutional capacity? (4) For the many scientists belonging to religious orders, how, if at all, did the rules and programs of their various orders influence their science?

* William Ashworth, 'Catholicism and early modern science', in D. Lindberg and R. Numbers (eds), *God and Nature: Historical Essays on the Encounter between Christianity and Science* (Berkeley, CA: University of California Press, 1986), chapter 5.

Varieties of Catholic scientific experience

The seventeenth century witnessed remarkable changes in the philosophy, methodology, and content of science. In a short list of the important developments one might include the establishment of the mechanical philosophy, the discovery that the laws of nature are mathematical, the acceptance of a new cosmology and its ramifications, an increasing reliance on experiment, measurement, and observation, and a growing tenor of anti-Aristotelian sentiment. Many Catholics played roles, large and small, in all facets of this revolution. What we would like to know is whether there was anything in Catholic theology that made such changes harder, or easier, to perceive or to accept. Did Catholic scientists have a view of God, nature, reason, process, Scripture, or authority that one can characterize as peculiarly Catholic, and did these views facilitate, or hinder, any aspect of the metamorphosis of science? In short, what difference did the Catholic faith make for the seventeenth-century scientist?

In order even to begin to answer such a question, we must look first at individual cases. Ideally one would like to consider a large number of Catholic scientists,... but since an exhaustive analysis is obviously impossible, I present instead... a brief discussion of five scientists in sequence. I did not choose these five at random, as a sociologist might wish, but rather selected those I consider to be the most important of the period, a choice that reflects my own conviction that the scientific revolution was perpetrated by great scientists and not by some kind of population pressure or genetic drift. For each of these five, we wish to know the extent to which scientific stances were influenced by religious views and vice versa. With all this in mind, we can then ask whether any common features have emerged that may be said to characterize the Catholic scientist in the seventeenth century.

Marin Mersenne (1588–1648) provides an interesting point of departure for this excursus, since not only was he instrumental in the rise of the mechanical philosophy, but his motivation was, at least initially, exclusively religious. Mersenne was alarmed in the 1620s by the growing threat of atheism (in its seventeenth-century sense of misbelief rather than disbelief). And for Mersenne the greatest danger lay in the philosophies of Italian naturalists such as Pietro Pomponazzi, Girolamo Cardano, and Giulio Cesare Vanini. Italian naturalism was considered dangerous to religion because it confused the natural with the supernatural and physics with metaphysics; essentially, it eliminated the boundaries between science and faith. Miracles, for example, were endangered by the naturalists, because in a world filled with sympathies and occult forces... anything could happen naturally. So Mersenne declared war not only on heterodoxy but on all occult philosophies, such as Hermetism, alchemy, and natural magic. ... And in order to preserve the realm of the miraculous and supernatural, Mersenne was driven to mechanize the natural. He reduced nature to an ensemble of phenomena that proceed according to natural law, recognizing that some kind of natural order is a prerequisite for a miracle that is contrary to nature. The mechanical philosophy, then, was Mersenne's instrument for preserving the realm of faith.

Having advocated a general philosophical position, Mersenne went further. He also maintained, and demonstrated by example, that the world of phenomena was worthy of study – that the laws of nature could be, and should be, determined by human observation and experiment. This attitude does not seem so surprising in the courtly Galileo, but Mersenne was a member of one of the most ascetic orders in all of France; as a Minim, he lived in perpetual Lent, practicing an extremely involved daily ritual. Yet Mersenne found it important and permissible to roll balls down planes, measure the rate of free fall, and determine the speed of sound. His example alone is sufficient to call in question the claim that personal observation of nature was uniquely associated with the rejection of religious authority advocated by Calvinism.

Many consider Mersenne's greatest contribution to science to have been the correspondence network he established and presided over, and it is interesting that here too his motivation seems to have been at least partly religious. Mersenne had no illusions about the ability of humans to penetrate very far beyond the observation of phenomena; he judged absolute knowledge of the essence of things to be beyond our powers. But the Christian does not need absolute knowledge to live a life of faith, nor does a scientist need to know causes in order to comprehend nature. So Mersenne concluded that the problems of science and faith could be worked out by gathering the opinions of intelligent men; by talking things over, a consensus of some kind could be reached that was good enough for all practical purposes. The idea that group activity could lay to rest the skeptical demon was most unusual; only Francis Bacon at this time was having similar thoughts (for quite different reasons), and Bacon never put such ideas into practice. Mersenne did, and in the process sowed the seeds for the social transformation of the scientific enterprise in the last half of the century.

Before leaving Mersenne, it is important to note that while he was beyond question a man of sincere Catholic faith, many of his theological views were quite untypical of the age. He supported Galileo when it was most unfashionable; he saw no theological problems in Gassendi's proposed revival of Epicurean atomism.... He was, in other words, tolerant beyond the norm of the times, whether Catholic or Protestant.

The religious beliefs of René Descartes (1596–1650) may have aroused the suspicions of Blaise Pascal, but Queen Christina had no qualms about the genuineness of his faith, for she converted to Catholicism in 1655 and gave the credit to Descartes. She was shrewder than many later critics, for Cartesian philosophy is quite inexplicable without God. The existence of God can be demonstrated by both causal and ontological arguments, and the proof of his existence is perhaps the most important link in the entire chain of deductions that gave rise to the Cartesian system. ... He is the guarantor of right reason and of the reliability of clear and distinct ideas. It is God's immutability that ensures the existence of laws and nature and necessitates the conservation of motion in the world. It is God's continuous presence that conserves those laws. So God is considerably more than first cause for Descartes.

Although Descartes believed that certain questions of theology could be answered by reason – the existence of God, the distinction of soul and body – he was very much opposed in general to rational theology. Matters of faith could be determined only by revelation; but matters of fact were the domain of reason and observation. Men abuse Scripture, he said, who use it to answer questions of reason, as do those who attempt to reason about faith. In truth, Descartes would have prefered to eliminate theology altogether, at least as represented by Scholasticism; it was an unnecessary middle ground between religion and philosophy, and the two realms of knowledge were better kept entirely separate.

Descartes had a similar disdain for theological arguments drawn from nature. The world was not for Descartes a collection of signs that demonstrated divine attributes or pointed the path to God. He rejected outright any doctrine of final causes, stating that whatever the purposes of God, they were too impenetrable to be discerned by mere observation of nature. His stance was quite similar to that of Bacon and rather different from that of most contemporary Christians, Catholic as well as Protestant.

Descartes's decision to ignore all theological questions except the ultimate one gave his work a flavor that could easily be disagreeable to more conventional philosophers and theologians. For all the attention he gave to God, his world of vortices was not noticeably Christocentric, a defect that Malebranche would later try to remedy. Descartes's assertion that doubt was a suitable vehicle for the acquisition of knowledge certainly came as a surprise to those weaned on Aquinas and seemed to many critics to undermine the very basis of religion (although Descartes himself did not apply his method to matters of faith). There are also passages in his work – particularly the posthumous treatises – that suggest dissimulation, as when Descartes asks us 'in our imagination' to move out of sight of the world God actually created five or six thousand years ago and to consider a new world where the laws of nature, acting on chaotic matter, produced a world very much like our own. Had Pascal been alive to read that, he would have had even more doubt about Descartes's sincerity.

But the most severe theological objection to Cartesianism would come from the implications of his theory of matter. Descartes's assurance that the worlds of faith and nature do not intersect ignored the reality of contemporary theology; and he failed to anticipate that, by equating matter with extension and denying the reality of secondary qualities, he had completely undermined the doctrine of the Eucharist. If accidents have no real existence, how could the accidents of bread remain while the substance was miraculously transformed into the body of Christ? Descartes was taken to task for this by Antoine Arnauld in the 'Fourth Set of Objections' to the *Meditations*; and, while Descartes attempted to defend the compatibility of his matter theory with transubstantiation, he was not noticeably convincing, and most of the ensuing problems faced by Cartesians would grow out of this one deviation from Scholasticism . . . .

At first glance Pierre Gassendi (1592–1655) seems to have traveled the same intellectual path as Descartes and Mersenne. Like his two contemporaries, he was adamantly opposed to Scholastic philosophy and Aristotelian science, and he bore a similar antipathy toward nature philosophies and alchemy. He too was profoundly affected by the skeptical revival of the period and was led ultimately to embrace a thoroughly mechanical philosophy of nature. But a closer look reveals that, while Gassendi may have started from the same point and reached some of the same conclusions, he really took an entirely divergent route, defended his conclusions quite differently, and held a view of the relation between science and religion that distinguished him markedly from both Descartes and Mersenne.

Gassendi was by inclination a Christian humanist in the tradition of Erasmus. When he began in the 1620s, along with Mersenne, to rebel against Scholasticism, his dissatisfaction did not lead him to a rejection of all authority, as it had Descartes; instead he instituted a search for a better ancient authority, which he ultimately found in Epicurean atomism. As a Catholic priest he readily saw the atheistic implications of this mechanical philosophy of antiquity, but he believed that, properly refurbished, atomism could be completely compatible with Christian theology – certainly more compatible than Aristotelianism. And so he tailored an atomic philosophy according to Christian guidelines. Since an infinite number of atoms was incompatible with the idea of a provident God (in such a universe, everything possible will happen), Gassendi determined that the universe of atoms is finite. He added to the human corporeal soul a separate, incorporeal mind that God creates *ex nihilo*. . . .

Gassendi thus ultimately arrived at a view of nature that was as mechanical as that of Descartes (although the two differed on important points, such as the possibility of a void and the divisibility of matter). But having gained a similar position, Gassendi defended it in quite a different way. It was evident to Descartes that matter was only extension; this conclusion could be rationally demonstrated. For Gassendi, by contrast, no fact of nature could be rationally demonstrated, and very few things, certainly not the nature of matter, were evident. Gassendi defended atomism because it was compatible with phenomena, whereas Aristotelianism was not. But it was one thing to say that atomism was plausible, quite another to say that we can learn precisely how the interaction of atoms causes sensible effects. Gassendi denied the possibility of gaining knowledge of such causes; only phenomena may be known, only effects may be observed, and the scientist need concern himself with nothing else.

Gassendi's strict scientific empiricism seems to have been the product of his fideistic theology. In Gassendi's view, reason was useless for religious purposes, since all religious truth comes by faith. Consequently he opposed all forms of rational theology, objecting strongly, for example, to Descartes's attempts at a rational proof of the existence of God. But if certainty comes only through revelation, then the study of the sensible world, beyond the pale of faith, is doomed to uncertainty. And so the scientist must be satisfied with the observation of phenomena, and he must resist the temptation to reason about causes that he can never know.

Gassendi's fideism led him to disagree with Descartes in one further respect. Descartes tried to insulate science from religion; Gassendi sought to remove the barriers. If our only certain truths are theological ones, it would be foolish indeed to ignore these when observing nature. And thus many of the premises of his atomic philosophy have an explicitly Christian origin. Consistent with this proposition is Gassendi's favorable opinion of arguments from design. Descartes rejected final causes out of hand, but Gassendi saw purpose in all of nature, and he suggested to Descartes that if he wanted to prove the existence of God, he should abandon reason and look about him; the order and harmony of the universe demonstrate God's existence and his attributes of goodness and providence. For Gassendi, the Two Books (nature and Scripture) were not to be kept on separate shelves. [. . .]

If Gassendi still resists categorization, Blaise Pascal (1623–1662) positively defies it. At once Christian pessimist, Cartesian rationalist, scourge of the Jesuits, mystic, skeptic, and experimental physicist, he remains an enigma impossible to resolve. He was clearly a man of different faces at different times . . . .

Perhaps Pascal's most novel insight was his realization that authority plays different roles in religion and science. In his remarkable scientific manifesto, the preface to the unpublished 'Treatise on the Vacuum', he asserted that in theology the authority of Scripture and the Fathers is the sole source of truth . . . . In those matters subject to reason and the senses, however, authority is totally useless, and we should 'pity the blindness of those who rely on authority alone in the proof of physical matters'. . . . Pascal must join Mersenne as another counter-example to the notion that Protestants were better able to disregard authority in scientific matters.

Given this clear demarcation of the provinces of reason and faith, it is not surprising that Pascal spurned the very premise of rational theology. His distaste for reasoners in religion was evident early in his career, when he pursued Sieur de Saint-Ange all the way to the archbishop, disputing his claim that even the Trinity could be demonstrated by reason. When the Jesuit Etienne Noël casually mentioned God in connection with the void in the barometer, Pascal replied – somewhat nastily – that the mysteries of the Divinity should not be profaned by scientific disputes. . . .

Pascal's feelings about religious arguments drawn from nature were more complex. Unlike Descartes, Pascal *did* believe that nature expressed the handiwork of the Creator .... But this latent theology of nature is nipped in the bud by Pascal's deeper conviction that God is not manifest, but irretrievably *hidden* in nature. 'God has hidden [truth] behind a veil', Pascal claims, and the study of nature will never lift that veil. If truth lay in nature, then anyone might discover it and know God. But for Pascal, such knowledge comes only through Scripture and grace. [...]

...There remains, however, one aspect of Pascal's career that refuses to bond with the rest: his foray into experimental physics. From the perspective of the historian of science his investigations of the barometer constitute a most brilliant achievement. But the Pascal we have portrayed had, quite simply, no business pursuing experimental science. His feelings about nature, even in his early years, should not have justified such concern as to whether air has weight or a void exists.... [T]he fact that jars is that Pascal *himself* became so engaged in this physical world, at the very time he was telling Mme Perier that tending to nature is idolatry, the worship of the creation rather than the Creator. There is of course no requirement, then or now, that a man live a life consistent with his beliefs, but I think it quite probable that Pascal felt the dissonance within and that he ultimately abandoned scientific inquiries because he could not justify them to himself. I make this point because it seems to set Pascal apart from the other Catholic scientists we have discussed. I think Descartes, Mersenne, and Gassendi were, to varying degrees, motivated in their science by religious factors. With Pascal such is not the case. Pascal was certainly a great Catholic and a great scientist. But if there is a causal thread that runs from the former to the latter, I fail to detect it.

Nicolaus Steno (1638–1686) is seldom mentioned in the same breath with Pascal or Descartes, yet he was arguably a more original thinker than either. His proposal that the surface of the earth contains the evidence of its own history – indeed, that an individual rock or fossil exhibits clues to the place and manner of its production – is as brilliant an insight as one can find in this marvelous century, and Steno's consequent unraveling of the geological history of Tuscany and his proposal of the organic origin of fossils shows that insight was coupled in his case with an extremely capable scientific methodology. Steno, in short, belongs on any list of great seventeenth-century scientists.

He belongs here, however, because he was a Catholic scientist, and he was so in a different sense from the others discussed. With Pascal or Descartes one never escapes the feeling that science and theology were in unstable equilibrium, that tensions existed between their worlds of faith and reason. With Steno one finds no hint of conflict. Raised a Protestant, Steno came to the Catholic faith quite deliberately, converting shortly after completing his treatise on the shark's head; and, wrapped in his newly acquired religion, he proceeded to write his great geological work, the *Prodromus*. It is true that his religious fervor would shortly carry him right out of the scientific arena, as he became priest, then archbishop (and soon it appears, a saint). But in those fruitful years between 1667 and 1670 he was equally at home in the spiritual and material realms, and his thoughts on certainty, nature, and Scripture are illuminating.

Steno had little use for the deductions of Cartesian rationalists, but he was also impatient with skeptics who argued that we can know nothing with certainty. There is indeed a great deal that is uncertain, admits Steno; with matter, to use his example, we do not know whether it is divisible or indivisible, whether it has other attributes than extension, or whether interstitial vacuums exist. Nevertheless, amid all this uncertainty we can still be sure that... in the case of rock strata, aggregates on top must have been formed after those beneath .... [...]

But by far the most arresting feature of Steno's geological thought is his failure to find any incompatibility between nature and Scripture. He was, after all, attempting to reconstruct the early history of the earth, a story which in his time was known *only* through the account in Genesis; but none of his findings, in his eyes, suggested disagreement with Scripture. Discussing the earliest phases of Tuscany's geological history, he points out that 'Scripture and Nature agree' that everything was once covered with water; as to how long it lasted, 'Nature says nothing, while Scripture speaks.' In other cases nature supplements Scripture, or neither provides an answer, but in *no* instance does nature suggest one thing and Scripture another.

Some scholars have interpreted this sanguinity as blindness and a tragic failing, and have lamented that the continued development of geology was held back for a century by Steno's reluctance to break away from a scriptural history of the earth. Such criticism misses the point. Steno would indeed have achieved towering greatness had he been able to anticipate every eighteenth-century development. But it is even more marvelous, and more instructive, that he was able to set down most of the principles of modern geology *without* departing from the traditional religious framework.

If there is anything to regret in Steno's career, it is that he was unable to hold on to his vision of the natural world as his theological interests expanded in the early 1670s. Already in 1673 he was extolling the unknowable over the world of observation, and in his last geological essay, the 'Treatise on Ornaments' of 1675, we see him reverting to a hermeneutic view of nature as a collection of signs and symbols, where each gem or stone evokes a religious truth or draws our minds to God. What this change tells us about Steno, or about science and religion in general, is difficult to ascertain. Some will think that Steno slid from the summit, but others will contend that he went on to better things. Not many geologists, after all, achieve sainthood.

What can we conclude from these portraits of five Catholic scientists – Mersenne, Gassendi, Descartes, Pascal, and Steno? Are they part of a larger picture? Do they, collectively, project something we can call a Catholic pattern? The answer, I think, is obvious; the term 'Catholic science', judging from these five individuals, has no meaning whatsoever. ... All were motivated by religious considerations, but each was led in a different direction, to positions that were mutually irreconcilable. Five is of course a tiny sample, but were you to enlarge it by including ... Marcello Malpighi, Giovanni Alfonso Borelli, Francesco Redi, and Evangelista Torricelli (and, of course, the implicitly present Galileo), the resulting overall picture would be even more confusing and a pattern even less evident.

The failure of a pattern to emerge is very significant. It means that nothing was inherently denied to the Catholic scientist by his personal faith. He could be, and was, rationalist, empiricist, skeptic, mechanical philosopher, mystic, natural theologian, atomist, or mathematizer. And the lack of a Catholic pattern considerably weakens the case for a Protestant one. To demonstrate this point, I suggest a thought experiment. Place our original Catholic five in a room with the Protestants Francis Bacon, Johannes Kepler, Christiaan Huygens, Robert Boyle, and Isaac Newton; add Galileo and Gottfried Wilhelm Leibniz for good measure; and divide the resulting dozen into two halves on any criterion *other* than religion. I doubt that anyone could produce in this manner a contingent exclusively of Catholics facing a side entirely of Protestants.

If Catholicism were just a personal creed, this discussion could end here, and we could conclude that the Catholic faith was rarely a hindrance, and often a considerable source of inspiration, for the architects of the scientific revolution. However, to be a Catholic in the seventeenth century was not just to subscribe to a creed; it was to be part of an institution, one of the most bureaucratically complex institutions of the era. The Catholic church could and did set policy on many matters of faith and conduct, and the true Catholic not only accepted these decisions but believed, as an essential ingredient of his faith, that the decisions of the church were necessarily correct. Because of the important role played by the church in determining questions of faith for the individual believer, it is vital that we now ask how the church, in its institutional capacity, reacted to the new developments of seventeenth-century science and what effect these reactions had on the Catholic scientific community.

Science and the institutional church

Religions are not, as a rule, novelty-seeking institutions, and it should come as no surprise that the Catholic church of the seventeenth century was decidedly reluctant to embrace many of the novel discoveries and reinterpretations of contemporary science. One could, after all, make the same statement about every other religious institution of the period, to say nothing of political and even educational institutions. Any religion more than a few hours old is going to take a dim view of radical theories that undermine its own dogma, and the Catholic church, as one of the most venerable and traditional of faiths, can hardly be faulted for resisting assured claims that the earth moves, that color is an illusion, that space is empty, and that in general modern scientists know more about nature than all the philosophers of antiquity.

The Catholic church, however, stands out from all other seventeenth-century institutions in that it not only criticized unwelcome ideas but was uniquely equipped to censor and repress such ideas and even to punish their advocates. This capacity for policing the faith was a legacy of the Counter-Reformation and two innovations of the Council of Trent – the Congregations of the Index and the Holy Office (Inquisition) – that were regularly used to identify heretical or dangerous ideas and ensure their containment and elimination. Moreover, since the church was both centralized and hierarchical, considerable pressure could be brought to bear on offenders by archbishops and bishops, generals of religious orders, and theological faculties at various universities. The resulting machinery of ideological suppression was formidable.

This apparatus, of course, was not established to deal with earth-movers or mechanical philosophers, but with those who would insinuate such Protestant heresies as consubstantiation or the priesthood of all believers into the mother church. Nevertheless, by the beginning of the seventeenth century several Renaissance philosophers of nature had become entangled in the machinery. ... Since this ... includes some of the most original minds of the Renaissance, and since their significance lies in their contributions to science and cosmology rather than theology, it seems appropriate to ask how the church found itself in what was, in hindsight, the regrettable position of adjudicating questions of natural philosophy.

The ground was laid by the desire at Trent to protect the faith from magic. Magic was anathema to mainstream Catholicism for many reasons. Natural magic could easily become supernatural magic, which involved trafficking with demons. Just as offensive, the supernatural often became natural in magical treatises, leading to explanations of the miraculous. Magic usually involved forms of divination, and the implication that the future is ordained by the lines on one's palm smacked of Calvinist predestination. Magic also gave an inflated role to the individual magus, which again suggested the taint of Protestantism. [...]

To exacerbate matters, when the works of these magical philosophers of nature were examined, other distasteful features also emerged, and these were included in the process of inquisition. Bruno (1548–1600), for example, was charged not only with defending magic and teaching that Moses and Christ were magi but with maintaining that stars have souls and that the number of worlds is infinite. Similarly Patrizi (1529–1597) was forbidden to advocate that there is only one heaven and that stellar space is infinite. These are doctrines that are not so much dangerous to the faith as they are anti-Aristotelian. By choosing to become the watchdog of Aristotelian philosophy as well as of Catholic theology, the church in 1600 was setting course for most of its confrontations with seventeenth-century science.

One of the roads down which the church was drawn led to the prohibition of heliocentrism and the condemnation of Galileo... [another] led to a rejection of alchemy, chemistry, and chemical medicine.

Alchemy was not included in the original church proscriptions against astrology and magic, perhaps because alchemy was in temporary eclipse at the time of the Council of Trent, perhaps because its foremost Renaissance champion, Paracelsus, was a Catholic (although certainly of a heterodox variety). A Paracelsian revival began around 1580, however, and most of the first wave of Paracelsians such as Joseph Duchesne and Oswald Croll were Protestants. Even Andreas Libavius, the principal early critic of Hermetic excess in chemistry, was a Lutheran, and as a result chemistry and chemical medicine in the first decades of the seventeenth century became an exclusively Protestant affair. Probably because of this, and certainly because Paracelsianism shared with magic the danger of tampering with the supernatural, the church tried to stamp it out. An early victim of this attempt at repression was the great physician Jan van Helmont (1579–1644). Helmont was one of the first Catholics to pick up the banner of Paracelsus, and as a result he suffered life-long persecution from ecclesiastical authorities. His treatise on the weapon salve and sympathetic cures was denounced by the Spanish Inquisition in 1625 as smacking of heresy. In 1630 Helmont was convicted by the Louvain theological faculty of perverting nature with magic and diabolical art. He later spent some time in prison, and proceedings against him were abandoned only in 1642. Consequently, he was able to publish nothing for most of his remaining years, and his significance became apparent only with the posthumous appearance of his works. [...]

The impact of this chemical persecution is not difficult to characterize: very few Catholics entered the field of chemistry, and except for the beleagured Helmont, there was no Catholic contribution to the discipline in the seventeenth century. Not that the science was mightily advanced by Protestants. But still, Libavius, Jean Beguin, Daniel Sennert, Niçaise Le Febvre, Nicolas Lemery, and Boyle – Protestants all – were at least able to make a start on the chemical revolution. No Catholic lent a hand. And it was not that chemistry in Protestant countries was free of criticism; many English opponents of Paracelsianism were even more vitriolic than the Jesuits. But in England neither side was suppressed, and out of the controversies emerged Boyle and the Newtonian school. In Italy nothing emerged at all. The Catholic failure to contribute to chemistry should give pause to those who would argue that bureaucratic interference, while regrettable, was only token interference, and that under the surface the search for truth went on.

The Catholic church's campaign against Cartesian philosophy is less notorious than its opposition to heliocentrism, but in many ways it is more illuminating, since it involved censorship at practically every level of the church hierarchy. Descartes, as is well known, was worried about the acceptability of his work in the 1630s; concerned about the Galileo affair, he abandoned plans to publish *Le monde.* Ironically, it was not Copernicanism at all, but his theory of matter, that eventually brought him into conflict with the authorities. As pointed out earlier, Descartes's denial of real accidents challenged the Thomistic explanation of the miracle of the Eucharist, which stipulated that the accidents of the bread remain while the substance is transformed into the body of Christ. Descartes attempted to reconcile his matter theory with transubstantiation in his reply to Arnauld, and all seemed well for a while. But after Descartes's death, when his philosophy was becoming increasingly popular in France, his views on the Eucharist were brought to the attention of Honoré Fabri (1607–1688), papal penitentiary in Rome, and an accomplished Jesuit scientist. As a result, Descartes's *Meditations* were placed on the Index in 1663. The previous year, Descartes had been condemned by the theological faculty at the University of Louvain, and the ecclesiastical pressure against Descartes began to build. In 1671, at the king's request, the archbishop of Paris requested his faculty to cease teaching Cartesian philosophy. The university of Angers took an even stronger stand in 1675, resulting in the condemnation and expulsion of the Cartesian Bernard Lamy. In 1677 the theological faculty at Caen also became officially anti-Cartesian and exiled several offenders. Even religious orders became involved. Nicholas Poisson, like Lamy an Oratorian, was forbidden by his superiors to write a biography of Descartes, and distribution of his Cartesian commentary was suspended. In 1675 the Benedictine Maurists forbade teaching of the Cartesian claim that matter equals extension. . . .

Did such persecution make any difference? Cartesian philosophy, unlike alchemy, continued to find support among Catholics. Probably the most serious result was personal distress, since most followers of Descartes seem to have been sincere Catholics and did not consider themselves heretics in the same class with Bruno. But occasionally the consequences of personal religious crisis manifested themselves in ways that are sad to witness. Jacques Rohault (1620–1675), one of the leading Cartesians in Paris in the 1660s, authored the most widely used Cartesian textbook of the last quarter of the century. But in his subsequent *Entretiens sur la philosophie,* which appeared after the archbishop's warning in 1671, he stated: 'I have no other principles than those of Aristotle. I recognize, as he does, privation, matter, and form. I agree with his general notions and his understanding of the words: substance, accident, essence, and quality.' It is, in its quiet way, an abjuration fully as tragic as that of Galileo.

Apart from these examples, and a few other skirmishes against atomism and the plurality of worlds, the church took little interest in the content of contemporary science, and most work passed unscathed through the censorship machinery. The threat of censorship and inquisitorial proceedings was always in the air, however, and it may be that the most devastating effect of church intervention in science was the dampening of the spirit of inquiry that had bubbled through Italian thought in the late Renaissance. This argument has been made before, although usually so overstated as to maintain that the church killed science in 1633 with one blow, which is of course nonsense. Nevertheless, it would be an overreaction to such hyperbole to assume that the flavor of Italian science was unaffected by the activities of the Curia. Cosmological discussion ceased except among the Jesuits ... and astronomy, with the single exception of Giovanni Borelli, was reduced to the making and using of telescopes. Medicine remained so traditional that Marcello Malpighi (1628–1694) lamented the repressive atmosphere at Bologna and was forced to turn to the Royal Society of London for encouragement and a forum for his discoveries. Even in physics, where the Italians excelled through the middle of the century, there was a noticeable reluctance to speculate about such things as the nature of matter or the significance of the vacuum. This philosophical apathy is most apparent in the activities and publications of the Accademia del Cimento, for which science became anonymous and philosophy invisible. ... In Italy after 1650, a hypothesis could hardly be found, and by 1700 science was struggling to survive.

But it is time we allowed an objection. Defenders of the institutional church who have read this far will be wondering if I have not overlooked an important point: if the church was so resistant to the content of contemporary science, why was it that so many Catholics in Holy Orders were engaged in various scientific investigations? ...

Science in the religious orders

The number of scientists within Catholic religious orders is impressive, as is the quality. ... [W]e can say with some assurance that taking vows was no barrier to becoming a scientist – in many cases a very good scientist indeed.

There is one order, however, that stands out from all others as the scientific order without rival in seventeenth-century Catholicism, and that of course is the Society of Jesus. I wish to focus the rest of my discussion on this remarkable group, for not only do we know a great deal about them, but the nature of Jesuit science raises many important questions concerning the influence of religious beliefs on scientific thought.

A complete bibliography of Jesuit scientific treatises of the seventeenth century would fill the space of this chapter. ... The Jesuits had a particular zest for experimental science; they were interested in every newly discovered phenomenon, from electrostatic attraction to the barometer to the magic lantern, and Jesuits played a major role in discovering many new effects on their own, such as diffraction and electrical repulsion. A recent history of early electrical science awarded the Jesuit order the honor of being the single most important contributor to experimental physics in the seventeenth century. Such an accolade would only be strengthened by detailed studies of other sciences, such as optics, where virtually all the important treatises of the period were written by Jesuits.

More significant than their numbers or their inventiveness is the fact that many of the Jesuits had a keen sense of the value of *precision* in experimental science – a sense that was not widely echoed by many of their more illustrious contemporaries. Riccioli's attempt to develop an accurate one-second pendulum by persuading nine Jesuit colleagues to count eighty-seven thousand oscillations over the course of a day, enabling him to identify an error of three parts in a thousand, is as amazing in its own way as Kepler's more famous dissatisfaction with an error of eight minutes of arc. It is instructive to remember that it was Riccioli – not Galileo, and not Mersenne, and certainly not Descartes – who first accurately determined the rate of acceleration for a freely falling body.

Another admirable feature of the Jesuit scientific enterprise was their appreciation of the value of collaboration. One might well argue that the Society of Jesus, rather than the Accademia del Cimento or the Royal Society, was the first true scientific society. Kircher (1602–1680), the impresario in Rome, was more than a match for Mersenne and Boulliau in Paris or Henry Oldenburg in London, in his ability to collect observations and objects from a worldwide network of informants. More important, Kircher published all this data in massive encyclopedias, which, together with similar efforts of Schott and Riccioli, were as vital as the early scientific journals in disseminating scientific information. If scientific collaboration was one of the outgrowths of the scientific revolution, the Jesuits deserve a large share of the credit.

Thus the Jesuits practiced science on a wide scale, were able (and often inspired) investigators, made many important discoveries and inventions, and encouraged the involvement of others. They do indeed seem to merit the praise they have frequently received. And yet when all this is said and acknowledged, there still remains the unavoidable feeling that Jesuit science was somehow seriously deficient. One does not appreciate this deficiency when reading *about* the Jesuit scientific achievements, but one certainly will by spending a few hours with a volume of Kircher, Scheiner, or Lana Terzi. It is hard to pinpoint what is wrong – indeed, as far as I know, no one has tried – but I wish to suggest what I see as the principal shortcomings.

One problem is that the Jesuit writers were, almost to a man, overly eclectic. ... But an extended encounter with a typical Jesuit treatise leaves the impression not of eclecticism but of a total lack of discrimination. Kircher, for example, in his *Mundus Subterraneus*, sandwiches descriptions of fossil fish between accounts of gems bearing the images of cities and stones in the shape of John the Baptist, and he can sustain such a mélange for hundreds of pages. There is no suggestion that some authorities might be more reliable than others; every fact or observation seems to be given equal weight. Even contemporaries who were equally fascinated by curiosities thought that Kircher was overly credulous, and Leibniz would later deride Kircher for believing tales of rocks sporting the features of Luther and for placing such fancies on the same ontological level as true figured stones.

A second deficiency of Jesuit scientific treatises, less obvious but more serious, is the lack of any philosophical superstructure holding together the facts being presented. There was surprisingly little interest in drawing conclusions about how nature operates. ... Riccioli (1598–1671), in a remarkably erudite discussion of new stars, lists fourteen possible explanations for their appearance, along with the strengths and shortcomings of each. This is eclecticism at its best. But he never chooses one, not even his own offering; he does not seem to care which one is correct, or even to realize that there can be only one true explanation, and that this one, when separated from the rest, will tell us something important about the cosmos. Riccioli and his fellow Jesuits present science as a game to be played or to be watched and admired, but the outcome of which is irrelevant. As a result, no Jesuit comes even remotely close to the stature of Descartes, Galileo, Pascal, or even Gassendi as a natural philosopher.

How is one to explain this... character of Jesuit science? How could so many intelligent scientists invest all that time and energy, become masters of the experimental method, discover all sorts of genuine natural effects, write all those magnificent treatises, and yet play such a small role in the essential developments of the scientific revolution? I am going to offer several possible explanations, and I wish to caution that these are conjectural and unproven, since the necessary supporting scholarship does not yet exist. I hope that in the near future we will see the rise of at least a cottage industry in the history of Jesuit science to test such conjectures, because until we finally solve the Jesuit enigma, we cannot be said truly to understand the role of religion in seventeenth-century science.

My first suggestion is that the Jesuits were handicapped by an emblematic view of nature, which they retained long after it had been discarded by other scientists. An emblematic worldview, which sees nature as a vast collection of signs and metaphors, was a staple feature of Renaissance thought, but in the seventeenth century Bacon, Descartes, Galileo, and their followers rejected the notion that everything in nature carries a hidden meaning. Nature instead was to be taken at face value and investigated on its own terms. The Jesuits, however, had become deeply committed to an emblematic view of the world, because in the late sixteenth century they had begun using emblems and images in their missionary program to attract the faithful....

...As long as one holds the view that nature is an elaborate hieroglyph, important only as a source of mystery and wonder, then the separation of true phenomena from false becomes secondary, if not irrelevant. Such a worldview produced enchantingly elaborate works of art and literature, but its dissolution was an essential feature of the revolution in science. The Jesuits were never able to abandon their emblematic world.

My second suggestion is that the Jesuits' reluctance to commit themselves to any one viewpoint or authority may be related to their probabilist stance in matters of moral theology. I do not pretend to understand the nuances of seventeenth-century casuistry, but the Jesuits were well known, indeed notorious, for their willingness to condone actions that other orders rejected as sinful. The Jesuit position, called probabilism, was that an action could be deemed moral if at least one respectable authority had judged it so, even if his opinion was less probable than that of authorities who denied its morality. The Jesuits found such a 'morally eclectic' position necessary if their missionary work was to be successful, but it meant that they became accustomed in theological questions to giving less probable opinions the same weight as more probable ones. And how interesting that they followed precisely the same course in their scientific writings! They recorded the opinion of every authority, provided he was a reasonable man, and they refused to reject a viewpoint merely because it was less probable than others. ... [I]t is my suspicion that the Jesuits embraced, consciously or not, a form of scientific probabilism that prevented them from taking any firm position, not only on matters of opinion but on questions of fact, and that this provided an insurmountable barrier to the development of any consistent natural philosophy.

A third factor that seems to have had an impact on Jesuit science was the legacy of fictionalism that the Jesuits inherited from Bellarmine, in the aftermath of the Inquisition's edict of 1616. Bellarmine, like Clavius before him, had subscribed to the opinion that astronomical systems were all fictional devices designed only to save the appearances of things, and that appearances have nothing to do with reality, which is the business of philosophers rather than astronomers. Fictionalism was common enough in the sixteenth century, but after Kepler the practice of using hypotheses that could not possibly be true fell rapidly out of favor. The Jesuits, however, were in an awkward position following the condemnation of Galileo, for they were the order charged with defending a geocentric cosmology. Initially they tried invective; as they grew more expert in science, they tried to disprove Copernicanism on scientific grounds; but by the time of Riccioli it had become evident to them that no system could be strictly proved or disproved by scientific means. How then to deal with the increasing superiority of heliocentric celestial mechanics? If the Jesuits had been realists, they would simply have proclaimed the system false, citing Scripture and church decrees, and refused to sanction the utility of an erroneous hypothesis. Such a position, while theologically quite sound, would have been very weak strategically; therefore Riccioli and other Jesuits instead revived the fictionalist stance, discussing heliocentric astronomy with great erudition and even considerable enthusiasm, but always with the caveat that it was merely a hypothesis, like dozens of other hypotheses that scientists adopted for convenience. The problem with fictionalism is that, like probabilism, it leads to excessive eclecticism and discourages the asking of larger questions. The Jesuits did take a position on Copernicanism – as mandated by Scripture and the Holy Office – but on matters where Scripture was silent, their fictionalist posture kept them from asking which hypothesis describes the way things really are. ... The goal was simply to save the phenomena. In the late Renaissance that was a laudable program. In 1670 it was not enough.

One final point should be made concerning the Jesuits' role as defenders of the church's stance on Copernicanism. Non-Jesuit Catholic scientists reacted in differing ways to the pronouncements on Galileo and heliocentrism. Descartes chose silence, at least temporarily, although he did not change his mind. Mersenne waffled, while Boulliau, to Descartes's amazement, went right ahead with the publication of a Copernican work.... Pascal ridiculed the pope for trying to change the facts.... These varying reactions make one point: that while most Catholics did genuinely believe that the church was one of the twin pillars of the faith, they were willing to admit that the church was occasionally wrong, or at least misguided. Such a tolerant attitude was denied the Jesuits; for them the church was never wrong, and its every pronouncement was to be accepted without question and defended without misgiving. The thirteenth Rule of Loyola is explicit: 'If we wish to be sure that we are right in all things, we should always be ready to accept this principle: I will believe that the white I see is black, if the hierarchical Church so defines it.' I think modern historians have failed to appreciate the significance of this point of faith and have misinterpreted many of the Jesuits' actions. Riccioli, Fabri, and others were not acting out of duty in defending geocentrism but out of *belief*. There is no reason to question the sincerity of such belief. ... This was not an easy task for the Jesuits; it is hard to see white as black, especially when others keep shouting 'White! White!' and the effort no doubt colored their views of other matters. I strongly suspect that the Jesuit proclivity for eclecticism, fictionalism, and probabilism, and their mistrust of the larger questions, were at least partially an outgrowth of a sincere attempt to accommodate a point of faith to a recalcitrant world of phenomena.

The Society of Jesus demonstrates, better than any other seventeenth-century case study, the difficulties that could ensue when religious concerns intruded into scientific affairs. The Jesuits were, after all, a missionary order; they were active in proselytizing for the church and keeping heretics at bay. It is not surprising that they would not, or could not, put aside such interests when contemplating the natural order. But when they allowed missionary concerns to shape their epistemology and view of nature, their scientific work was inevitably affected. It seems that the very factors that made the Society such a successful religious order, and set it apart from all others, also figured strongly in Jesuit scientific work, isolating it irretrievably from the main currents of the scientific revolution.

Chapter Eight
Science in Seventeenth-Century England

8.1 Andrew Cunningham, *Harvey and blood circulation**

It is not currently possible to reconstruct the precise sequence of experiments which led Harvey to discover the circulation of the blood, for the book he published in 1628 about the discovery was a carefully worked-out argument, not a laboratory notebook. He called his book *An Anatomical Exercise on the Motion of the Heart and Blood in Animals*, and wrote and published it in Latin. But we can listen here to the breathless way he describes the moment when it occurred to him that the blood circulated. Harvey has been arguing that the blood must pass from the right side of the heart to the left via the lungs (the so-called 'lesser' circulation):

> Truly when I had often and seriously considered with myself and long turned over in my mind how great the quantity of blood was – evident partly from the dissection of living animals for experiment's sake, and from the opening up of blood-vessels and from many ways of investigating, partly from the symmetry and magnitude of the ventricles of the heart and of the vessels which go into it and out from it (since Nature, making nothing in vain would not have allotted that greatness proportionately to those vessels, to no purpose), partly from the beautiful and careful construction of the valves and fibres, and from the rest of the fabric of the heart, and likewise from many other things – viz., the abundance of blood passed through was so great, and the transmission was done in so short a time, that I realised that the fluid from the ingested food could not supply it: and indeed that we should have the veins empty, quite

* Andrew Cunningham, 'William Harvey: the discovery of the circulation of the blood', in R. Porter (ed.), *Man Masters Nature: 25 Centuries of Science* (London: BBC Books, 1987), pp. 73–6.

> exhausted, and the arteries on the other hand burst with too much intrusion of blood, unless the blood did pass back again by some way out of the arteries into the veins, and return to the right ventricle of the heart.
>
> I began to bethink myself whether it had a sort of motion as if in a circle, which afterwards I found to be true, and that the blood was thrust forth and driven out from the heart through the arteries into the flesh of the body and all the parts, by the beating of the left ventricle of the heart... and that it returns through the veins into the vena cava, and to the right ear of the heart...

What settled it all for Harvey, it seems, was the presence of little 'doors' (membranous flaps) in the veins: such things do not occur in the arteries, so what purpose could there be for them in the veins? Fabricius had actually discovered these 'little doors', but had thought their purpose was to slow the outward flow of the nutritive blood. Faced with his new dilemma about the route of blood, Harvey investigated them anew and discovered that they allowed only one-way flow – that they were *valves*. Their role had to be to prevent the blood, returning through the veins to the heart, from slipping back.

Thus did Harvey discover the circulation of the blood. But he had not been trying to discover whether the blood circulated, nor had he initially even been trying to confirm a hunch that it circulated. Harvey was engaged in a quite different enterprise, with its own goals and purposes. This was one reason why the idea that blood circulated dawned on him only slowly and reluctantly. But there was another reason too. Harvey was a willing follower of Aristotle in his whole approach. Aristotle the philosopher and investigator of 'the animal' provided Harvey with his method, his logical tools, his topic, his goals. Indeed Harvey conducted his anatomy in a philosophical spirit: like a good philosopher he was looking for *causes*, trying to give accounts which explained why 'the things themselves' are as they are. The problem that arose from Harvey's devotion to Aristotle was that Harvey could not be totally happy with his new discovery, because he could not find out the most important thing about it: he could not discover *why* – for what purpose – the blood circulated. For a philosophical anatomist, as Harvey was, this was the greatest disappointment: he had to settle for 'speculating' and 'likely reasons'.

Harvey's discovery was not greeted with immediate belief. Many people thought it was absurd, and others took it as a threat to their understandings of how the body worked. Controversy went on for years. John Aubrey, the gossipy antiquarian, heard about it from Harvey himself:

> I have heard him say, that after his book of the *Circulation of the Blood* came out, that he fell mightily in his practice, and that 'twas believed by the vulgar that he was crack-brained; and all the physicians were against his opinion, and envied [had ill-will towards] him; many wrote against

> him. With much ado at last, in about 20 or 30 years time, it was received in all the universities in the world; and...he is the only man perhaps that ever lived to see his own doctrine established in his life-time.

Harvey's discovery of the circulation of the blood was an accidental by-product of his attempt to revive and reinstate in seventeenth-century England the anatomical enterprise that Aristotle had conducted in Athens twenty centuries before. This greatest of modern discoveries about the body was made by a man trying to do anatomy like an ancient Greek philosopher. Only an Aristotelian anatomist could have thrown up the particular set of phenomena which required the hypothesis of the circulation of the blood to explain them: and only an Aristotelian anatomist did. Even though in Harvey's day there were non-Aristotelian and even anti-Aristotelian philosophers, none of them was asking the kind of questions, and of the right object in nature ('the animal'), which could have led them to think that the blood might be circulating. Such questions made sense only to Aristotelians. The discovery of the circulation of the blood was exclusively (and necessarily) the product of an Aristotelian way of thinking and investigating.

All this may seem a bit strange. For we would instinctively expect this great discovery to have been a product of a deliberate attempt to release men's minds from their bondage to the teachings of the 'ancients', and in particular from the teachings of Aristotle. Given that this discovery was made in seventeenth-century England, we would also expect it to be associated with the 'mechanistic' way of thinking then associated (in different forms) with the work of René Descartes, Robert Boyle and Isaac Newton. 'Mechanistic' explanations of natural phenomena are ones which seek to explain events primarily in terms of the motion of the smallest particles of inert matter: it is the approach of which we are the heirs, and which underlies our modern science. We are so imbued with this way of thinking, that the temptation is for us to reconstruct this discovery so that it accords with a mechanistic view, by (for instance) stressing the one instance of measurement that Harvey uses in his argument, and the analogy he makes between the action of the heart and that of a pump. And, because it seems so obvious to us that the blood circulates – it seems to us that it was a fact just waiting to be discovered – we would expect Harvey to have been the final person in a tradition of people *looking for* evidence and proof of the circulation. But all this would be wrong. For Harvey, the body did not act like a machine in any way: it was alive, and all its activities and functions were those of life. For him the beat of the heart was a vital, not a mechanical, phenomenon, and the blood itself was alive.

In his own day Harvey's approach, though only newly brought back into practice, was very old-fashioned. He looked to Aristotle as his primary authority, and had no time at all for such people of his own day as Descartes, who were promoting the 'mechanical philosophy'. Indeed Harvey's anatomical practice was as old-fashioned as his politics. Throughout his life he was a loyal supporter of the monarchy, never wavering even in the darkest days of the Civil War and Commonwealth. He dedicated his book announcing the discovery to the King, his patron and employer, Charles I. It was not just a matter of flattery that Harvey's opening words there describe the king in his kingdom as being like the heart in the body: 'Most serene King', he wrote, 'the heart of animals is the foundation of life, the ruler of each of them, of the microcosm it is the sun from which all *vegetatio* derives, all vigour and strength emanates. The king is equally the foundation of his kingdoms, and of his own microcosm he is the sun, of the State he is the heart from which all power emanates, every favour originates.' It is clear that when, as a loyal Aristotelian, Harvey first chose to study the heart in the animal, he was also, as a loyal monarchist, choosing to study it as the king of the animal body.

8.2 Catherine Wilson, *The invisible world of the microscope**

[...] In 1608 the spectacle-maker Hans Lipperhey of Middelburg applied (unsuccessfully, for his priority was already contested) for a thirty-year exclusive right of manufacture for a set of lenses mounted in a tube. The resulting telescope, quickly transmitted westward into England and southward into France, Germany, and Italy, achieved instant celebrity through the series of sensational observations of the moon and planets made by Galileo in 1609–10 and reported in his *Sidereus nuncius*.

The compound lens system qua microscope insinuated itself far more gradually. There were several reasons for this. First, the magnification of small objects, as opposed to distant ones, was not a novelty and could be accomplished when desired with a single lens; second, the microscope was not presented as an agent of a revolution in physics and natural philosophy, as the telescope was by Galileo; third, being able to see far was obviously useful in affairs of the world – in navigation and warfare especially – but being able to see small was not. The technology was transferable. Galileo had used magnifying lenses for small objects even in the period of the *Sidereus nuncius* and later copied the design of the compound instrument, and Cornelius Drebbel designed a microscope based on the principle of the Keplerian telescope, consisting of two convex lenses, and exhibited it in Paris in 1621 or 1622. But microscopes were still something of a rarity in 1625.

That less contention was attached to the question of the true inventor of the microscope than to the question of the true inventor of the telescope shows that it was the habilitation of the instrument, the creation of a context for it in which magnification had a purpose, that required originality and purposiveness. It is appropriate here to acknowledge the role played by

---

* Catherine Wilson, *The Invisible World: Early Modern Philosophy and the Microscope* (Princeton, NJ: Princeton University Press, 1995), pp. 74–5, 81–8.

Galileo's early promotion of (in contrast to Bacon's near indifference to) artificial optics as a means of achieving knowledge. The book that popularized the microscope, Hooke's *Micrographia*, did not appear until more than half a century after the *Sidereus nuncius*. It then both demonstrated the scope of microscopy – its application to the plant, animal, and mineral worlds, and to human artisanship – and supplied a moral-theological-practical justification for it. [...]

A lens produces a curved image; either the edges are out of focus or the center is. This can be compensated for by using a curved surface to mount the specimen, but then not all of its parts will be equally enlarged. By making the lens opening smaller, one achieves some improvement, but again at a cost of light. Chromatic aberration is a consequence of the fact that different wavelengths of light are differently refrangible; the red rays arrive in a different location from the blue rays and produce colored halos or fringes around the edge of the image. Newton had argued that chromatic aberration could not be reduced on account of the dispersiveness of all transparent bodies. The aberration can be corrected by using two kinds of glass of different refractive indices; but the theory of the achromatic refracting telescope was published only in 1758, and its manufacture was not perfected until much later. Finally, not one ray of light but a multitude depart from every point on the observed object and are focused differently; this is responsible for the blur known as 'spherical aberration'. Elliptical and hyperbolic lenses, as Descartes saw, will also eliminate spherical aberration, but these are exceedingly difficult for a technician to grind and were beyond the reach of seventeenth-century opticians.

Aberration was a far more serious problem for compound than for single microscopes. Yet the compound microscope had certain advantages, in addition to its larger field. The lenses were not so minute and difficult to handle: among the single lenses made for the Royal Society by John Mellin in about 1680, for example, was one only one twenty-fifth of an inch across. The compound instrument could be constructed with a stand and made to sit on a table, leaving the hands of the viewer free to manipulate the specimen, and it could be fitted with interchangeable objective lenses to give varying powers of magnification in conjunction with the ocular lens. It could be attractively decorated, as Hooke's tooled and gilded leather-covered instrument was. Though Leeuwenhoek used silver and sometimes gold for the body of his microscopes, their beauty was and is recherché compared with the gleaming proportions of the larger instruments. The compound microscope did not have to be positioned so close to the eye, which many observers found uncomfortable.... But Hooke testified to the superiority of the single lens 'to those whose eyes can endure it', observing that the colored fringes that disturb the image in the compound microscope are avoided with the single lens. There were other inconveniences experienced with single-lens microscopes: the short focal length of high magnifiers meant that the object had to be placed extremely close to, or even touching, the lens, dirtying it or sometimes scratching it. Holding the apparatus up to the light was tiring, and mounting the specimen by, for example, getting it to stick on a pin was more trouble than having it lie flat. For all these reasons, an optically inferior device might be preferred. [...]

The range of designs of the earliest microscopes, the commercial varieties of which all issued from a few well-known workshops, contrasts with the progressive standardization the instrument underwent as trial and error proved some designs more functional than others and as aesthetic considerations became less and less relevant to opticians in the design of scientific instruments. Price is correct to emphasize that instruments experience a symbolic-decorative phase before utilitarian values gain the upper hand entirely. One might accordingly suspect that... the scientific community in the seventeenth and eighteenth centuries was more interested in designing and possessing optical instruments than in using them to explore the world.

...But in fact an explosively wide range of investigations was made in the fifty years between 1640 and 1690. Leeuwenhoek looked at semen, blood, milk, bone, hair, spittle, the brain, sweat, fat, tears, sap, salts and crystals, protozoa and parasites, sponges, mollusks, fish, spermatozoa and embryos, pores and sweat, and muscle fibers. Grew specialized in plant parts; Malpighi, the most methodical, studied the kidneys, the lungs, the gall, the brain, fat, and bone marrow, as well as chicken embryos and the fine anatomy of insects, particularly the silkworm. Swammerdam, in his short career, was drawn mainly to insect anatomy and metamorphosis. [...]

It has been observed that, in the pre-Newtonian phase of the Royal Society, by contrast with its eighteenth-century phase, living subjects occupied more attention than celestial mechanics; Boyle even went so far as to say that the contrivance of the eye of a fly or a man's muscles was superior to the sun and the system of the heavenly orbs. It was not simply a question here of the prestige of the mathematical sciences after the *Principia* and Newton's ascendancy, but one of positive avoidance. There were theological reasons in the earlier period for preferring to concentrate not on lumps of matter in infinite spaces but on exquisite contrivances.

This interest was amply repaid: *Micrographia* is a captivating book. Its author, Hooke, who was apparently the brains behind the experimental, mathematical, and mechanical competence of Robert Boyle, the great rhetorician of the mechanical philosophy, was first employed directly by him when Boyle went to Oxford in 1655. In 1662 he became the paid curator to the infant Royal Society, and acquired other titles in due course: city surveyor, Cutlerian Lecturer in Mechanics, and Gresham Professor of Geometry. He took over the making of microscopical observations from Christopher Wren in 1661, the society having ordered him to present and publish observations on moss, vinegar, bark, blue mold, and spiders' eyes. He brought his microscope to Royal Society meetings, mainly during 1663 and 1664, and on those occasions invited the members to compare his 'schemes' or drawings of those subjects with their appearance through the microscope.

These drawings were composites, not records of individual observations. 'I never began to make any draught', he says, 'before by many examinations in several lights, and in several positions to these lights, I had discovr'd the true form.' For Hooke, these illustrations were not meant to be direct reproductions of a momentary optical experience, but rather an improvement on momentary witnessing that would give the general form stripped of the idiosyncrasies of the individual specimen or observation.

Hooke's objects of study were insects, especially fleas, lice, flies, moths, and bees; plant material; mold and fungus; inorganic objects, such as snowflakes and stones; man-made objects, such as needles, razors, cloth, and paper; animal parts, such as hair, fish scales, and the sting of the bee; and transitory phenomena, such as sparks struck from flint and the colors of thin films of mica. Observation 18 is the famous examination of cork with its 'little boxes or cells' without passageways between them, which Hooke supposed to be an excrescence on the tree's bark. *Micrographia* was meant, it has been argued, as an offering to the Royal Society's sponsor, Charles II, as evidence of the accomplishments of the society. It upheld, as John Harwood points out, the society's claims to be directly concerned with 'things', to be dedicated to an ideal of objectivity, and if what *Micrographia* delivered was not precisely useful knowledge, the society had at least shown itself capable of producing something other than talk and socializing. The book was vastly more entertaining than the artificial-language-manual-cum-encyclopedia that the society had recently sponsored in John Wilkins's dinosauric *Essay towards a Real Character*. But Hooke's most prescient statements are deeply buried in the observations, where they are often tangential to the ostensible subject matter; Hooke used his assignment to communicate some of his freest speculations on physics in general. Into this category fall his remarks on colors and diffraction in observation 58, his statements about combustion, heat, and respiration in the section on charcoal (observation 16), his remarks about the cause of petrification (observation 17), and the aforementioned discovery of 'cells' in cork. Yet the associated illustrations were not the most striking or memorable, and the prize for visual interest must go to the animal and vegetable sections. Newton, whose eye was for the serious, seems to have found most interesting the observations on plate colors and diffraction.

The reception of the book was highly positive: it was much sought after, and print runs sold out repeatedly. A summary in German appeared in 1667 with copies of the most popular engravings, and the *Journal des savants* reviewed it in 1668, lamenting that few people could read English. The *Journal des savants* praised the illustrations and included a copy of the magnificent louse foldout. Huygens wrote to Robert Moray of the Royal Society that he was delighted to have received the book at last, that he admired the care with which it had been drawn and engraved, the revelations of mechanics and geometry in the work of nature, and the theoretical studies of colors. Leibniz had also heard of it and tried persistently for years to obtain a copy, receiving one finally in 1678.

8.3 Steven Shapin, *Boyle's house of experiment**

[Robert] Boyle had laboratories at each of the three major residences he successively inhabited during his mature life. From about 1645 to about 1655 he was mainly in residence at the manor house of Stalbridge in Dorset, an estate acquired by his father, the first Earl of Cork, in 1636 and inherited by his youngest son on the earl's death in 1643. By early in 1647 Boyle was organizing a chemical laboratory at Stalbridge, perhaps with the advice of the Hartlibian circle, whose London laboratories he frequently visited. Late in 1655 or early in 1656 he removed to Oxford, where his sister Katherine, Lady Ranelagh, had searched out rooms for him in the house of the apothecary John Crosse, Deep Hall in the High Street. He was apparently able to use Crosse's chemical facilities, and his own rooms contained a pneumatic laboratory, where, assisted by Hooke, the first version of the air pump was constructed in 1658–1659. During his Oxford period Boyle also had access to a retreat at Stanton St. John, a village several miles to the northeast, where he made meteorological observations but apparently did not have a laboratory of any kind.

Boyle was away from Oxford for extended periods, staying sometimes at a house in Chelsea, sometimes with Katherine in London, and sometimes with another sister, Mary Rich, Countess of Warwick, at Leese (or Leighs) Priory in Essex. In 1666 he had Oldenburg look over possible lodgings in Newington, north of London, but there is no evidence that he ever occupied these. And he periodically stayed at Beaconsfield in Buckinghamshire, possibly at the home of the poet Edmund Waller. But Oxford remained his primary residence and experimental workplace until he moved into quarters with Katherine at her house in Pall Mall in 1668. This was a house (actually two houses knocked into one) assigned to Lady Ranelagh by the Earl of Warwick in 1664. It stood on the south side of Pall Mall, probably on the site now occupied by the Royal Automobile Club. Although luxury building in this area was proceeding apace in the Restoration, at the time Boyle moved in Pall Mall still retained a rather quiet and semi-rural atmosphere. During the 1670s Boyle's neighbors included Henry Oldenburg, Dr. Thomas Sydenham, and Nell Gwyn.

Boyle's laboratory in Katherine's house was probably either in the basement or attached to the back, and there is some evidence to suggest that one could obtain access to the laboratory from the street without passing through the rest of the house. The unmarried Boyle seems to have dined regularly with his sister, who was a major social and cultural figure in her own right, living 'on the publickest scene', and who entertained his guests at the family table. He remained there until his death in 1691, which closely followed Katherine's. [...]

---

* Steven Shapin, 'The house of experiment', *Isis*, 79 (1988), pp. 379–80, 382–8.

The threshold of the experimental laboratory was constructed out of stone and social convention. Conditions of access to the experimental laboratory would flow from decisions about what kind of place it was. In the middle of the century those decisions had not yet been made and institutionalized. Meanwhile there were a variety of stipulations about the functional and social status of spaces given over to experiment, and a variety of sentiments about access to them.

To the young Robert Boyle the threshold of his Stalbridge laboratory constituted the boundary between sacred and secular space. He told his sister Katherine that '*Vulcan* has so transported and bewitched me, that as the delights I taste in it make me fancy my laboratory a kind of *Elysium*, so as if the threshold of it possessed the quality the poets ascribe to *Lethe*, their fictions made men taste of before their entrance into those seats of bliss, I there forget my standish [inkstand] and my books, and almost all things'. The experimenter was to consider himself 'honor'd with the Priesthood of so noble a Temple' as the 'Commonwealth of Nature'. And it was therefore fit that laboratory work be performed, like divine service, on Sundays. (In mature life Boyle entered his Pall Mall laboratory directly after his morning devotions, although he had apparently given up the practice of experimenting on the Sabbath.) In the 1640s he told his Hartlibian friends of his purposeful 'retreat to this solitude' and of 'my confinement to this melancholy solitude' in Dorset. But it was said to be a wished-for and a virtuous solitude, and Boyle complained bitterly of interruptions from visitors and their trivial discourses. [. . .]

Many contemporary commentators remarked upon the ease of access to Boyle's laboratory. John Aubrey wrote about Boyle's 'noble laboratory' at Lady Ranelagh's house as a major object of intellectual pilgrimage. 'When foreigners come to hither, 'tis one of their curiosities to make him a Visit.' This was the laboratory that was said to be 'constantly open to the Curious, whom he permitted to see most of his Processes'. In 1668 Lorenzo Magalotti, emissary of the Florentine experimentalists, traveled especially to Oxford to see Boyle and boasted that he was rewarded with 'about ten hours' of his discourse, 'spread over two occasions'. John Evelyn noted that Boyle 'had so universal an esteeme in Foraine parts; that not any Stranger of note or quality; Learn'd or Curious coming into England, but us'd to Visite him'. He 'was seldome without company' in the afternoons, after his laboratory work was finished.

But the strain of maintaining quarters 'constantly open to the curious' told upon him and was seen to do so. As an overwrought young man he besought 'deare Philosophy' to 'come quickly & releive Your Distresst Client' of the 'vaine Company' that forms a 'perfect Tryall of my Patience'. Experimental philosophy might rescue him 'from some strange, hasty, Anchoritish Vow'; it could save him from his natural 'Hermit's Aversenesse to Society'. When, during the plague, members of the Royal Society descended upon him in Oxford, he bolted for the solitude of his village retreat at Stanton St. John, complaining of 'ye great Concourse of strangers', while assuring Oldenburg that 'I am not here soe neere a Hermite' but that some visitors were still welcome. Even as John Evelyn praised Boyle's accessibility, he recorded that the crowding 'was sometimes so incomodious that he now and then repair'd to a private Lodging in another quarter [of London], and at other times' to Leese or elsewhere in the country 'among his noble relations'.

Toward the end of his life Boyle took drastic and highly visible steps to restrict access to his drawing room and laboratory. It is reported that when he was at work trying experiments in the Pall Mall laboratory and did not wish to be interrupted, he caused a sign to be posted on his door: 'Mr. Boyle cannot be spoken with to-day.' In his last years and in declining health, he issued a special public advertisement 'to those of his ffriends & Acquaintance, that are wont to do him the honour & favour of visiting him', to the effect that he desired, 'to be excus'd from receiving visits' except at stated times, '(unless upon occasions very extraordinary)'. Bishop Burnet said that Boyle 'felt his easiness of access' made 'great wasts on his time', but 'thought his obligation to strangers was more than bare civility'.

That obligation was a powerful constraint. The forces that acted to keep Boyle's door ajar were social forces. Boyle was a gentleman as well as an experimental philosopher. Indeed, as a young man he had reflected systematically upon the code of the gentleman and his own position in this code. The place where Boyle worked was also the residence of the son of the first Earl of Cork. It was a point of honor that the private residence of a gentleman should be open to the legitimate visits of other gentlemen. Seventeenth-century handbooks on the code of gentility stressed this openness of access: one such text noted that 'Hospitalitie' was 'one of the apparentest Signalls of *Gentrie*'. Modern historians confirm the equation between easy access and gentlemanly standing: 'generous hospitality was the hallmark of a gentleman'; 'so long as the habit of open hospitality persisted, privacy was unobtainable, and indeed unheard of'. And as the young Boyle himself confided in his *Commonplace Book*, a 'Noble Descent' gives 'the Gentleman a Free Admittance into many Companys, whence Inferior Persons (tho never so Deserving) are... excluded'. Other gentlemen knew who was a gentleman, they knew the code regulating access to his residence, and they knew that Boyle was obliged to operate under this code. But they did not know, nor could they, what an experimental scientist was, nor what might be the nature of a different code governing admittance to his laboratory. In the event, as Marie Boas wrote, they might plausibly come to the conclusion that Boyle 'was only a virtuoso, amusing himself with science, [that] he could be interrupted at any time. ... There was always a swarm of idle gentlemen and ladies who wanted to see amusing and curious experiments.' When, however, Boyle wished to shut his door to these distractions, he was able to draw upon widely understood moral patterns that enabled others to recognize what he was doing and why it might be legitimate. The occasional privacy of laboratory work could be assimilated to the morally warrantable solitude characteristic of the religious isolate.

8.4 Simon Schaffer and Steven Shapin, *Making experimental knowledge**

[...] Boyle's and Hooke's air-pump was, in the former's terminology, an 'elaborate' device. It was also temperamental (difficult to operate properly) and very expensive: the air-pump was seventeenth-century 'Big Science'. To finance its construction on an individual basis it helped greatly to be a son of the Earl of Cork. Other natural philosophers, presumably as well supplied with cash as Boyle, shied away from the expense of building a pneumatic engine, and a major justification for founding scientific societies in the 1660s and afterwards was the collective financing of the instruments upon which the experimental philosophy was deemed to depend. Reading histories of seventeenth-century science, one might gain the impression that air-pumps were widely distributed. They were, however, very scarce commodities. ... Boyle's original machine was soon presented to the Royal Society of London; he had one or two redesigned machines built for him by 1662, operating mainly in Oxford; Christiaan Huygens had one made in The Hague in 1661; there was one at the Montmor Academy in Paris; there was probably one at Christ's College, Cambridge, by the mid-1660s; and Henry Power may have possessed one in Halifax from 1661. So far as can be found out, these were all the pumps that existed in the decade after their invention.

Without doubt, the intricacy of these machines and their limited availability posed a problem of access that experimental philosophers laboured to overcome. Less obviously, the control of access to the devices that were to generate genuine knowledge was a positive advantage. The space where these machines worked – the nascent laboratory – was to be a public space, but a restricted public space, as critics like Hobbes were soon to point out. If one wanted to produce authenticated experimental knowledge – matters of fact – one had to come to this space and to work in it with others. If one wanted to see the new phenomena created by these machines, one had to come to that space and see them with others. The phenomena were not on show anywhere at all. The laboratory was, therefore, a disciplined space, where experimental, discursive, and social practices were collectively controlled by competent members. In these respects, the experimental laboratory was a better space in which to generate authentic knowledge than the space outside it in which simple observations of nature could be made. To be sure, such observations were reckoned to be vital to the new philosophy and were judged vastly preferable to trust in ancient authority.

---

* Simon Schaffer and Steven Shapin, *Leviathan and the Air-Pump: Hobbes, Boyle and the Experimental Life* (Princeton, NJ: Princeton University Press, 1985), pp. 38–9.

Yet most observational reports were attended with problems in evaluating *testimony*. A report of an observation of a new species of animal in, for example, the East Indies, could not easily be checked by philosophers whose credibility was assured. Thus all such reports had to be inspected both for their plausibility (given existing knowledge) and for the credibility and trustworthiness of the witness. Such might not be the case with experimental performances in which, ideally, the phenomena were witnessed together by philosophers of known reliability and discernment. Insofar as one insisted upon the foundational status of experimentally produced matters of fact, one ruled out of court the knowledge-claims of alchemical 'secretists' and of sectarian 'enthusiasts' who claimed individual and unmediated inspiration from God, or whose solitary 'treading of the Book of Nature' produced unverifiable observational testimony. It is not novel to notice that the constitution of experimental knowledge was to be a public process. [...]

8.5 Lawrence Principe, *Boyle as alchemist**

[...] Cataloguing Boyle's alchemical pursuits requires first that we be able to identify alchemical activities accurately, and especially to distinguish them satisfactorily and consistently from more properly *chemical* activities. Admittedly, in some seventeenth-century situations, the distinction is neither warrantable nor valuable, but if we wish to use the label 'alchemy' to describe a set of activities and beliefs (and we should be able to do so) we need to be clear and consistent about its referents. Texts can be differentiated based on their motives and theoretical assumptions, but practical processes by themselves present a greater problem. The very same process (for example, the calcination of lead) may be properly alchemical or chemical (i.e. non-alchemical) depending upon the intent, goals and interpretation of the operator. Occasionally it has been all too easy to classify experimental activities as 'alchemical' or 'chemical' based upon preconceived notions of the experimenter, usually revolving around his received 'credibility' or 'modernity'. For negotiating this difficult terrain, I believe it is safe and legitimate to call endeavours aiming at traditional alchemical desiderata – the Philosophers' Stone, the extraction of mercuries and sulphurs, the alkahest, etc. – as alchemy, although it does not follow that alchemy should be restricted to such topics.[1]

Boyle himself draws a distinction in the middle of a troublesome area – transmutation. Boyle's belief in metallic transmutation has been recognised

---

* Lawrence Principe, 'Boyle's alchemical pursuits', in M. Hunter (ed.), *Robert Boyle Reconsidered* (Cambridge: Cambridge University Press, 1994), pp. 92–7, 100–2.

1 In correspondence with Lawrence Principe, the following article was suggested for further discussion on the distinction between alchemy and chemistry: William R. Newman and Lawrence M. Principe, 'Alchemy vs. chemistry: The etymological origins of a historiographical mistake', *Early Science and Medicine*, 3 (1998), pp. 32–65.

for many years, but only in a *non-alchemical* sense.... But what has not been hitherto recognised is Boyle's belief in *two separate types of transmutation* – one mechanical, the other properly alchemical. Boyle clearly differentiates mechanical transmutation from alchemical projection later on in the *Origine of Formes*:

> ...there may be a real Transmutation of one Metal into another, even among the perfectest and the noblest Metals, and that effected by Factitious Agents in a short time, and, if I may so speak, after a Mechanical manner. I speak not here of Projection...because, though Projection includes Transmutation, Transmutation is not all one with Projection, but far easier than it.

The 'projection' to which Boyle refers is the traditional alchemical method of transmuting metals by *projecting* (i.e. casting on, from *proiecere*) a minute quantity of the Philosophers' Stone upon a crucible full of molten lead or hot mercury, thereby converting it in a few minutes into pure gold.

Boyle's belief in the reality of alchemical transmutation (by projection of the Philosophers' Stone) is made abundantly clear in his unpublished *Dialogue on the Transmutation of Metals*. This text recounts a debate set at a 'Most Noble Society', quite possibly the Royal Society, concerning the possibility of transmutation by projection. The 'Anti-Lapidist' party, headed by a character named alternately as Erastus or Simplicius, argues that alchemical theories are unclear and undemonstrated, that the accounts of transmutation are unreliable and that the entire idea of the Philosophers' Stone and transmutation by projection is unbelievable. On the other side, Zosimus, heading the 'Lapidist' party, counters that alchemical theories may well be deficient, but alchemical claims do not transgress the bounds of possibility, and that there are sufficient accounts of transmutations and related phenomena to credit them. Arguments and counter-arguments are presented, until Pyrophilus enters and recounts his first-hand experience with an 'Anti-Elixir' which when projected upon molten gold successfully transmuted it into a base metal. (Boyle published this conclusion to the unpublished *Dialogue* anonymously in 1678 as *An Historical Account of a Degradation of Gold*.)...

Boyle's belief in the Philosophers' Stone led him to try to make it for himself. Early in his career in March 1646/7, upon receiving in pieces the chemical furnace he had ordered, Boyle lamented to his sister Katherine, that 'I see I am not designed to finding out the Philosophers' Stone, I have been so unlucky in my first attempts at Chymistry'. Of course, this mention might be dismissed as youthful enthusiasm or a joking complaint. After all, in the apologetic preface to his 'Essay on nitre', written in the mid-1650s and published in 1661, Boyle protests that he laboured at 'Chymical Tryalls... without seeking after the Elixir, that Alchymists generally hope and toyl for, but which they that know me know to be not at all in my aime'. In spite of this protestation, many less obvious instances leave no doubt about

Boyle's continuing attempts to confect the Stone. One of these appears in brief intimations in *On the Unsuccessfulness of Experiments, The Sceptical Chymist* and more fully in *Origine of Formes*. In the last, Boyle describes a powerful solvent he calls his *menstruum peracutum* which can dissolve gold and transmute a portion of it into silver. He uses this process to show the enormous changes which can be effected on stable bodies, and deploys it, at least overtly, to expound his corpuscularian hypothesis. But his careful and deliberate choice of expressions and vocabulary creates an allusive sub-text, accessible to readers conversant in alchemical literary styles, which reveals the original goal of the process as the preparation of the Philosophers' Stone. For example, when remarking upon the dissolution of gold in his solvent, he writes that it is 'wont to melt as it were naturally... without Ebullition (almost like Ice in luke-warm water)'. This seemingly unremarkable simile is actually a canonical phrase of the alchemical literature reserved exclusively for the dissolution of gold in Philosophical Mercury, the first crucial step in confecting the Stone. [...]

Even the casual observer of alchemy knows that when the preparation of the Philosophers' Stone is mentioned, the topic of secrecy cannot be far distant. Secrecy is a crucial element of alchemy – all alchemists command their students to keep silence, and they themselves shroud their knowledge in an obscure metaphorical double-talk which continues to deter and confound readers. In a fragment, possibly from the *Dialogue on Transmutation*, Boyle complains that

> Divers of Hermetic Books have such involv'd Obscuritys that they may justly be compared to Riddles written in Cyphers. For after a Man has surmounted the difficulty of decyphering the words & terms, he finds a new & greater difficulty to discover ye meaning of the seemingly plain Expression.

Many historians have been at pains to distance Boyle from the secrecy of earlier traditions, and to portray him as a major figure in the promotion of open communication and the public validation of knowledge. Although these analyses do have some merit, they are not applicable to all of Boyle's work, and in particular they are not representative of his alchemical pursuits. There is, in fact, no shortage of evidence attesting to Boyle's commitment to *secrecy* in alchemical matters, for when dealing with alchemy Boyle adopted much of the secretive style of traditional alchemical writers. I have found that Boyle employs no fewer than ten different codes, ranging from simple word-, letter-, or number-replacements, to more complex nomenclators and ciphers, throughout his papers and correspondence. Significantly, these concealment techniques occur only in the context of traditional alchemical pursuits....

Boyle is often purposefully obscure even in published writings. I have already mentioned the corpuscularian text and alchemical sub-text of the gold process. In addition, several works, including *Sceptical Chymist* and the *Usefulness of Experimental Philosophy*, use the principle of dispersion – a traditional alchemical method of concealment whereby dissevered parts of a single process are scattered disconnectedly through large stretches of unrelated text. . . .

Boyle's remarkable paper 'Of the incalescence of quicksilver with gold' should be mentioned in this regard. This paper which appeared in the 21 February 1675/6 issue of *Philosophical Transactions* describes a specially prepared mercury which grew hot when mixed with gold, contrary to the normal behaviour of quicksilver. But Boyle is so evasive about describing this mercury that the publication is far more frustrating than informative. He never explains how the mercury is prepared, but only recounts its effects. Contrary to the norms of Royal Society discourse, and certainly contrary to the prevailing depictions of Boyle's commitment to open communication, Boyle flatly refuses to answer any questions regarding the mercury.

Boyle believed that this incalescent mercury held great power in reference to the preparation of the Philosophers' Stone. Boyle's interest in this one substance alone spanned nearly forty years, from its first preparation in 1652 to its inclusion in a memorandum (*ca*. 1691) listing the most important items to be discharged before his death. He explicitly connects this mercury with the alchemical sources, writing that he found in alchemical texts 'some dark passages, whence I then ghess'd their knowledge of it, and in one of them I found, though not all in the very same place, an Allegorical description of it'. In a typically alchemical manner, he excuses his secrecy by citing the 'political inconveniences that may ensue if it prove to be one of the best kind and fall into ill hands'. [. . .]

The incalescent mercury paper should be read not as a source of information, but as a request for it from the alchemical community. Boyle was apparently unsure of how to prepare the Philosophers' Stone from the mercury. The solution was to obtain direct contact with knowledgeable alchemists rather than wandering in the uncertainty of obscure alchemical texts. He had already questioned 'several prying Alchymists' who 'of late years travelled into many parts of *Europe* to pry into the Secrets of Seekers of Metalline Transmutations' regarding this substance. Boyle addresses the paper to as large an audience as possible by publishing it in English and Latin in parallel columns – a publishing practice both unprecedented and unrepeated in *Philosophical Transactions*. He unambiguously indicates the alchemical nature and importance of the paper by connecting his results to alchemical work published previously under allegorical disguise. Boyle then demonstrates his acceptance of alchemical rules by committing himself to secrecy, and culls out responses to the paper from non-adepts by refusing to accept questions. Finally, having targeted his audience and proven his communion with them, he asks the elite adepti of the alchemical community for further information – hoping that he

> may safely learn... what those that are skilful and Judicious enough to deserve to be much considered in such an affair, will think of our Mercury... The knowledge of the opinions of the wise and skilful about this case, will be requisite to assist me to take right measures in an affair of this nature. And till I receive this information, I am obliged to silence.

[...] Traditional alchemy always held close links with the metaphysical and the supernatural; thus, as Boyle became more involved in traditional alchemy, and increasingly convinced of its claims, he would have found such connections in alchemical texts. He probably began his alchemy–spirit realm link with the unanimous testimony of the adepti that alchemical knowledge came by God's revelation. Alchemists considered the Philosophers' Stone as a *donum Dei*; it could not be prepared solely by the sweat of the laborant, but required cooperative revelation by God. Biblically, God's messages often came through angels or other spiritual means, thus these beings could likewise be the imparters of alchemical truths. Indeed, Boyle's early alchemical collaborator George Starkey claimed in a letter to Boyle (26 January 1651/2) that his knowledge of the secrets of alchemy was revealed to him in a dream by a good spirit [*Eugenius*] sent by God. It was partially this sacred nature of alchemical knowledge which demanded secrecy from the alchemists, as illicit communication of such secrets would (to use an alchemical phrase) cast the pearls of divine revelation before the swine and, in the words of Elias Ashmole, 'render one *Criminall* before *God*, and a *presumptuous violator* of the *Caelestiall Seales*'. This same consideration may likewise have helped provoke Boyle's secrecy. The indiscretion of open alchemical communication might be accounted the sin against the Holy Spirit (through Whom the knowledge had been selectively imparted) which Boyle so feared. [...]

It has often been noted that accounts of witchcraft and magic were valuable because, if proven, they would demonstrate the existence of spirits against the assaults of atheists.... Alchemy, if its objects had the power to summon rational spirits, could provide a superior weapon. Whereas witchcraft was notoriously difficult to demonstrate either in its operation or its effects, the red powder of projection and its products could be demonstrated (though not prepared) by anyone. Further, if spirits were involved in alchemical change, the success of a spirit-assisted transmutation would provide an instance of immaterial, incorporeal substances inducing changes in material, corporeal substances. This would silence those who doubted that an incorporeal God could affect matter and that an immaterial human soul could actuate the human body. Both points were on Boyle's mind, as many fragments among the Boyle Papers present arguments refuting those who deny the action of the incorporeal on the corporeal. In reverse, if, as Boyle suggested, angels were in fact attracted by the Philosophers' Stone by 'congruities or magnatisms capable of involving them', this fact would be an instance of the incorporeal being affected by the corporeal. [...]

Boyle was concerned, especially in his later years, about the possible misapplication of the New Philosophy, which, by succeeding in banishing occult explanations, might be used to deny the existence of the spirit realm as well. Boyle was acutely aware that, though the 'tumultuous Justlings of Atomical Portions of senseless Matter' could explain the origin of forms and qualities, they could not explain the origin of the world – although they could describe the composition of a vinous spirit, they could not describe the composition of a rational, incorporeal spirit. The concern over the New Philosophy's possible tendency to atheism is well-known, and underlies much of Boyle's *Christian Virtuoso*. Indeed, the preface of that work addresses the inapplicability of mechanical views to 'Incorporeal and Rational Beings' while asserting the importance of studying them. In sum, Boyle found himself in a potential dilemma: his conspicuous success in applying mechanical explanations to the world resulted in a tension with his theological commitments, as these new explanatory models threatened to distance spiritual agents (including God) from all contact with the natural world, possibly rendering them superfluous. Boyle's late-career view of alchemy would have helped bridge the gap between experimental philosophy and theology. Alchemy complemented and corrected the system of natural philosophy that Boyle supported. Alchemy was the interface between the rational, mechanical functioning of the corporeal world and the supra-rational, miraculous workings of the spiritual world. [. . .]

8.6 Rob Iliffe, *The 'Principia': Newton's authorial role**

By autumn 1687, Newton's masterpiece was public. As far as we can tell, the account of the Vestal ceremonies and his analysis of idolatry remained an entirely private interest, though his disdain for Roman Catholicism became known outside Cambridge when – in April and May 1687 before Judge Jeffreys – he represented the University in efforts to prevent James II imposing a Benedictine Monk as MA in Sidney Sussex College. Apart from this, Newton had revived his alchemical and scriptural studies, and found time to become a Member of Parliament for Cambridge University in 1689. Nevertheless, he was also planning both an edition of his optical work, and a second edition of the *Principia*. To this end, he set about collecting accounts of various errors in the 1687 text, and worked in harness with a number of selected acolytes. Amongst these, the Swiss mathematician Fatio de Duillier soon became a favorite. In April 1690, Fatio went to Holland with a list of corrections and additions, a copy of which he left with Huygens. As a confidant of Newton, Fatio was well

* Rob Iliffe, 'Is he like other men? The meaning of the *Principia mathematica*, and the author as idol', in Gerald Maclean (ed.), *Culture and society in the Stuart Restoration* (Cambridge: Cambridge University Press, 1995), pp. 170–3, 175–6.

situated to function as a conduit for the views of Huygens and Leibniz, and he wasted little time in reporting Newton's private beliefs to these scholars.

From February 1690, Fatio let it be known that Newton wanted to entrust him with the proposed second edition, but Newton was still stalling at the end of 1691. Yet the project looked like being realized in early 1692, when Fatio informed Huygens that Newton believed in the 'Newtonian' *prisca*; that October Newton supplied him with a new list of errata and addenda. Fatio had designs of his own, and still claimed to Leibniz in 1694 that Newton believed his own theory of gravity was the one he favored over all other 'mechanical' accounts. However, David Gregory had noted as far back as December 1691 that 'Mr. Newton and Mr. Hally laugh at Mr. Fatio's manner of explaining gravity'. Gregory was an avid disciple of the author of the *Principia* from an early stage, and was amply rewarded when Newton's backing helped him beat Halley to obtain the Savilian Chair of Astronomy at Oxford in 1691. In May 1694, Newton let him see 'The Original of Religions' as well as a number of textual changes which Gregory was told would be incorporated in some way into the second edition.

The range of Newton's use of the ancient snippets was vast. He was to show that his mathematical tools were known to the Ancients, and told Gregory that 'the[ir] resolved locus [was] a Treasury of Analysis', and much of the work centered around restoring lost books such as Euclid's *Porismata* (dealt with in Book VII in the work of Pappus). Other remarks recorded Newton's claim that 'a continual miracle is needed to prevent the Sun and the fixed stars from coming together through gravity', and that 'the great eccentricity in Comets in directions both different from and contrary to the planets indicates a divine hand: and implies that the Comets are destined for a use other than that of planets'. Jupiter's and Saturn's satellites could take the place of other planets (such as Earth), 'and be held in reserve for a new Creation'.

All this agreed very well with the doctrines of the Ancients, and especially Thales. More radically, Newton now argued that 'the philosophy of Epicurus and Lucretius is true and old, but was wrongly interpreted by the Ancients as atheism'. In July, Gregory confirmed that Book Three of the *Principia* would show that the Egyptians taught the Copernican system, and that a text explaining 'the authentic design of the Ancients' would be appended to the entire work, in which 'the errors of the moderns about the mind of the Ancients are detected'. Gregory was now confident that his 'Notes on the Newtonian Philosophy' would be published, as he had proposed to Newton after their meeting in May. In any case, Newton himself had the new Classical Scholia ready for publication. As Gregory's notes reveal, Newton had shifted his attention to the ancient atomists, while still incorporating significant pieces from Macrobius's *Somnium Scipio* concerning the mysteries of the Pythagorean philosophy. Newton's style in these Classical Scholia did not suggest that they were speculations, and he forcefully asserted that the Ancients 'were aware of' and 'believed in' the true physical system of the world.

This was a complex display of hermeneutic analysis. Newton claimed to decipher what the Ancients meant when they spoke in code, because he had rediscovered the truths underlying their mysteries. His treatment of the Ancients' work also allowed Newton to state publicly his views on issues which otherwise would lead to unnecessary disputes. So to Proposition Nine, he proposed a new scholium which described the cause of gravity, by explaining what the Ancients taught about the new phenomenon:

> Thus far I have explained the properties of gravity. But by no means do I consider its cause. However I will say in what sense the Ancients theorized about it. Thales held that all bodies were animate, inferring this from magnetic and electrical attractions... He taught that everything was full of Gods, and by Gods he meant animate bodies.

Likewise, Newton agreed with Cudworth that the Ancients had designated atoms 'hieroglyphically' as 'monads', referring to physical and not mathematical indivisibles. If this *prisca sapientia* in its prelapsarian pristinity was what the Ancients had believed, then the *Principia* was also true, and one needed the great work to understand the Classical remains. But how was the ordinary mortal to understand the *Principia*?

While releasing evidence of his private interests to people like Fatio and Gregory, Newton was also issuing instructions on how to approach the public text. These linked strategies enabled him to exercise imperious control over the way the book was to be read and understood. Authority came from his status as its author, and rival accounts of its meaning had better beware in case Newton or his allies ruled that these were misunderstandings. Such proprietorial concerns had nearly forced him to shelve the whole project as early as 1686 when he was told that Hooke was claiming priority both for the discovery of universal gravitation and of proof that planets in elliptical orbits followed the inverse-square force law. [...]

The *Principia* gained in stature amongst a wider but still restricted audience as Newton consolidated links with various allies and moulded himself into a public figure, becoming Warden and then Master of the Mint in the late 1690s. But it was his private life which attracted the interest of other scholars. Droves of foreign visitors began to visit. The Marquis de l'Hôpital asked John Arbuthnot in 1696 'does he eat & drink & sleep? Is he like other men?'; he was 'surprized when the Dr. told him he conversed chearfully with his friends, assumed nothing & put himself upon a level with all mankind'. Yet Newton suggested at various times that his aim in writing was to satisfy only a select band of scholars. This elite readership was always the mathematically literate, and in the early 1690s he composed a draft conclusion to a proposed edition of the *Opticks* suggesting that he had not been able to publish in the *Principia* his belief that Nature 'observes the same method in regulating the motions of smaller bodies wch she doth in regulating those of the greater' because

> This principle of nature being very remote from the conceptions of Philosophers I forbore to describe it in that Book leas[t It] should be accounted an extravagant freak & so prejudice my Readers against all those things wch were ye main designe of the Book: & yet I hinted both in the Preface & in ye book it self... but the design of yt book being secured by the approbation of Mathematicians, [I have] not scrupled to propose this Principle in plane words.

[...]

The post-*Principia* Newton moulded himself into a *fin de siècle* priest of nature, although the extensive religious significance of the text remained generally unknown to the public until the 'General Scholium' and its footnotes on idolatry were published in the second edition of 1713. His treatment of the *Principia*, and his management of the dissemination of its truths among the mathematically illiterate, imitated what he had discovered was practiced amongst the learned Ancients. Newton's private life remained necessarily elusive, for his publishing and self-fashioning strategies required that he and his disciples appeal to this morally unpolluted sanctum for authoritative accounts of both his and the *Principia*'s true meaning. Rather than assuming (along with contemporaries) that achieving direct access to Newton was also to gain entry to some real or privileged meaning of the text through the mind of 'genius', historians should perhaps view the deployment in any given setting of discourses invoking privacy as strategies to speak authoritatively about truth and meaning....

8.7 Betty Jo Dobbs, *Newton: philosopher by fire**

Isaac Newton studied alchemy from about 1668 until the second or third decade of the eighteenth century. He combed the literature of alchemy, compiling voluminous notes and even transcribing entire treatises in his own hand. Eventually he drafted treatises of his own, filled with references to the older literature. The manuscript legacy of his scholarly endeavor is very large and represents a huge commitment of his time, but to it one must add the record of experimentation. Each brief and often abruptly cryptic laboratory report hides behind itself untold hours with hand-built furnaces of brick, with crucible, with mortar and pestle, with the apparatus of distillation, and with charcoal fires: experimental sequences sometimes

---

* Betty Jo Dobbs, *The Janus Faces of Genius: The Role of Alchemy in Newton's Thought* (Cambridge: Cambridge University Press, 1991), pp. 1, 5–13.

ran for weeks, months, or even years. As the seventeenth-century epithet 'philosopher by fire' distinguished the serious, philosophical alchemist from the empiric 'puffer' or the devious charlatan or the amateur 'chymist', so may one use the term to characterize Isaac Newton. Surely this man earned that title if ever any did. [. . .]

My studies since 1975 have yielded hints that Newton was concerned from the first in his alchemical work to find evidence for the existence of a vegetative principle operating in the natural world, a principle that he understood to be the secret, universal, animating spirit of which the alchemists spoke. He saw analogies between the vegetable principle and light, and between the alchemical process and the work of the Deity at the time of creation. It was by the use of this active vegetative principle that God constantly molded the universe to His providential design, producing all manner of generations, resurrections, fermentations, and vegetation. In short, it was the action of the secret animating spirit of alchemy that kept the universe from being the sort of closed mechanical system for which Descartes had argued. . . .

. . . Newton stood at the beginning of our modern scientific era and put his stamp upon it irrevocably. He may be seen as a gatekeeper, a Janus figure, for one of his faces still gazes in our direction. But only one of them. Like Janus, who symbolized the beginning of the new year but also the end of the old one, Newton looked forward in time but backward as well. . . .

I do not assume the irrelevancy of Newton's pursuit of an ancient, occult wisdom to those great syntheses of his that mark the foundation of modern science. The Janus-like faces of Isaac Newton were after all the production of a single mind, and their very bifurcation may be more of a modern optical illusion than an actuality. Newton's mind was equipped with a certain fundamental assumption, common to his age, from which his various lines of investigation flowed naturally: the assumption of the unity of Truth. True knowledge was all in some sense a knowledge of God; Truth was one, its unity guaranteed by the unity of God. Reason and revelation were not in conflict but were supplementary. God's attributes were recorded in the written Word but were also directly reflected in the nature of nature. Natural philosophy thus had immediate theological meaning for Newton and he deemed it capable of revealing to him those aspects of the divine never recorded in the Bible or the record of which had been corrupted by time and human error. By whatever route one approached Truth, the goal was the same. Experimental discovery and revelation; the productions of reason, speculation, or mathematics; the cryptic, coded messages of the ancients in myth, prophecy, or alchemical tract – all, if correctly interpreted, found their reconciliation in the infinite unity and majesty of the Deity. In Newton's conviction of the unity of Truth and its ultimate source in the divine one may find the fountainhead of all his diverse studies.

. . . Mathematics was only one avenue to Truth, and though mathematics was a powerful tool in his hands, Newton's methodology was much broader than that implied by the creation of mathematical models, and Newton's goal was incomparably more vast than the discovery of the 'mathematical principles of natural philosophy'. Newton wished to penetrate to the divine principles beyond the veil of nature, and beyond the veils of human record and received revelation as well. His goal was the knowledge of God, and for achieving that goal he marshaled the evidence from every source available to him: mathematics, experiment, observation, reason, revelation, historical record, myth, the tattered remnants of ancient wisdom. With the post-Newtonian diminution of interest in divinity and heightened interest in nature for its own sake, scholars have too often read the Newtonian method narrowly, selecting from the breadth of his studies only mathematics, experiment, observation, and reason as the essential components of his scientific method. For a science of nature, a balanced use of those approaches to knowledge suffices, or so it has come to seem since Newton's death, and one result of the restricted interests of modernity has been to look askance at Newton's biblical, chronological, and alchemical studies: to consider his pursuit of the *prisca sapientia* as irrelevant. None of those was irrelevant to Newton, for his goal was considerably more ambitious than a knowledge of nature. His goal was Truth, and for that he utilized every possible resource. [...]

Perhaps the most important element in Newton's methodological contribution was that of balance, for no *single* approach to knowledge ever proved to be effective in settling the epistemological crisis of the Renaissance and early modern periods. Newton had perhaps been convinced of the necessity of methodological balance by Henry More, who had worked out such a procedure within the context of the interpretation of prophecy. Since every single approach to knowledge was subject to error, a more certain knowledge was to be obtained by utilizing each approach to correct the other: the senses to be rectified by reason, reason to be rectified by revelation, and so forth. The self-correcting character of Newton's procedure is entirely similar to More's and constitutes the superiority of Newton's method over that of earlier natural philosophers, for others had certainly used the separate elements of reason, mathematics, experiment, and observation before him.

But Newton's method was not limited to the balancing of those approaches to knowledge that still constitute the elements of modern scientific methodology, nor has one any reason to assume that he would deliberately have limited himself to those familiar approaches even if he had been prescient enough to realize that those were all the future would consider important. Because his goal was a Truth that encompassed not only the 'mathematical principles of natural philosophy' but divinity as well, Newton's balancing procedure included also the knowledge he had garnered from theology, revelation, alchemy, history, and the wise ancients. It has been difficult to establish this fact because Newton's papers largely reflect a single-minded pursuit of each and every one of his diverse studies, as if in each one of them lay the only road to knowledge. . . . In only a few of his papers may one observe his attempt to balance one apparently isolated line of investigation with another.

The characteristic single-mindedness reflected by each set of Newton's papers has led to the modern misunderstandings of Newton's methodology, for study of any one set may lead to a limited view of Newton's interests, goals, and methods, and the papers have all too often been divided up into categories that mesh more or less well with twentieth-century academic interests. . . .

. . . Newton's alchemy had almost always been considered the most peripheral of his many studies, the one furthest removed from his important work in mathematics, optics, and celestial dynamics. Most students of Newton's work preferred to ignore the alchemy, or, if not to ignore it, then to explain it away as far as possible. But one may, perhaps pardonably, remain unconvinced that a mind of the caliber of Newton's would have lavished so much attention upon any topic without a serious purpose and without a serious expectation of learning something significant from his study of it. Indeed, working one's way through Newton's alchemical papers, one becomes increasingly aware of the meticulous scholarship and the careful quantitative experimentation Newton had devoted to alchemical questions over a period of many years. Clearly, *he* thought his alchemical work was important. So one is forced to question what it meant to him: if Newton thought alchemy was an important part of his life's work, then what was that life's work? Was it possible that Newton had a unity of purpose, an overarching goal, that encompassed *all* of his various fields of study? . . .

In certain ways Newton's intellectual development is best understood as a product of the late Renaissance, a time when the revival of antiquity had conditioned the thinkers of Western Europe to look backward for Truth. Thanks to the revival of ancient thought, to humanism, to the Reformation, and to developments in medicine/science/natural philosophy prior to or contemporary with his period of most intense study (1660–84), Newton had access to an unusually large number of systems of thought. Each system had its own set of guiding assumptions, so in that particular historical milieu some comparative judgment between and among competing systems was perhaps inevitable. But such judgments were difficult to make without a culturally conditioned consensus on standards of evaluation, which was precisely what was lacking. . . . As a consequence, Western Europe underwent something of an epistemological crisis in the sixteenth and seventeenth centuries. Among so many competing systems, how was one to achieve certainty? . . .

But Newton was not a skeptic, and in fact his assumption of the unity of Truth constituted one answer to the problem of skepticism. Not only did Newton respect the idea that Truth was accessible to the human mind, but also he was very much inclined to accord to several systems of thought the right to claim access to some aspect of the Truth. For Newton, then, the many competing systems he encountered tended to appear complementary rather than competitive. The mechanical philosophy that has so often been seen as the necessary prelude to the Newtonian revolution probably did not hold a more privileged or dominant position in Newton's mind than did any other system. The mechanical philosophy was one system among many that Newton thought to be capable of yielding at least a partial Truth. [...]

Not only was Newton's goal a unified system of God and nature, it was also his conviction that God *acted* in the world. Though Newton avoided most hints of pantheism and though his Deity remained wholly 'other' and transcendental, Newton had no doubt that the world was created by divine fiat and that the Creator retained a perpetual involvement with and control over His creation. The remote and distant God of the deists, a Deity that never interacted with the world but left it to operate without divine guidance, was antithetical to Newton. Newton's God acted in time and with time, and since He was so transcendent, He required for His interaction with the created world at least one intermediary agent to put His will into effect. Just such an agent was the alchemical spirit, charged with animating and shaping the passive matter of the universe. [...]

Chapter Nine
Scientific Academies across Europe

9.1 Mario Biagioli, *Princely patronage of the Accademio del Cimento**

[...] As noticed by several historians, [Prince] Leopold [de' Medici] never provided his academy with a legal charter. He called it into session or suspended its activity whenever he desired. He set its experimental agenda, paid for the experimental apparatus from his own purse, and tended to draw his academicians from mathematicians and philosophers who were already on the Medici payroll. It seems that the very name of 'Accademia del Cimento' was a retrospective invention connected to the publication, in 1667, of the *Saggi* – a book presenting a selection of experiments conducted at the (by then) defunct academy. Finally, the academy was neither formally established nor disbanded. It began to meet around 1657, slowed down its activities after 1662, and stopped convening after 1667 when Leopold became cardinal and moved temporarily to Rome. As one academician remarked, the academy was nothing more than an expression of the 'prince's whims'.

The Cimento's status as an unofficial academy was, I think, a direct result of Leopold's participation in it. A prince of Leopold's rank could easily taint his image by working together in an *official* context with his subjects (some of whom were of quite low social background). Things were made even more complicated by the Cimento's commitment to experiments, that is, to a practice involving the use of mechanical devices. People of high social status could *observe* such activities only within settings that provided appropriate 'status shields'. An appropriate etiquette had to be followed. For instance, in the Bolognese public anatomy lessons analysed by Ferrari,

---

* Mario Biagioli, 'Scientific revolution, social bricolage, and etiquette', in Roy Porter and M. Teich (eds), *The Scientific Revolution in National Context* (Cambridge: Cambridge University Press, 1992), pp. 26–32.

a secret compartment was built into the anatomy theatre so that 'authorities, ladies or other persons' could watch the mechanically-connoted dissection without being seen. People of higher social status could be polluted by much less. Pope Gregory XV could not participate openly even in a semi-private academy gathered in the Vatican Palace by the cardinal nephew to listen to orations on biblical subjects. As reported by an observer, the pope 'participated' only *in incognito*, 'remaining retired in a small chapel' attached to the cardinal's room where the gathering took place.

Leopold controlled the possibilities of status-pollution in various ways. First, as shown by the preface to the *Saggi*, he tried to make sure he would be perceived as a princely supervisor rather than an active hands-on participant. Second, he presented the academy as something belonging to his private sphere. In fact, a prince could display himself naked to his servants in the privacy of his bath, but he could not do so in a more public space. Therefore, the participants to the Cimento could not become 'academicians' in the sense of being members of an official corporate body. Leopold's status required them to be his 'scientific servants'. In fact, they were not allowed to display their association to the academy by using titles such as 'Accademico del Cimento'. They had no official relationship with Leopold except as his subjects.

The same issues of status that made Leopold keep the academy as a fully private enterprise prevented him from entering into scientific disputes. In fact, disputes belonged to people who had an axe to grind – like members of the ignorant and self-interested lower classes. The academy's vocal commitment to the experimental method – one that led to accurate descriptions of experimentally (re)produced effects rather than to the explanation of their causes – was not only a result of Leopold's desire to keep clear of possible conflicts with theologians: it reflected the politeness of the philosophical etiquette to which he was bound by his own status.

By having his 'academicians' perform and describe experiments rather than seek their causes, Leopold made sure that the activity of the Cimento would not lead to status-tainting disputes. For analogous reasons, Leopold was exceedingly cautious of having himself invoked as the judge in scientific disputes. When that happened – as with Huygens and Fabri on Saturn's rings – he passed the matter to his academicians. They were instructed to perform careful experiments and, without passing any final judgement, to report what their experiments (based on *models*) suggested about the tenability of the contenders' claims (which the Cimento considered only as *hypotheses*).

Similarly, in the *Saggi* (the text through which the Cimento 'went public' in 1667) Leopold made sure that the academy's activity was represented as having unrolled as smoothly as possible, undisturbed by internal disputes. The frequently strong tensions and explicit disagreements recorded in the academicians' private correspondence were made invisible in the *Saggi*. Moreover, the book was written in a collective voice. No voice of any academician, except that of the Secretary who wrote the report, is ever made explicit. On several occasions, the text's cautiousness to avoid signs of individual authorship went so far as to try to 'objectify' the narrative by adopting the passive of the third person singular ('it was taken', 'it was seen', 'it was thought', 'it was provided').

There is more to the connection between the book's voice and Leopold's status than his attempt to present the academy as a haven of consensus and a space where only non-disputational knowledge was produced. For instance, it is significant that Leopold did not try to publicize his academy by opening up its meetings to many qualified visitors (as . . . London's Royal Society tended to do), but by distributing (without selling) an elegantly illustrated book presenting a selection of its experiments.

By doing so Leopold was probably trying to kill two birds with one stone. Through the *Saggi*'s textual strategies he managed to efface himself sufficiently from the academy's activities to preserve his princely status and yet not enough to delegitimize the academy's results. Unlike Robert Boyle and the Royal Society who bound themselves to certify knowledge through 'competent' and 'open' witnessing managed through a fairly intricate etiquette, the Cimento's results were presented as credible simply because they had been certified by somebody of Leopold's status.

Because of Leopold's effaced but effective presence, the *Saggi* did not need to reproduce the names of the witnesses and experimenters nor any other specific circumstantial information about the execution of the experiments. In general, the *Saggi* presented neither complete reports of individual experiments nor the 'typical' experiment, but a collage-like narrative composed by various narrative segments taken from different experiments – a procedure that would have not met the strict requirements of what Shapin has called Boyle's 'literary technology'.

In a sense, the present and yet invisible Leopold was the *incognito* certifier of the academy's work. But, because the *Saggi* did not mention any academician in particular, the credit for the work of the academy fell by 'default' on the prince. Leopold became the author *in absentia* – the only way in which he could be an author and enhance (rather than jeopardize) his image. The Cimento's unnamed academicians resemble Boyle's technicians studied by Shapin. They were indispensable as *workers*, but were not legitimate enough to 'make knowledge', that is, to be *authors*. However, unlike Boyle, Leopold did not utilize the academicians' involvement in the experiments to blame them for possible failures. This was not a result of Leopold's good nature but of his very high social status. No embarrassing failure could be represented in a princely experimental narrative. To Leopold, any such accident was equivalent to an embarrassing etiquette blunder at court.

Unlike Boyle, Leopold had the writer of the *Saggi* give his subjects full credit for having *performed* the experiments. However, despite the apparent differences, there is an underlying similarity between Leopold's and Boyle's textual strategies. For instance, in Boyle's case, the assistants were represented as nameless and unable to produce knowledge because it was the patron who had to be presented as the author. It was Boyle who had the status and credibility necessary to 'make knowledge'. The assistants 'collaborated' with him only in the sense that they took care of mechanical tasks that could not be dealt with by somebody of Boyle's status. Leopold's case was different and yet structurally homologous to Boyle's. Having a higher status than Boyle, Leopold was bound to a lower threshold of pollution. Consequently, he could not present himself as participating in scientific activities as much as Boyle could. It was because of this that Leopold's academicians received more credit than Boyle's assistants. However, giving some credit to the academicians did not deprive Leopold of authorship. In fact, because his academicians–subjects were kept nameless and because Leopold was the ultimate source of the academy's credibility, the credit reverted to the prince. Leopold was an author in the only way he could be one: *in incognito*. The peculiar voice of the *Saggi* provided a skilful solution to a problem of etiquette: to allow Leopold to fashion himself as author while remaining unpolluted by the status-tainting features of the knowledge-making process.

The namelessness of Leopold's academicians may reflect the same dynamics of princely power-image that led to the replacement of Cassini's name on the medal celebrating his astronomical discoveries with the 'learned men whom the King maintained at the Observatory'. More generally, one may suggest that the anonymity of some of the early published works of the Académie des Sciences should not be seen just as a sign of allegiance to a Baconian ideal, but may be seen also as reflecting the logic of the image of its royal patron. Symmetrically, the Royal Society's acceptance of the individual authorship of their members may also have reflected the English king's lesser involvement and interest in that institution. That the Royal Society as an institution could dissociate itself from the views of its individual members may be read also as a sign of its relative independence from the king. Because of the relative 'distance' between the king and the Society, his image was not directly at stake in its members' printed work. This estrangement of the royal patron, I think, made it possible for the fellows to emerge as individual authors. For individualism to emerge, the absolute monarch had to go.

This analysis has mapped out some of the specific status dynamics that governed Leopold's relationship with the Cimento and has suggested how similar dynamics may have framed the notion of scientific authorship in other European scientific institutions, but it has not addressed explicitly what Leopold's motives may have been in gathering and sponsoring an experimental academy. Practical usefulness and national technological development do not seem to have been a high-priority concern for Leopold – a junior prince of a small and increasingly agricultural state. Nor could spectacle have played much of a role in Leopold's decisions. Although the activities of the Cimento were rhetorically presented as the heritage of

Galileo's science, they could not compare with the spectacularity of his discoveries. . . .

The Cimento's activity suited Leopold's image because its experimental discourse (like that of Boyle's experimental philosophy) presented a non-contentious type of knowledge – one appropriate to a prince of post-Reformation Italy. Moreover, the Cimento's activities were quite well suited to the codes of courtly *sprezzatura*. No sweat-inducing machines (like air-pumps) but plenty of elegant glasswork (frequently destroyed during the experiments with truly aristocratic nonchalance) populated its experimental space. Moreover, several of the 'experiences' were not produced in the academy's various meeting places, but observed in the field. They resembled botanizing and collecting (or courtly *conversazioni*) as much as laboratory practices.

Academicians used the artillery pieces of the fortress of Livorno (not operated by themselves but by low-class gunners) to prove Galileo's claims about the parabolic trajectory of projectiles. They stayed up at night somewhere in Palazzo Pitti (probably in courtly late-night gatherings like those described in the *Book of the Courtier*) watching the formation of elegant ice crystals in water containers. On other nights, they busied themselves determining the speed of sound by watching the flare of faraway guns and measuring the time it took to the sound of the firings to reach them. Therefore, the experiments of the Cimento were not part of a 'laboratory life'. Rather, they resembled those courtly activities – like dancing or fencing – that characterized the daily life of a prince. The meetings of the Cimento did not demarcate a modern professional space but took place in the private sphere of the prince. As shown by the sometimes very frequent and unscheduled meetings, its activities were closely connected to the varying schedule of Leopold's daily life. He called its meetings as he would have called a hunting expedition.

That the Cimento did not meet regularly and never received a legal charter was not a sign of Leopold's casual attitude (on the contrary, he was exceedingly orderly and a great organizer) but rather of his status. A fixed schedule and a statute would have been an intolerable restriction of his freedom. To engage in something like the Cimento, Leopold had to keep it as something completely private. It needed to be something that was perceived to be completely his own, something he could fully control and display his power and status by controlling.

At the same time, because of its private character, this setting would provide him with an 'informal' space in which he could participate in (and legitimize) the academy in ways which would have been impossible if the Cimento had happened to be an official institution. In fact, had the Cimento been an official body, Leopold would have had to behave according to the strict etiquette that regulated the public life of a prince like himself – an etiquette that would have prevented him from mingling with technicians. The textual strategies of the *Saggi* reflected similar concerns:

they conveyed Leopold's familiarity with (and yet distance from) the Cimento. Those techniques were the textual analogues of the informal formality that surrounded the activities of the academy.

These considerations suggest a relationship between privateness and publicness, participation and distance. Leopold's example indicated that a junior prince could participate in scientific activity only if this was presented as a strictly private enterprise. The case of Louis XIV suggested that an absolute monarch's public involvement in science could be only accompanied by his not participating in the scientific activities he was publicly legitimizing and supporting. In fact, Louis XIV visited the Académie at the Observatory only once, in 1682, during a purely ceremonial event. From the available evidence, it does not seem that Charles II – the king of England who chartered the Royal Society in 1662 – ever visited that institution.

Leopold's concern with fitting his academy into the codes of country life is confirmed by his attention to the literary style of the *Saggi*. First of all, it is telling that he selected Count Lorenzo Magalotti (a sophisticated *virtuoso* and *letterato*) rather than Giovanni Borelli (his most brilliant but not-so-polished academician) as secretary of the academy and writer of the *Saggi*. Then, when the manuscript of the *Saggi* was finally completed, Leopold made sure it was reviewed for the elegance of its style as much as for the accuracy (and religious orthodoxy) of its scientific content. Also, Leopold must have been concerned with the potential tediousness of the *Saggi* since – halfway through the text – Magalotti felt compelled to include a disclaimer about the necessary dullness of the description of the experiments. . . .

. . . Although the type of scientific patronage embodied by the Cimento was very different from Cosimo II's patronage of Galileo and from Louis XIV's establishment of the Académie des Sciences, it shared in the image and power dynamics that framed those other two cases. Everything in the Cimento (methodology, 'research programme', experimental activity, organization, etiquette and public representation) was closely tied to the necessities of Leopold's status. In short, the Cimento was not the toy of a rich, provincial, junior prince tired of the occupations of court life. Whatever our evaluation of the science done by the Cimento, it exemplifies the structural features of Italian princely patronage of science at the end of the seventeenth century. Because Italy's political scenario (from the small size of the states to the princes' power-image dynamics) could not accommodate the institutional formats that were developing in other European countries, and because a discipline like astronomy was becoming less spectacular than it had been at the beginning of the century, there were fewer options for social bricolage between the resources of individual scientific practitioners and those of the princely patrons. The Cimento and the other informal academies developed around Roman cardinals and prelates were among the few venues within that scenario.

9.2 M. Hunter, *The usefulness of science in seventeenth-century England**

Restoration scientists were obsessed by the usefulness of their studies. As Thomas Sprat put it in his *History of the Royal Society*, contrasting the new science with the sterile scholastic philosophy: 'While the Old could only bestow on us some barren Terms and Notions, the New shall impart to us the uses of all the *Creatures*, and shall inrich us with all the Benefits of *Fruitfulness and Plenty*.' In the Restoration, as in the Interregnum, many shared Bacon's conviction that the advancement of learning had suffered in the past not least through the dissociation of 'Speculative men' and 'men of experience': 'it were to be wished (as that which would make Learning indeed solid & fruitfull) that *active* men would or could become writers', so that natural philosophy might directly ameliorate human life.

Just how literally such remarks are to be taken has been the subject of some controversy. ... Bacon not only placed a high value on science's utilitarian benefits: he also held up the advancement of knowledge in the mechanical arts as a model to intellectuals. ... But it is clear that most of those concerned with science had a strong sense of the essential primacy of intellectual considerations in the pursuit of knowledge about the natural world. Though anxious to see science applied, they were aware that too immediate a stress on practical considerations obscured the general principles on which these were based. This was Bacon's own view: despite his hopes for the amelioration of life and his belief in the fundamental unity of understanding with utility, he advocated a judicious balance of intellectual and practical ends, knowing that 'experiments of Light' were to be preferred to 'experiments of Fruit' when priorities clashed. [...]

There is a real danger of reading the appeal to utility too narrowly and assuming that it referred exclusively to practical, everyday needs. Many defenders of the utility of the Royal Society meant... to protest the merits of improved knowledge of the natural world rather than utilitarian applications. [...]

It is even possible that the stress on utility had an element of public relations, emphasis on likely tangible benefits being intended to justify broader intellectual concerns to a hostile public. For it is essential to remember that the new science was attacked as trivial and unimportant, contrary to the claims of its propagandists. [...]

This suggests artificiality, and the stimuli to intellectual activity are certainly distorted by overstressing practical considerations. But scientists nevertheless felt strongly that it was in their power to improve human life, either by applying theory or by bringing Baconian method to bear on

---

* M. Hunter, *Science and Society in Restoration England* (Cambridge: Cambridge University Press, 1981), chapter 4.

industrial and other techniques. ... Perhaps the best testimony to this is the project for a great collaborative 'History of Trades' that the Society espoused, which took up Bacon's notion of what was in effect the technological counterpart of his projected natural history. Information about technical processes was to be collected for its value in its own right and as a potential source of data for scientific hypotheses, while, through collation and comparison, it was also hoped that improvements noted in one area could be introduced in others. [...]

The History of Trades programme had creditable success in stimulating careful descriptions of industrial practices. ... One was dyeing, ... another was tanning. ... Other matters considered included salt-making, the production of alum and the brewing of cider, on which a large amount of research was published by the Society in 1664. Mining technology also received extensive attention, not least in articles in early volumes of the *Philosophical Transactions*. In each subject an attempt was made to check information and collate reports from different experts and hence to produce as authoritative an account as possible.

Such efforts had their agrarian counterpart in the concern to improve farming techniques of the so-called 'Georgical Committee', one of the groups set up to consider areas of the Society's interests in 1664. The intention of this was (in the words of a report of its meetings) 'the composing of a good History of Agriculture and Gardening, in order to improuve the practise thereoff'. [...]

Related to this was a great enthusiasm for silviculture, and perhaps the epitome of this early collaborative activity was Evelyn's *Sylva, or a Discourse of Forest-Trees and the Propagation of Timber in His Majesties Dominions* (1664). This resulted from extensive researches by various Fellows of the Society in response to a request to help improve the timber supply from the Commissioners of the Navy. Though Evelyn's name is solely attached to the work, his primary task was to supervise the disparate work of others and to present it in an acceptable form with his 'exquisite pen'. On its publication the book was a success, selling over a thousand copies in less than two years, 'a very *extraordinary* thing in *Volumes* of this bulk', as Evelyn was assured by booksellers. It reached a second edition in 1670 and a third in 1679, and by this time it was making a substantial profit for its publishers. ... *Sylva* was a useful compendium of information on all aspects of planting and it inspired the accumulation of more, some of which was incorporated in successive editions. More important, it certainly stimulated afforestation. [...]

Much attention was also paid to new inventions, either propounded by the Society's Fellows or originating elsewhere but brought to the Society for its approval and encouragement. In 1664 it was even suggested that the Society be given the duty of inspecting all proposed mechanical devices to see if they were 'new, true, and useful'. [...]

Equally important, intellectually fertile Fellows like Hooke and Petty were encouraged to devise machines and gadgets, and not only in the

Society's earliest years but thereafter a number of fruitful inventions came from scientific circles. [. . .]

There was also sustained interest in theoretical problems of practical significance – notably the longitude, attacks on which continued without any real success throughout the period. . . . A related development was the establishment in 1675 of a Royal Observatory at Greenwich, where John Flamsteed became the first Astronomer Royal. [. . .]

Scientists also tried to purvey theoretical knowledge in a practical form. Perhaps the best example of this is Petty's *Discourse . . . Concerning the Use of Duplicate Proportion* (1674). . . . In it, Petty tried to bring Galilean physics to a wider public, illustrating the value of calculations in shipping, artillery and building. [. . .]

All in all, these efforts to find ways in which intellectuals could serve current needs show an impressive attempt at relevance. Underlying them is a concern to get to grips with the problems of the contemporary economy. The bid to describe and improve agricultural and industrial practices manifested a sense of the value of improved production. Timber supply was a real crux due to pressure from the combined demand of industry and shipping. Comparable importance could be attached to dyeing processes and salt extraction, while butter and cheese were 'the most considerablest manufacture In England'. Above all, the close involvement of men of science in maritime affairs shows their conviction that intellectuals should contribute to this critical area of national life. [. . .]

It was increasingly clear that intellectuals could not effect improvements in practical matters as easily as they often supposed. Moreover those concerned with applied knowledge were increasingly isolated in their technological work even when frequenting scientific circles. . . . The younger generation of scientists was apparently less prone than the older to combine science and technology. [. . .]

Of the reasons hitherto suggested for the decline in technology, perhaps most persuasive has been the argument that, after its hopeful beginnings at the Restoration, science became increasingly elitist between 1600 and 1700. [. . .]

Though aloofness undoubtedly existed, it was present from the start of the Restoration, balancing the impulse against it. The History of Trades was itself symptomatic of a realignment in European intellectual life, of the conviction growing since the Renaissance that intellectuals should involve themselves in practical matters traditionally considered beneath them: a gentleman was, after all, partly defined as a man who did not soil his hands by manual labour. It is striking to find aristocrats and statesmen studying industrial processes in the early Royal Society and tempting to see this as a shift towards closer co-operation between intellect and economic life with wide implications. Yet the residual force of the older view is clear. . . . Intellectuals often valued 'secrets' on technological topics and were loath to vulgarise them by public dissemination. Much has been made of Bacon's attack on secrecy and his call for the free dispersal of knowledge: this assault on traditional attitudes which impeded scientific co-operation has rightly been seen as one of his most critical messages to the seventeenth century. But the change in attitude that he advocated could not be effected overnight. [...]

The problem was perhaps severest in subjects like shipbuilding where technical data could be regarded as arcana of state, but similar attitudes are to be found even in men like Boyle. ... On the other hand, the success of *Sylva* may have indicated how improvements could be purveyed among the educated class in the provinces with which the Royal Society was in contact, whereas... influence on the more numerous lower social orders was much harder to gain. The possibility of using a network of prominent landowners to spread innovation seems to have occurred to the Royal Society's organisers, echoing a more general feeling that in agriculture the example of the socially elevated was a useful agent of change. [...]

But the whole question of elitism may be misconceived in implying that changing approaches to technology resulted from internal developments in the scientific community, rather than from the reaction of scientists to external factors that they met in trying to make themselves useful. The most striking feature of their technological programme was arguably the gap between ambition and achievement, and this most easily explains why attitudes changed as they did, and why different people reacted in different ways. ... Of all the projects so far mentioned, however, few had more than slight effects, and we must now consider why the results were so disappointing compared with the good intentions we have chronicled.

The History of Trades and the projected account of agriculture well illustrate the need to separate activity and results. Scientists devoted much energy to such projects, but it is anachronistic to value the information they recorded for its historical interest even when it was never published. In its contemporary setting, the outcome was less impressive. [...]

As manifestations of the encyclopaedic impulse which was to culminate in the eighteenth century they are not without significance, and they certainly helped teach intellectuals about practical matters, which some saw as valuable at the time.... Though intellectuals supposed that they could easily and quickly master any craft and suggest improvements overlooked by ignorant artisans, this was not necessarily the case. ... It early became obvious that many processes were not easily described in words or even diagrams, since the operation had to be carried out precisely to be successful: 'the practick part' was just not susceptible to this kind of treatment. [...]

Inventions entailed comparable problems, since, despite the optimistic salesmanship of their protagonists (often echoed in modern books), they were not always as feasible as was claimed.... Intellectuals were also bad at judging economic factors, though sophistication grew during the Restoration. ...In the overpopulated agrarian economy of seventeenth-century England there was little need for the labour-saving gadgets of scientific enthusiasts, so they were never taken up. This explains why nothing came of John Evelyn's suggestion that the 'sembrador' – a Spanish machine that ploughed, sowed at an equal depth and harrowed all at once – be introduced into England, whatever its success in as depopulated a country as Spain. Lack of demand probably also accounts for the failure of Charles Howard's mode of making tannin without bark. The price of bark was falling rather than rising so there was little incentive to adopt his method, which was slower than that usually employed since tannin had to be extracted from the branches where it was less concentrated.

Another difficulty arose in naval matters. Even when the theoretical knowledge of intellectuals gave them an advantage over others – as in doing experiments on fluid displacement and making calculations about the most efficient form of hull design – the technical problems involved were too great for them to master. The results were therefore disappointing, even if they sometimes challenged accepted views. ... One should therefore treat sceptically the favourite suggestion of intellectuals as to why their proposals were not generally adopted – the ignorance and conservatism of craftsmen, shipwrights and farmers. ... The findings of economic historians do not suggest a general opposition to technical innovation, and scientists' most extreme denunciations must be rejected. The received picture of the agricultural history of the era – based on the record of men like Aubrey – is of a steadily accelerating acceptance of improvements like the use of root crops and new fodders and the growing of specialised cash crops in areas in which they were economically advantageous. In industry there were comparable developments: it has even been argued that the habit of undertaking small-scale adjustments necessitated by the replacement of wood by coal (itself a crucial adaptation) gave English artisans the attitudes which helped bring about the Industrial Revolution. On the whole the interest of intellectuals like those of the Royal Society followed rather than led such changes. Modifications in processing alum and copperas were both merely reported to the Society, while the agricultural inquiries naturally revealed innovation rather than led to it. Even when intellectuals made investigations these were often irrelevant to practice, so that improved techniques noted by travellers abroad were never applied although desperately needed at home. [...]

The large and diffuse audience for such information and the slowness with which it could be reached – let alone change effected – contrasted markedly with the tightly-knit European intellectual community. Here the efficiency of networks of communication like Oldenburg's correspondence popularised scientific discoveries and theories almost overnight, ensuring swift and widespread acclaim for their inventors. [...]

It is therefore not surprising that scientific discovery had attractions for intellectuals that technological improvement lacked, resulting in an inevitable preference for theoretical over practical pursuits. It is easy to oversimplify this process, since the utilitarian Baconian preoccupations of science meant that the two were never exclusive. But this, more than anything, explains the Royal Society's withdrawal from technological improvement, while the growing complexity of scientific knowledge also encouraged this dissociation to an extent not precedented in the Interregnum. Intellectual life was satisfying whereas, both now and earlier, attempts to improve technology ended in repeated disappointment.

It is hence revealing that scientists who aspired to a more significant role than that of professor did not go to technology to find it. Instead, they moved out of science into positions of power where they could use their talents in the more traditional world of public affairs.

9.3 P. B. Wood, *Apologetics and ideology in Sprat's 'History'**

The method described in the *History* was concerned almost exclusively with the cooperative compilation of natural histories, with the primary aim of utilitarian benefit. The Society's design, according to Sprat, was to 'make faithful *Records*, of all the Works of *Nature*, or *Art*, which...come within their reach', to rid learning of its 'confused heap of vain, and useless particulars', and to disentangle the facts known about nature from general theories. The collection of natural histories consisted of two parts, the preliminary collection of data, and the subsequent collective judgement of the Fellows concerning the facts. The activities of Henry Oldenburg typify the initial stage; Oldenburg's 'universal intelligence' network and his publication of the *Philosophical transactions* were designed to gather and disseminate information suitable for inclusion in the comprehensive natural histories planned by the Society. Collective effort was of paramount importance at this stage due to the volume of information required, and because such collective effort ensured that the materials gathered would not be subject to the bias of individuals. The subsequent validation of the information compiled by these means required the use of experiment which, it was claimed, was the ultimate judge of the truth of matters of fact. All reports to the Society were, if possible, verified by experiments, which were repeated 'till the whole *Company* [were] fully satisfi'd of the certainty and constancy; or, on the otherside, of the absolute impossibility of the effect.'

In 1661 Robert Boyle had published two essays concerning unsuccessful experiments in his *Certain physiological essays*, arguing that many experiments, especially chemical ones, were not repeatable due to the impurities of the substances used, the numerous species of particular substances, and other unknown contingencies. Boyle went on to argue that even observations were variable, pointing to the variation of the compass and to examples drawn from anatomy. This sceptical critique of experiment and observation, coming from one of the Society's most eminent natural philosophers, was

---

* P. B. Wood, 'Methodology and apologetics: Thomas Sprat's *History Of The Royal Society*', *British Journal for the History of Science*, 13 (1980), pp. 6–8.

no doubt seen as undermining the method described in the *History*, given the fundamental role assigned to experimentation. Consequently, Sprat included a reply to 'an *Objection*... rais'd by the *Experimenters* themselves'. While admitting that in many chemical experiments the number of contingencies involved rendered it unlikely that effects were repeatable, he argued that by a more scrupulous method of experimentation, using the same kinds of ingredients, the same proportions, and observing the same circumstances in the execution of the experiment, the same effects would always follow. However, his reply begged the question at issue, since Boyle had claimed that there were difficulties in the very preparation of the ingredients and in following more than once exactly the same procedure in the execution of an experiment. Sprat failed to meet Boyle's arguments by locating the problem in the experimenter rather than in nature, since Boyle had himself been largely concerned with the practical difficulties of the experimentalist. The rejoinder to Boyle, although inept, illustrates the demands, imposed by the *History*'s apologetic ends, of presenting the method as unproblematic and of giving the illusion of assured success.

Naturally experimentation was a significant source of information in its own right, apart from its role of verifying the reports of other natural philosophers. According to Sprat no certain art of experimentation existed; that is, unlike logic or rhetoric, experimentation was largely a process of trial-and-error. He stated that 'true *Experimenting* has this one thing inseparable from it, never to be a *fix'd* and *settled Art*, and never to be *limited* by constant Rules'. Furthermore, he claimed that the members of the Society designedly experimented *at random* to prevent the premature development of theories: 'they have made the raising of *Rules* and *Propositions*, to be a far more difficult *task*, than it would have been, if their *Registers*, had been more *Methodical*'. Thus the Society, on the *History*'s account, aimed to 'heap up a mixt Mass of *Experiments*...', and 'they confin'd themselves to no order of subjects'. It is instructive to compare Sprat's description of the method of natural history with Hooke's 'Method for making a history of the weather', included in the *History* as a model example of the Society's achievements. In contrast to the *History*'s picture of random fact-gathering, Hooke's 'method' provides a schema for the systematic recording of observations, including correlations reflecting casual connections between the weather and other phenomena. Moreover, the aim of his 'method' was to further theoretical analysis:

> ...these [observations]... may be registered so as to be most convenient for the making of comparisons, requisite for the raising *Axioms*, whereby the Cause or Laws of Weather may be found out...

Rather than being a designed attempt to prevent theoretical inferences, as Sprat had characterized the natural histories of the Society, Hooke's 'method' was designed to facilitate such inferences, including a column devoted to 'general Deductions, Corollaries or Syllogisms, arising from the comparing the several *Phaenomena* together'. Hooke's short tract was in fact an example of the kind of natural history envisaged by Bacon as the basis for the application of the inductive method.

Hooke's own prescriptions for natural historians were outlined in detail in his 'A general scheme or idea of the present state of natural philosophy', written in 1665 or 1666. ... What emerges from his discussion is that the collection of natural histories is to take place within a well-defined context of methodological and metaphysical presuppositions. Ideally, the naturalist must be well acquainted with all systems of philosophy in order to guess 'at the Solution of many Phenomena almost at first Sight'. The naturalist does not engage in random fact-gathering, for his queries and hypotheses guide his investigations. Hooke required the naturalist to know the laws of mechanics because they were the basic laws of nature. Moreover the histories of trades were of particular importance since they furnished a stock of mechanical models for causal explanations. As Oldenburg had written to Spinoza: '...we [the Royal Society] spend much time in preparing a history of the mechanical arts, feeling certain that the forms and qualities of things can best be explained by mechanical principles...'. Thus the significance of the programme for the history of trades was determined, in part, by the relationship it bore to the mechanical philosophy. The method of Hooke's consummate natural historian, then, is in contrast to that outlined in the *History*. This contrast further illustrates how Sprat's aims of popularization and recruitment led to a simplification and distortion of the procedures of those like Hooke who were actually engaged in the technicalities of research. [...]

9.4 A. Stroup, *Scientific utility at the Académie Royale des Sciences**

Like the academicians, Pontchartrain and Bignon respected knowledge for its own sake, and they tried to protect and direct the republic of letters, including the Academy of Sciences. They guaranteed the survival of the Academy by continuing to fund it, and some of their measures at least temporarily improved institutional morale. They also repaired the Academy's image by increasing membership and printing books and articles by academicians, hence creating an illusion of institutional vigor. But in reality funding was reduced, many academicians did not attend meetings, and several publications represented old research.... To survive, the Academy had to enhance the king's *gloire*, and elegantly produced books filled with discoveries obtained under royal auspices did just that.

The Academy and its protectors had still another strategy, one which would appeal to a patron who had no head for scientific theories, but who

* A. Stroup, 'Utility as a goal', *Transactions of the American Philosophical Society*, 77/4 (1987), chapter 6.

wanted better maps, better health, and better military equipment. By showing that their work could benefit king and kingdom in such practical ways, academicians could demonstrate to their royal patron the value of the Academy. This strategy was consistent with their own notions about natural philosophy and mathematics, for in common with other savants of the age, academicians believed that theoretical research would have practical consequences.

The minutes and publications of the 1690s show that the Academy pursued applied science in many forms. Mathematicians and astronomers focused their utilitarian attentions on navigation and cartography. In his audiences with Louis XIV, Cassini repeatedly assured the king that astronomical work would improve these spheres of knowledge so important to a ruler. Data collected at the Observatory, by Jesuits in the Far East, or by Cassini's provincial and foreign correspondents were all used to correct the longitude of major cities, as the map on the Observatory floor witnessed.

Botanists and chemists sought to improve medicine, and the minutes and official histories of the Academy are filled with discussions of remedies. In 1694, for example, Charas presented papers on medicaments, especially opium, and in the following year Homberg proposed a way of improving cataract operations.

Manufacturing, horology, and calendrical reform also found a place in the Academy's interests. Homberg studied dyes, and Morin de Toulon's report on porcelain is a chemist-mineralogist's study of a process in which European royalty invested heavily during the seventeenth and eighteenth centuries. Cassini reviewed a proposal for reforming the calendar, and Varignon took an interest in improving clocks.

The Academy also had an educational function, realized not only through its publications but also by its room of machines at the Observatory. This was open to the public for the study of natural philosophical apparatus such as the burning mirror; models of industrial, military, and agricultural machines; astronomical instruments; and the map of the moon prepared by Patigny under Cassini's direction.

The Academy had begun under Colbert an ambitious project of mapping the entire kingdom, but funding lapsed in the 1680s for the extension of the meridian and other surveying necessary for the task. Although Pontchartrain did not recommit the treasury to this project until after the reorganization of 1699, he did support a less well known cartographical project in which the Academy played a role. This was the *Neptune françois* published in 1693, a collection of maps which surveyed the European coasts from Trondheim in Norway to the straits of Gibraltar. [...]

The Academy's interest in technology is epitomized by its studies of machinery, which date back to Colbert's protectorship. Throughout the century academicians and their paid associates designed and tested apparatus, and outsiders also submitted inventions to the group, hoping for its approval. [...]

Some of the inventions the Academy considered in the 1690s had a decidedly military purpose, as when Amontons presented a design for flexible, cheap, and lightweight pontoons, or Dalesme demonstrated the recoil of a cannon. A visitor to the Observatory in 1696 came away impressed with the usefulness of the *salle des machines* to military engineers and teachers of fortification and navigation. ... Several members had military or naval experience. ... These academicians were no armchair strategists but rather active campaigners. The king's wars affected the Academy not only by reducing its funding, but also by drawing the attention of members to military technology and tactics.

The Academy also provided cover for clandestine activities on behalf of the crown. Thus the Dutch instrument-maker Hartsoeker not only supplied telescope lenses and presented theoretical papers to the Academy, but also assisted French diplomatic initiatives for peace with Holland. [...]

The Academy served the king's civil needs as well. A principal concern since the 1670s was the supply of water to the palace at Versailles. Academicians had surveyed for aqueducts under Colbert and Louvois, and from this work the savants developed books on surveyor's levels and hydrography. ... Even the yearly measurements of rainfall at the Observatory were justified and perhaps motivated by concern for maintaining water levels at Versailles.

The Academy and its three ministerial protectors were serious about utilitarian goals, but a bald summary of accomplishments in this category perhaps overemphasizes the extent and nature of these investigations during the 1690s. By comparison with previous years, the Academy's practical research was smaller in scope and piecemeal. Under Pontchartrain it lacked both the emphasis on theoretical underpinnings that had characterized the Academy's technological research under Colbert, and the commitment to massive projects of public works which both Colbert and Louvois had sponsored. Furthermore, the Academy's utilitarian activities cost less under Pontchartrain. ... Finally, the number of inventions submitted to the Academy for approval also declined.

But the Academy was not the sole representative of the crown's practical expectations from science, for Pontchartrain and Bignon supplemented the Academy's technological functions by forming a sister-society. This was the Compagnie des arts et métiers, which also figures in the treasury payments of the 1690s. [...]

The Compagnie's minutes were inserted in the Academy's own minutes to emphasize the connection between Compagnie and Academy. Until 1699 the Compagnie formed a separate, allied group with strictly technological concerns. [...]

The mission of the Compagnie was a source of dispute. Des Billettes envisaged a group which would take over some of the functions of the Academy, such as reviewing inventions, either as a permanent subdivision of the Academy or as a separate entity. He and his colleagues hoped also to compile an encyclopedia of arts and crafts, and the three interviewed artisans, sketched tools, workshops, and mills, and read treatises on technology. But their energies were from the start focused on one particular craft, printing, because of the inclusion of Anisson, director of the Imprimerie royale, and Philippe Grandjean, a type-founder, in their Compagnie. Anisson, Bignon, Pontchartrain, and the king wanted a new typeface for the royal printing house, something distinctive that was scientifically designed for maximum legibility and beauty.

Disappointed at this narrowing of their charge, Truchet, Des Billettes, and Jaugeon rationalized it on the grounds that printing served all the other arts, and they set out to pursue the subject exhaustively by also examining the ancillary arts of binding, goldworking, tanning, papermaking, type-casting, and typesetting. … Tensions between the savants and the royal printer grew, for the Compagnie des arts et métiers did not wish to be a mere consultant to the Imprimerie royale, while Anisson reasonably feared that its research would become so diffuse as to undermine work on his type-face. But once they had designed the typeface – the famous *romain du roi* – the savants' own apprehensions were confirmed, for although their treatise on printing was finished by 1704, it was never published. Having served Louis's *gloire* with the typeface, the savants were incorporated into the Academy where they again took up their work on the encyclopedia of arts and crafts.

Judging from such records as have been identified, the Compagnie was costly. … Without complete figures only imprecise comparisons are possible, but the crown certainly spent more on the Compagnie than on the research of the Academy during the same years. Thus the Academy's decreased expenditures on utilitarian activities under Pontchartrain do not signify his preference for pure over applied science, but rather result from his experiment with a separate technological and consultative institution, which ultimately was incorporated into the Academy, where some savants thought it belonged in the first place. … Thus from 1691 through 1699 the Academy's technological interests accounted for 10.3 percent of the research budget. […]

Utilitarian interests not only were an end in themselves but also had become fashionable in the late seventeenth century. Amateurs collected machines and intricate toys and created a market for books on mechanics. Thus, the editor of a dictionary had paid artisans in the 1680s to explain their technical vocabulary, just as members of the Compagnie were to pay them in the 1690s to reveal their methods. Knowledge of the arts and crafts and an understanding of complicated apparatus amused gentlemen and attracted savants, and royal funding reflected not only the utilitarian but also the fashionable impulse.

Academicians and members of the Compagnie shared ties of friendship, knowledge, and patronage. Since patronage was contingent on practical results, to advertise specific achievements helped ensure the survival of the two institutions. One result was that the fates of the Academy and the Compagnie were tied to the Imprimerie royale, the linchpin in the ministerial policy of publicizing scholarly activities. Although the savants did not entirely accept the primacy of printing, Pontchartrain and Bignon emphasized it in justifying, in very different ways, the existence of the Academy and the Compagnie.

A second result was that the members of these learned societies shared the frustration of seeing their scholarship channeled into a narrower role than they had expected. Savants appreciated knowledge primarily for its own sake, although they also anticipated practical benefits. But the king had more liking for concrete results than for theory, and saw utility primarily in terms of royal *gloire*. Thus the crown tended to treat the savants as technical consultants to itself or to the Imprimerie royale, whereas the savants saw themselves as benefiting knowledge, medicine, technology, and cartography in a broader sense.

Seventeenth-century scientists were savants with useful knowledge. This utilitarian potential helped justify subventions to the Academy as well as to its sister-society, even in the troubled 1690s. But the incentives were insufficient to ensure that the Academy and the Compagnie receive support at Colbertian levels. The crown supported scholarship, whether for its own sake or for more utilitarian motives, partly because it could do so at relatively little cost. It is the irony of Pontchartrain's protectorship that inadequate funding undermined the morale of the royal scholarly societies he governed, but that part of his success in preserving them was due to their relatively small budgets at a time of general financial exigency.

Chapter Ten
The Reception of Newtonianism across Europe

10.1 Larry Stewart, *The selling of Newton**

In the past decade the role of science in the early eighteenth century has come in for close scrutiny and increasing debate. There is specifically one rather large and problematic issue, that is, the relationship between science and technology in England in the first half of the eighteenth century when, it is generally agreed, the Industrial Revolution had not yet made any discernible impact. There are those historians who have insisted that the Newtonian natural philosophy had nothing whatever to do with the mechanical creations and innovations of artisans and craftsmen. This may be understandable because Newtonian science has come to be regarded as fundamentally mathematical and experimental – and not even comprehensible, except in the broadest terms, to the Augustan virtuosos. This has often created the version of science as a purely rational and cerebral activity distanced from and above technology, a science unsullied perhaps by the grime of mechanics' hands. One might speculate on the ideological origins of such a universe, but it seems that one can at least see that such a version of events is determined in part by the question that proposes a direct causal relation between cerebral science and rank technology. The argument evidently is that, if one cannot find the historical evidence that establishes a precise link between Newton's interparticulate forces and the partial vacuum of the Savery engine, then one must conclude that no relationship existed.

But historical associations are never quite so simple. One could easily demonstrate that the Newtonian natural philosophy was deliberately propagated among men whose interests tended to be more practical than philosophical. It is not to make any grand ideological claim to state that

* Larry Stewart, 'The selling of Newton', *Journal of British Studies*, 25 (1986), pp. 178–92.

Newton's science found its way into the commercial coffeehouses of London in the early eighteenth century where it encountered an eager and receptive audience. Public lectures were the vehicle by which an otherwise esoteric and incomprehensible mathematical natural philosophy was made intelligible to a wider public than could have been expected to have read the *Principia*. Indeed, in 1735 the schoolmaster, lecturer, and instrument maker Benjamin Martin made this specific objection against some of the more famous explications of Newtonianism. Henry Pemberton's *View of Sir Isaac Newton's Philosophy* may have been 'too expensive for the Purse of the Publick in general', as was Newton's *Principia* itself. But John Clarke's *Physics* (Martin meant the translation of Samuel Clarke's commentary on Rohault) was defective in 'Order and Method', while even the highly popular *Physico-Theology* of William Derham gave 'a delightful view of Nature broke all to Pieces'. It was this situation, in part, that authors and lecturers like Martin sought to correct. For example, one of the most prolific of the eighteenth-century lecturers, John Theophilus Desaguliers, endeavored to cover in his courses what he regarded as the essential topics of natural philosophy. These regularly included mechanics, hydrostatics, and Newton's laws of motion as well as his experiments in optics. An undated prospectus, but certainly after 1716, Desaguliers's 'Course of Mechanical and Experimental Philosophy' included not only hydrostatics, the properties of air, and optics but also 'explanations of Newton's law of gravity and a model of an Engine for raising Water by Fire'. A similar prospectus of 1725 made it clear that 'any one, though unskill'd in Mathematical Sciences, may be able to understand all those Phaenomena of Nature, which have been discovered by Geometrical Principles, or accounted for by Experiments'.

While the origins of the public scientific lectures are uncertain, shortly after the Revolution of 1688 they were on their way to becoming an industry. Following a lead set by chemical demonstrators in the late seventeenth century, the Reverend John Harris gave mathematical lectures at the Marine Coffee House in Birchin Lane, behind the Royal Exchange, in a series established by the brewer Charles Cox 'for the public good'. After almost a decade Harris was succeeded by James Hodgson, who had recently left the employ of the astronomer Flamsteed to join with the elder Francis Hauksbee, the instrument maker, in lectures on experimental philosophy. Hodgson continued to lecture with Hauksbee and at the Marine until 1709 or 1710, having obtained an appointment at Christ's Hospital Mathematical School at the end of 1708. The lectures at the Marine were then continued by Humphry Ditton, one of the most interesting and underestimated of the Newtonian polemicists until his early death in the spring of 1715.

The connection between Christ's and the Marine Coffee House suggests that education, whether formal or popular, was characterized by a deliberate aspect of utility. Christ's Hospital Mathematical School was specifically designed to prepare boys in the art of navigation in order to fit them to become apprentices to the captains of ships. We know that, when Hodgson initially became involved with the elder Hauksbee in 1704, the courses were advertised to be given at Ayers's Writing School in St. Paul's Churchyard. This may represent the origins of Hodgson's friendship with Thomas Ayers, writing master and teacher of navigation, whom Hodgson nominated to fellowship in the Royal Society in 1707. Ayers's practical interests were indicated by his early activities in the society when he discussed some of the numerous technical proposals and claims made by the Huguenot Denis Papin. Furthermore, Ayers was concerned in the agreement signed with Sir James Lowther in 1715 to exploit the Newcomen engine, a matter in which several of the Newtonians had an interest.

These associations merely represent a glance at a world in which the lecturers as entrepreneurs of science found that they had a commodity that merchants as well as the aristocracy found attractive. Toward the end of the reign of Queen Anne the result was the emergence of a veritable industry of science, joined by Desaguliers from Oxford and William Whiston, sent down from Cambridge for unorthodox religious views that he adamantly refused to conceal. Few of these lecturers, whose numbers continued to grow throughout the first half of the century, stuck to the niceties of experimentalism. On the contrary, many effectively sold their lectures as practical and useful. When the Whig pamphleteer Richard Steele organized in 1712 a series of entertainments, including science, in the institution that he called his 'Censorium', he explicitly proposed that 'All the works of Invention, All the Sciences, as well as mechanick Arts will have their turn'. What Steele and the lecturers were doing was acknowledging that their audience was no longer limited merely to royalty or aristocracy. This was confirmed in 1719 by Desaguliers, who had then become involved with Steele's enterprise. In that year Desaguliers advertised a course of experimental philosophy to include an improvement on the Savery engine 'of the greatest Use for draining Mines, supplying Towns with Water, and Gentlemens Houses'. It is undoubtedly the case that such an approach was very successful. By 1734 it is estimated that he was engaged in his 121st course, some consisting of 300 experiments. [...]

The importance of such entrepreneurs of science as Desaguliers or Harris lies not merely in their roles as proponents of Newtonian philosophy. Equally significant were the claims that such individuals made for those who attended the lectures and became well versed in natural philosophy and experimentalism. The Newtonians made the most of their opportunities in a world that made public support of their lectures and subscription to their books just as possible as private patronage had once been limited. That is not to say that private patronage was insignificant. On the contrary, in some circumstances it was very much alive, as it was for Desaguliers, who cultivated the support of James Brydges, the first duke of Chandos. Yet this too led into the world of entrepreneurs and financiers in the warrens and alleys around the Royal Exchange. It was in these quarters, as in the Marine in Birchin Lane, that opportunities abounded in the flood of projects that very nearly swamped early eighteenth-century commerce. [...]

...There is no question that those whom we have known largely as scientific lecturers spent ample time trying to secure patents for one kind of invention or another. For example, in 1720 Desaguliers, along with Daniel Niblet, a coppersmith, and William Vream, an instrument maker, obtained a patent for a device that used the heat of steam to dry various substances from hops to gunpowder and turpentine. William Vream had formerly been employed by the elder Francis Hauksbee, who died in 1713, and had more recently been associated with Desaguliers. Niblet maintained his alliance with Desaguliers at least until 1736, when they were still assessing improvements to the steam engine. The younger Francis Hauksbee, who had carved out a position for himself as an instrument maker, lecturer, and, ultimately, curator of experiments for the Royal Society, attempted in 1728 to secure a patent for 'a new Method of Applying the Centrifugal Force of a Body Moving in a Curve Line to the Moving of all the Various kinds of Mechanical and Hydraulical Engines or Instruments'. Unfortunately, like most patent applicants in the early part of the century, Hauksbee was as vague as he could be and still maintain his credibility. [...]

The move in the early eighteenth century from private patronage to the public of the coffeehouses represents a venture into commerce for the entrepreneurs of science. They were not shy in responding to the needs of the patrons they encountered. In part this explains the emergence of a great number of practical inventions from those who were otherwise active in public lectures. Besides Desaguliers and Watts, there was the younger Francis Hauksbee, who, with Benjamin Robinson of the York Buildings Company, obtained a patent in 1728 'for preserving the Planks and Sheathing of Ships sailing to the East and West Indies'. Within two months Hauksbee obtained yet another patent, this time with Benjamin Lund, a Quaker merchant of Bristol, for 'a method for the more advantagious manufacturing of Copper Oare & extracting Silver from Copper and mixing of Copper Calamy & Charcoale together for making Brass'. Their patent was immediately challenged by the Bristol Brass Company, especially their claim to be able to extract silver from the copper ores, although the crown upheld the patent.

It was this web of connection among a wide range of companies that provided the philosophical lecturers the chance to exploit their talents as inventors. This was especially true of the water-supply companies and those engaged in extractive industries such as mining, where knowledge of the latest steam devices or of the methods of assaying and smelting of ores might be in great demand. The utilization of the Savery and Newcomen inventions by partnerships and joint stocks is a good example. This was even the case among the groups created to export knowledge of the Newcomen engine to Europe at a time when the aftermath of the South Sea Bubble was making it difficult for joint-stock companies to operate with their customary abandon. Several of the individuals concerned were connected with Desaguliers, who was, as his advertisements for his lectures tell us, actively demonstrating the principles of steam technology. It was Desaguliers who recommended Joseph Fischer von Erlach to his former student the Dutch experimentalist s'Gravesande in 1721 for a proposed engine to be built for the landgrave of Hesse. In September 1720, the engineer John O'Kelly, about whom very little is known, arrived in Liege to construct an engine to drain the coal mines with his partners Canon Wanzoulle, Baron d'Eynstatten d'Aubay, and Lambert van den Steen. O'Kelly too seems to have had some connection with Desaguliers, for, when the partners had a falling out and O'Kelly was in danger of being imprisoned for debt, it was Desaguliers's patron, the duke of Chandos, who intervened with the secretary of state, Lord Carteret, once he had obtained details of the case. Such connections were important even in international affairs. [...]

The fleeting glimpses we catch of scientific lecturers near the Royal Exchange or in the coffeehouses of country towns tell us a little of the dynamism of Newtonian England. In the early eighteenth century there was emerging the kind of social and economic structure that made the manipulation of nature not only possible but also desirable. Who better to do so than those who claimed to know nature best? Who better to encourage them than improving landlords and financiers who were spurred to act by a financial revolution that promoted exploitation? What is striking about this attitude, which some found to be reprehensible and corrupt, was that the disruptions of the financial markets wrought by the South Sea Bubble in 1720 merely caused many to pause momentarily before engaging in the next project.

It needs to be recognized that landowners were no more immune to the promises of self-styled engineers than were the early industrialists. Witness the constant ventures of Sir James Lowther of Whitehaven, the earls of Dudley, or the earl of Hopetoun and Archibald Grant in Scotland to exploit the resources of their estates or of land they leased on a speculative basis – in some measure by employing those who were familiar with the world of the London natural philosophers and public lecturers. Natural philosophers such as Desaguliers were effective in marketing their knowledge as the foundation for improvement. Because of this, improvement – as an eighteenth-century spectacle – has an epistemological foundation closely linked to the success of Newtonian natural philosophy. This is not the same thing as claiming Newtonianism produced industrialism. But nor were these aspects of eighteenth-century culture in effective quarantine from one another. Newtonian science, and especially the manner in which it was so vigorously sold, fits nicely into the power relations of a society that was just constructing the incipient framework of capitalism that was to make the Industrial Revolution possible. This should not, however, be reduced to a necessary foundation or to an ideological one. Rather, the relationship between natural philosophy and the emergence of an industrial England was a subtle one that was tied to the forces of land as much as to the marketplace, to the interests of the aristocracy who scoured their estates as to those views of radical materialists or heterodox theologians.

If we are to approach the meaning of Newtonianism for the early eighteenth century, we have to recognize that Newton's followers created from his natural philosophy a social movement of many facets within the intellectual and political circumstances of late Stuart and early Hanoverian England. Perhaps Newton would not have recognized such a creation, although it is doubtful that anyone kept awake by the noise and smoke of the Mint could have missed what was happening. It was not Desaguliers's position as chaplain to the duke of Chandos that helped to establish the acceptability of the technologically oriented lectures. Such lectures were obviously successful beyond the social world of the aristocracy and would have been so without either Desaguliers or Chandos. Both were responding to the circumstances that helped to engineer the transformation that found expression in the Industrial Revolution. What is remarkable about the lecturer and his lordship was that they were aware of the small role they had in the larger scheme that they tried time and time again to orchestrate. Obviously, natural philosophers were not immune to the blandishments of commercialism. If this led them to project technical success that they failed to deliver, we must not ignore the involvement of Newton's apostles with incipient industrialism.

10.2 Simon Schaffer, *Newtonianism**

Making sense of Newtonianism: the interests of interpreters

Newtonianism possesses its literary classics, the *Principia* and the *Opticks* (1704). The former is a mathematical treatise on mechanics and astronomy. Its central techniques involve ways of representing change of quantity by geometrical and dynamical models. Instantaneous changes in variable quantities become incremental variations in line lengths. The effects of continuous forces are made equivalent to sums of instantaneous impulses deflecting moving bodies from a rectilinear path. The time a body takes when moving round some point from which a single force is exerted on it is computed by the area swept out by the line linking the body to the point.

Each technique possessed its own history. In the *Principia* the work of Newton's predecessors such as Christiaan Huygens and Johann Kepler was transformed. Kepler's laws of planetary motion were generalised as motions in a central-force field. New phrases appeared; in analogy with Huygens's 'centrifugal force', 'centripetal force' referred to actions tending

* Simon Schaffer, 'Newtonianism', in R. C. Olby, G. N. Cantor *et al.*, *Companion to the History of Modern Science* (London: Routledge, 1990), pp. 613–18.

towards the force-centre. With these tools, the treatise presented an analysis of the motions of planets and moons, tides and comets. Newton's work in mathematics later appeared in texts on analysis, first published in 1711, and on general mathematics, edited in 1707 by William Whiston, Newton's successor at Cambridge. Newton's achievement was swiftly seen as a compelling account of natural motion and its quantitative description.

Newton's book on light and colour, the *Opticks*, was assembled during the 1690s from earlier material. The treatise wore an experimental dress: it contained detailed protocols for trials with prisms, mirrors and lenses, and, as a coda, a set of queries on fire and light, matter theory and experimental philosophy, which grew in length in the successive editions from 1704 to 1717. One way of speaking about this achievement was to make the *Principia* into a programme for mechanics and astronomy and to read the *Opticks* and its associated texts as a charter for a new experimental philosophy. Both works contain some passages which present strictures on method, including a passage on experimental method first published at the end of the *Opticks* in 1706; a 'General scholium' added to the *Principia* in 1713; and some 'Rules of reasoning', as they came to be known, prefaced to the *Principia*'s third book. Newton's epigones published extracts from some other texts, including significant letters to Robert Boyle, the Anglican priest Richard Bentley and others of Newton's contemporaries, principally concerned with optics, matter theory and natural theology. The Newtonian corpus was formed out of this rather uneven mixture of documents.

To chart the uses to which this legacy was put, it is necessary to consider the local interests served by its application. These interests help clarify why some writers won Newton's blessing while others attacked his system. Examples from optics and chemistry show how this worked. Many eighteenth-century natural philosophers held that Newtonianism included loyalty to a model of light as minute, rapid particles of matter affected by strong short-range forces between these particles and common matter. This was a doctrine which they claimed to find in Newton's *Opticks* and in some of his earlier papers on light and colour, composed in the 1670s and reprinted in the 1780s. Yet the doctrine of the corpuscular character of light was rejected by several natural philosophers. There was no period when Newton's writ ran throughout Europe. The greatest of mid-century academicians, Leonhard Euler, espoused a wave theory of light and denied that all matter gravitated equally. Others behaved differently. While Thomas Young also developed an alternative model of light, based on vibrations in a space-filling ether, he claimed a good Newtonian warrant for his own views. ... Such examples illustrate the more general thesis that interpretation of the Newtonian enterprise has been consequent upon the interests of interpreters. We must try to understand these purposes in order to explain why a particular version of Newton's meaning was given, and to examine who it was who shared this interpretation.

Present-day historians and philosophers have their interests too. They have discussed Newtonianism in a variety of ways. Newton is treated as the legislator of modern sciences and Newtonianism as a *language*, a vocabulary and a grammar which most scientists began to use to speak about their world. His texts are combed for his views on method and his celebrated discussions of the significance of mathematics, the use of experiment, the suspicion of hypothesis and the relation between force and matter. Newtonianism becomes synonymous with 'classical' physics. A second form uses Newton to classify the variations in eighteenth-century natural philosophies. It attends to the obvious authority of Newton in the Century of Enlightenment. In this version of history, selected researchers are identified as 'Newtonian' if their views conform to those which the historian attributes to Newton. If they do not, then the natural philosophers in question can be seen as 'anti-Newtonian' or even as victims of 'misreadings' of the master's dogma. Neither approach accounts for the interpretative practices of Newton's audiences.

It is a notable feature of the development of western sciences since the Renaissance that they have formed well-defined groups of collaborating enquirers. The outward and visible forms of the scientific enterprise, such as the performance of experiments and their repetition by other credited colleagues, shared stories about the origin and progress of scientific specialities and disciplines, heroes and villains, provision of training through textbooks, laboratories and academies, networks of publication and informal communication, are all accompaniments of such groups. Much discussion has been devoted to ways in which such groups can be defined. What gives them coherence? What function do they play in research, and how do they come into being and pass away?

Each historiography answers differently. The first suggests that the local ecology of groups of co-workers is a less significant feature of scientific life than the intellectual assumptions upon which all such collaboration depends. 'Newton' becomes the name of an heroic mind which promulgated these assumptions and gave them decisive authority. Yet a wide variety of social settings can be invoked to see how Newton's authority emerged and became powerful. The period following Newton's death witnessed the emergence of scientific societies across Europe, modelled on the Paris Académie des Sciences. Members of these academies strenuously developed and debated the implications of the mathematics, mechanics and astronomy which Newton constructed. At the same period, promoters of lecture demonstrations and public shows expanded their appeal and the range of instruments and experiments. Significant commercial activity in London and elsewhere was accompanied by deliberate efforts to base the claims of speculative entrepreneurs upon Newton's natural philosophy. 'Newtonian' natural philosophy began to appear, in textbooks by writers such as Willem s'Gravesande (1688–1742) and John Theophilus Desaguliers (1683–1744), and in more popular volumes such as Voltaire's *Éléments de la philosophie de*

Newton (1738) and *Il Newtonianismo per le dame* (1737) composed by Francesco Algarotti. This implies that the classification of eighteenth-century sciences into those of a 'mathematical' and those of an 'experimental' emphasis, and the division of Newton's legacy into traditions of commentary upon the *Principia* and upon the *Opticks*, can provide a preliminary model for an exploration of Newtonianism. Furthermore, as the disciplinary map of natural philosophy changed during this century, so did the various senses of Newtonianism.

Newtonian powers: eighteenth-century experimental philosophies

Between 1690 and 1727 Newton and his colleagues worked hard to establish agreement about the sense to be given to the terms of his natural philosophy and its implication for contemporary culture, religious and political. Several platforms were available for participants in these debates. At the Royal Society in London, where Newton became President in 1703, experimenters such as Francis Hauksbee (*c.* 1666–1713) and his successor Desaguliers were employed in staging shows to illustrate Newtonian principles. Command over this institution was important. Hauksbee's work on electricity, light and pneumatics, initially conceived as a development of the research programme initiated by Robert Boyle, was very rapidly absorbed by Newton into his account of the relationship between force and matter. Trials performed by Hauksbee from 1705 were reported in subsequent editions of Newton's *Opticks*.

Desaguliers' role was equally decisive. From 1710 he continued the lectures on experimental philosophy delivered at Oxford by John Keill. As demonstrator at the Royal Society from 1714, Desaguliers' shows played a key role in the responses Newton organised to challenges from Continental critics. Visitors were able to witness expert and authoritative replications of Newton's optical trials. It was Desaguliers, for example, who in 1714 revived the label of 'crucial experiment' for one such trial using two prisms. The Newtonian group held that this demonstrated that light could be analysed into truly 'primitive rays' which could not then be made to change colour upon further refraction. The experiment appeared in idealised form at the front of the 1722 Paris edition of the *Opticks*. Such texts, including the translation of s'Gravesande's textbook on Newtonian philosophy which Desaguliers produced in 1721, accompanied and reported the celebrated experimental work. Experimental techniques shown at the Society and elsewhere in London were transmitted to France, Holland and Italy. Popular texts such as those of Voltaire and Algarotti appeared as a result of these contacts and the consequent emergence of audiences for this form of experimental philosophy. Desaguliers and Newton provided their performances with a commentary which stipulated the right way of interpreting the public trials. In the process, a Newtonian vocabulary was formed, involving passive particles, short-range attractive forces and active principles such as the causes of electricity, light and chemical fermentation. Newton's queries to the *Opticks* became a charter for a mature research programme.

The establishment of Newton's authority within the universities was also a significant aspect of this process. Networks of patronage and influence were decisive. At Newton's Cambridge, powerful figures such as Richard Bentley aided the careers of men who played significant roles in Newtonian debates, such as Whiston, Samuel Clarke and Roger Cotes. Political and religious controversy accompanied this process. Newton was treated by some of his allies as an exemplary member of the Whig establishment, while others saw his programme as threatening and divisive. Several of those Cambridge men who supported Newton were accused of heresy and became deeply suspect to Tories and High Churchmen, who perceived that Newton's natural philosophy seemed to attribute force to matter. The critics charged that Newtonians revelled in conceptual obscurity and offered a path to rationalist deism and free thought. In 1710 Whiston was expelled from the University. Clarke was accused of denying Anglican orthodoxy. Newton's closest colleagues found themselves enrolled in the defence of Newton's mathematics, cosmology and theology against critics such as the philosopher and divine George Berkeley, whose *Analyst* (1734) contained a brilliant assault on the foundations of Newtonian mathematics. Responses to Berkeley from mathematicians such as James Jurin and Colin Maclaurin provided opportunities for the development of Newton's programme, as did the contemporary war waged between the British natural philosophers and the allies of G. W. Leibniz, a contest which touched on issues of metaphysics, natural philosophy and religion.

Presented in polemical replies to major domestic and foreign critics, Newtonianism was a system born in conflict and a language produced for specific pugnacious purposes. The debate on the cause of gravity illustrates this formation. In the *Principia*, Newton seemed to eschew all causal talk about forces; having given their mathematical measure, he offered few indications of his account of the principles which allowed them to arise from and affect matter. The question was discussed with the Cambridge men: in the 1690s, Bentley gave sermons which enlisted Newtonianism in the defence of religion, Roger Cotes replaced Bentley as editor of the *Principia*'s second edition (1713), and Clarke translated the *Opticks* into Latin in 1706, fought Leibniz on Newton's behalf in 1715–1716, and transformed a celebrated treatise on Cartesianism, Jacques Rohault's *Traité de physique*, into a Newtonian textbook for Cambridge students (1697).

It seemed important to provide an answer to the question of the cause of forces such as gravity, since Newton faced charges from critics such as Huygens and Leibniz that an uncaused force which acted through empty space on the centres of bodies was no better than a 'miracle' or an 'occult' quality. If these accusations could be sustained, then the moral propriety of Newtonian natural philosophy would be undermined. But Newton's statements on the cause of gravity needed careful management to make them into an adequate response. Newton and his allies asked whether a force like gravity had a material cause. In passages in the *Principia* and in conversations during the 1690s Newton answered in the negative, since all matter gravitated, and any matter in space would disturb motion. Cotes told Newton in 1712 that this must mean that gravity was essential to matter, and published this view in his preface to the *Principia*'s second edition (1713). Newton and Clarke disagreed. Newton informed Bentley in 1693 that gravity could not be essential to matter, and reiterated this view in a query added to the *Opticks* in 1717. Samuel Clarke preached in 1706 that gravity must be infused into matter by God. He told Leibniz in 1715 that this did not mean that force was a 'miracle'. Arguments between Newton and his readers, whether friendly or hostile, spawned several different answers to the Newtonian question of force.

Contemporary with the contest with Leibniz, for example, Newton published statements which indicated that some fluid medium might indeed provide the cause of forces. This principle might be electrical. Newton helped himself to Hauksbee's experiments to sustain this view. Furthermore, Newton carefully changed passages in the 1713 *Principia* and the 1717 *Opticks* to allow some ethereal fluid which might be present in space and inside bodies and which could produce these forces. As presented in Britain in the first decades of the eighteenth century, Newtonianism banned certain phrases from natural philosophical discourse. No fluid made of common matter filling all space could be mentioned. It must not be said that gravity, or any force, was inherent in and essential to matter. It was allowable to remain silent on the cause of gravity. It was also permissible to tell stories about some rare fluid, itself susceptible to short-range forces, and to use electrical, chemical and optical experiments to provide support for such a substance.

This discourse provided natural philosophers with some important resources: a range of fluids evinced in demonstration experiments with electrical machines, air pumps and other devices, mathematical techniques for estimating forces and their effects, a theological claim about the first cause of all natural change and the sense that mathematical sciences might not need to state the real causes of the forces which they described. None of these resources were unique to Newton's allies. Others found them useful too. Newton's resources were applied to problems of medicine, moral philosophy and political economy. [...]

10.3 John Brooke, *Religious utility of science**

In his *Dictionary of chemistry* (1789), the entrepreneur James Keir (1735–1828) reported that a new spirit was abroad: 'the diffusion of a general knowledge,

* John Brooke, *Science and Religion: Some Historical Perspectives* (Cambridge: Cambridge University Press, 1991), pp. 152–63.

and of a taste for science, over all the classes of men, in every nation of Europe'. Even allowing for exaggeration, that growing appetite for science contrasts with the leaner fortunes of certain religious institutions. Fifty years earlier in England it had not been uncommon to hear Anglican clergy bewailing a widespread notion that Christianity had been discredited. The contrast is striking. Between 1660 and 1793 the scientific world established itself with more than seventy official scientific societies (and almost as many private ones) in urban centers as far removed as St. Petersburg and Philadelphia. In France alone there were thirty. The established churches, however, often perceived themselves to be in danger, both from dissenting religious movements and from a ground swell of rationalism and ridicule. But what connections were there between the popularization of science and rationalist movements that threw Christian theologians on the defense? ...

Attacks on the power of the Christian churches, and of the Roman Catholic Church in particular, were launched by deists, who denied the authority of doctrines supposedly derived from revelation; by materialists, who denied a duality between matter and spirit; and by agnostics, who, like David Hume (1711–76), argued that it was impossible to know anything about the nature of God that need affect human conduct. The extent to which each of these critiques drew on the resources of science will be our principal concern.

Historians are in no doubt that there were connections between reverence for science and irreverence toward religion, especially among representatives of Enlightenment culture in France. [...]

An attitude of respect for the sciences, but condescension toward orthodox religion, was usually sustained by extolling the power of human reason. Newton's gravitational theory, with its solution to the problem of planetary motion, symbolized what the human intellect could achieve. Science was respected not simply for its results, but as a way of thinking. It offered the prospect of enlightenment through the correction of past error, and especially through its power to override superstition. An English deist, Matthew Tindal, saw such untapped wealth in human reason that he suggested in 1730 that men ought to be judged, and would be judged by the Almighty, according to the use they had made of it. The most rational religion, as Joseph Priestley (1733–1804) later argued, was the one that would prevail, once all religions could compete on equal terms. A common grievance among eighteenth-century rationalists was that established churches were privileged only through arbitrary political power. [...]

The religious utility of science

Any suggestion that science was valued in the eighteenth century only, or even principally, as an antidote to 'priestcraft' would be false. As propagandists for science found new social contexts in which to assert its utility, it was practical utility, in the sense of solving technical problems, that usually came to the fore. A classic example would be the determination of longitude at sea, a problem kept in the limelight by the association of mercantile interests with the safety of ships. One of Newton's popularizers, William Whiston, was instrumental in petitioning the English Parliament in 1714 to offer incentives for a viable solution. As repositories of specialized knowledge, scientific academies and societies were increasingly consulted. The Royal Society was approached in the 1770s when the British government wished to know whether a blunt or pointed lightning conductor would be the more effective. By opening new avenues for patronage, the scientific societies often provided a forum in which the rhetoric of utility could yield new dividends. The chemist William Cullen (1710–90) addressed himself to problems in agriculture, bleaching, and salt purification to please aristocratic patrons with whom he conversed at the Philosophical Society of Edinburgh. In England, the Lunar Society of Birmingham gave Joseph Priestley access to the wealth of entrepreneurs such as Matthew Boulton, James Watt, and Josiah Wedgwood. In return, he would offer advice on the effects of different gases in the steam engine or analyze clay samples for Wedgwood's pottery.

Studies of eighteenth-century scientific societies have, however, shown that the concept of utility carried broader meanings than the solution of technical problems. Even in a town such as Manchester, at the heart of Britain's industrial revolution, the men who gathered at the Literary and Philosophical Society (founded 1781) justified their allegiance to science with a multiplicity of arguments. The promise of technical application was dangled before manufacturers, even if relatively few benefited. The misfortune, declared one member, Thomas Henry, 'is that few dyers are chemists, and few chemists dyers'. But science was also commended as polite knowledge, a means of cultural expression particularly congenial to the medical membership. An acquaintance with natural knowledge could be urged as one mark of a gentleman. As founder of the Derby Philosophical Society in 1783, Erasmus Darwin (grandfather of Charles) announced that his society would seek 'gentlemanly facts'. Science was also promoted as rational entertainment, suitable for the youth of a burgeoning town who might otherwise be seduced by tavern or brothel. A 'relish for manly science' was considered by one member of the Manchester society, Thomas Barnes, to be second only to religion in cultivating a mind likely to succeed in business.

Scientific knowledge could be prized for its supposed objectivity, in salutary contrast to the warring factions of political or religious parties. A dispassionate quest for objective knowledge was proclaimed a virtue in its own right, as by Fontenelle, the perpetual secretary of the Paris Academy of Science, whose eulogies of deceased members invariably praised their selfless commitment to human welfare. In the Manchester society, science was commended as a profession and an integral part of a reforming ideology that placed higher value on intellectual attainment than noble birth or inherited wealth. Finally, science was considered useful as a means of theological instruction. It gave content to arguments for God's power and foresight.

This belief that science had religious utility was widespread in the eighteenth century, particularly in Protestant countries. Certainly the God inferred from nature was not always the God of Christian orthodoxies; but in the minds of many it was. And, for them, the marvelous adaptations in the organic world underlined the wisdom of the God in whom they already believed. Christian, as well as anti-Christian, values could assist the popularization of science. Indeed, popular lecturers on science in Britain habitually displayed the powers of fire and electricity as impressive, almost theatrical, testimony to divine power. On a more elevated plane, the popularization of Newton's philosophy in England was due, in part, to the sermonizing of Anglican divines who, like Samuel Clarke (1675–1729), insisted that what was commonly called the course of nature was 'nothing else but the will of God producing certain effects in a continued, regular, constant, and uniform manner'.

When Robert Boyle had drawn up his will in July 1691, he had bequeathed fifty pounds annually for sermons to prove the Christian religion 'against notorious infidels, viz. Atheists, Theists, Pagans, Jews, and Mahometans', without descending to controversies among Christians themselves. From the sermons of the first Boyle lecturer, Richard Bentley, it is apparent that certain features of Newton's science could fit the bill. The ambiguities in Newton's philosophy of nature ... were systematically resolved in favor of a God active in both nature and history. Against materialist options, Bentley rejoiced that gravitational attraction was not an innate property of matter. He also argued that a mechanistic account of action-at-a-distance would require the emission of an effluvium from the attracting body and projected in every direction – a concept which he found repugnant. He also capitalized on Newton's point that gravity operated from the center of bodies, its magnitude depending on mass, not surface area. This seemed to exclude mechanical causation based on the pressure of corpuscles. For, if gravity could penetrate to the center of a solid body, it had to be a power 'entirely different from that by which matter acts on matter'. That was the opinion of Clarke who, like Bentley, inferred that the world depended 'every moment on some Superior Being'.

For Bentley, as for Newton, the elegance of the inverse-square law was proof of the divine mathematician, as was the fact that the planets had gone into closed orbits. Against those who derived the universe from a chance collision of atoms, Bentley had another Newtonian device. Committed to the ultimate unity of all matter, Newton had argued that the density of a body reflected the amount of empty space it contained. This porosity of matter was confirmed by the transmission of light corpuscles through as solid a material as glass. Such was the range of densities that most bodies must contain more space than matter. The implication, seized by Bentley, was the paucity of real matter in the universe. There were so few atoms compared with the dimensions of space that, given a uniform distribution, the odds were against any two colliding, let alone enough to make a world.

Bentley's arguments for the religious utility of science had political as well as philosophical dimensions. In the belief that atheism would encourage social instability, he was assuming that rational argument would be an effective counter. In other more specific ways, the political circumstances in England during the last decade of the seventeenth century created the space for Bentley's polemic. Following the Revolution of 1688, when William of Orange displaced James II from the throne, those clergy who sought to justify their allegiance to a new king could argue that a special divine mandate must take priority over a divine right to rule. Low Church Protestants warmed to the notion that God had willed the expulsion of James, whose Catholic sympathies had produced unwelcome effects in as cloistered an environment as Cambridge.

During the 1680s the fear of popery had been a prominent theme in the thinking of those who would call themselves Whigs. For John Locke, the prospect of a Catholic monarch, with a threatening allegiance to a foreign power, had meant the abandonment of 'the whole kingdom to bondage and popery'. Newton himself had shared Locke's opinion that Catholicism could only appeal to uninquiring heads and unstable minds. Yet the beast had reared its head in Cambridge just as he was completing his *Principia* (1687), and in so ugly a manner that he had sacrificed his privacy to slay it. The occasion was a mandate from James II, asking Cambridge University to confer the degree of Master of Arts upon Alban Francis, a Benedictine monk, without requiring him to take the usual oath to uphold the Anglican faith. If James could catholicize his army, this looked like the thin end of the wedge in a bid to control the universities. Newton had encouraged the university to resist the royal will and was one of its representatives when the royal fury demanded an explanation.

Newton's science could be invested with political meanings because it was so easy to construct parallels between God's intentions for nature and for society. Newton's emphasis on the freedom of God's will, as disclosed in nature, could be useful when justifying the intervention of God's will in human affairs – even in the removal of a king. And, since God's activity in nature was mediated by physical laws, one could also argue for the moderating effect of Parliament in constraining an overzealous monarch. It would, however, be misleading to imply that such analogies provide the only explanation for the popularization of Newton's science. Prominent among those who promoted it were Huguenot émigrés from France and Scotsmen who, like the mathematician Colin Maclaurin (1698–1746), appreciated the quality of Newton's intellectual achievement. Among the first teachers of Newton's science at the University of Oxford were the high churchmen John Freind and John Keill, the latter having followed his mentor, David Gregory, from Edinburgh.

It would also be misleading to imply that Newtonian divines such as Richard Bentley and Samuel Clarke were representative of a dominant orthodoxy within the Anglican Church. At least eighty percent of Anglican clergy in the period 1689–1720 were high churchmen who looked to parliamentary legislation rather than intellectual argument to suppress the atheist threat. By their fellow clergy, the Boyle lecturers were often perceived as heretics rather than upholders of orthodoxy. And not without reason. Newton privately disbelieved in the doctrine of the Trinity. His successor, William Whiston, lost his Cambridge chair by going public on that very issue. Bentley and Clarke, in the perception of their critics, were tarred with the same brush. To argue for the religious utility of the latest science was hardly a passport to preferment.

The fact that certain Anglican clergy assisted the popularization of Newton's science must not be allowed to conceal another important fact – namely that theological contentions were sometimes to the fore in critiques of Newtonian philosophy. This was especially true on the Continent of Europe where Leibniz, wielding a different metaphysical theology, would have no truck with Newtonian concepts. Leibniz was a Lutheran philosopher who, in wishing to do justice to the claims of theology, was also motivated by the desire to see Europe's Protestant churches united. Writing to his Catholic employer in 1679, he had even claimed that his philosophy would be accessible to the Jesuits. Rejecting the Cartesian definition that matter consisted only in extension, he insisted that every true substance had its own individuality, its own 'soul'. With the claim that he was restoring a concept of substantial form, he implied that he was creating space for a Catholic piety that had been compromised by Descartes.

During the second decade of the eighteenth century, Leibniz became embroiled in controversy with Newton's advocate, Samuel Clarke. Their exchange, which amounted to five letters each, began with a declaration from Leibniz that natural religion in England appeared to be in a state of decay. Writing to Caroline, princess of Wales, in November 1715, he observed that 'many will have human souls to be material; others make God himself a corporeal being'. Having cast aspersions on Newton and his followers, Leibniz excited a swift rejoinder from Clarke, who published the ensuing correspondence in 1717, following Leibniz's death. Theological issues dominated the debate, which was, however, colored and intensified by other matters. Antipathy had grown between Newton and Leibniz because of a priority dispute over the calculus. Indeed, Newton once complained that Leibniz's critique of his gravitational theory had been designed to persuade the Germans that he (Newton) had lacked the wit to invent the new technique. National pride and personal jealousies were also involved. Having become court philosopher to the House of Hanover, and since on the death of Queen Anne there was a Hanoverian succession to the English throne, Leibniz cherished the hope that he would gain an official position at the new English court. For Newton and his supporters this was not a delightful prospect. Leibniz, for his part, was concerned that his royal pupil, Caroline, was falling under the spell of her Newtonian tutor, Samuel Clarke. The attitude that he struck may also have had deeper political roots, in that his aversion to systems of nature that stressed God's unrestrained power may have sprung from their use in sanctioning absolute powers claimed by an earthly sovereign – notably Louis XIV, whom he had long recognized as a threat to the peace of Europe, and of the German States in particular. In 1683 he had satirized Louis's military expansionism with references to Mars, the God of War.

An antipathy to voluntarist theologies is evident in Leibniz's remark that a secure foundation for law is to be found not so much in the divine will as in His intellect, not so much in His power as in His wisdom. Justice would be established 'not so much by the Will as by the Benevolence and Wisdom of the Omniscient'. The theological reasoning that underlay this position was concerned with the grounds on which God could be praised for His work in creation. If nature was deemed to be good merely because it was the product of God's will, this provided weaker grounds than if God could be shown to have deliberately structured the world according to standards of goodness, beauty, and wisdom that were independent of His will.

In the context of natural philosophy this meant that the world must reflect the rational constraints by which divine wisdom had been guided in the process of creation. Accordingly, theological arguments were pitted by Leibniz against the Newtonian vacuum:

> To admit a vacuum in nature is ascribing to God a very imperfect work. I lay it down as a principle, that every perfection, which God could impart to things without derogating from their other perfections, has actually been imparted to them. Now let us fancy a space wholly empty. God could have placed some matter in that space: Therefore there is no space wholly empty: Therefore all is full.

If there were a vacuum, there was no good reason why God should have stopped creating rather than enriching His work with further variety. Newton's atoms were also repudiated. It involved an unacceptable breach of continuity to suppose that at a certain stage of division one would suddenly encounter units not further divisible. The gravitational force was rejected as unintelligible. Because Newton had failed to explain its agency in mechanical terms, it had to be either an occult quality or a perpetual miracle. Leibniz had already set his face against perpetual miracles in his criticism of Nicolas Malebranche, who had responded to the Cartesian problem of how mind could interact with body by postulating a continuous intervention of God. Leibniz had preferred an alternative solution, sometimes called 'psychophysical parallelism', according to which mental states correlated with physical states, but with each the result of independent causal chains, bound only by a preestablished harmony. There was nothing to be gained, in physiology or physics, by requiring constant miracles.

In Leibniz's theology there were grounds for praising God if His creation could be shown to be the best of all *possible* worlds. But such a world could hardly require the cosmic repairs urged by Newton, Bentley, and Clarke. Miracles, Leibniz insisted, were to supply the needs of grace, not to remedy second-rate clockwork. The controversy raged, too, on the nature of space and time, with Leibniz ridiculing Newton's reference to space as the sensorium of God. If the deity were possessed of organs, He could not be spirit but must be a corporeal Being. To crown his objections, Leibniz appealed to his principle of preestablished harmony. Once the motions of celestial bodies had been fixed, and living organisms created, everything that followed had been 'purely natural, and entirely mechanical'.

There was irony in this. For Leibniz's desire to uphold a category of the *purely* natural, the *entirely* mechanical, only confirmed the impression in English circles that it was his philosophy, not Newton's, that would benefit the deists. Leibniz would consider that a false impression because he took pains to distinguish between his kingdom of nature and kingdom of grace. The latter was the province of human minds, which were causally independent of the material world and not naturally destructible. Even the kingdom of nature was an empire within an empire. The natural philosopher could provide his mechanical explanations for the workings of an organism, but this need not exhaust the manner in which God's creatures were to be conceived. They also embodied His purposes. Teleological and mechanical explanation belonged to different modes of analysis; they were not mutually exclusive. Even the behavior of inanimate objects illustrated a purposiveness enshrined within nature. Leibniz was particularly impressed by the science of optics, which showed that light always seemed to travel by the 'easiest' path. In the last analysis he believed, as strongly as the Newtonian divines, that science had a religious utility. It gave content to talk of perfection in the construction of the universe. His complaint was that Newton's universe was not perfect enough. [...]

Chapter Eleven
Science in the Scottish Enlightenment

11.1 Paul Wood, *University science in eighteenth-century Scotland**

[...]

II

Despite the loss of the Scottish Parliament in 1707, Edinburgh boasted the most complex institutional structure of all the Scottish burghs during the eighteenth century. As the centre of the legal system and the Kirk, Edinburgh nurtured a number of different professional communities and, since it also continued to serve as the Scottish base for most of the landed classes, the town could therefore support a highly diverse, and sometimes specialized, set of clubs and societies. Thus although the university figured large in the cultural landscape of the Athens of the North, it did not dominate its civic environment like the colleges of Glasgow and Aberdeen did in their respective locales. This meant that itinerant lecturers, for example, were able to cultivate an audience for natural philosophy in Edinburgh, whereas this was not, as we shall see, necessarily the case elsewhere in Scotland.

While the other Scottish universities were comparatively slow to jettison the traditional system of regenting, Edinburgh was the first to adopt fixed professorial chairs in 1708, as part of Principal William Carstares' campaign to make his college competitive with the Dutch universities that he so greatly admired. In place of the old scholastic curriculum taught by the regents, Edinburgh now offered four core arts classes, given by the Professors of Humanity, Greek, Logic and Metaphysics, and Ethics and Natural

* Paul Wood, 'Science, the universities, and the public sphere in 18th-century Scotland', *History of Universities*, 13 (1994), pp. 101–2, 104, 106–20.

Philosophy, to which students could add courses of their choice offered by the other professors. Thanks to this new teaching regime, the four arts professors were, in effect, guaranteed a relatively comfortable income (although their salaries were not generous by the second half of the century); but for those teaching subjects like mathematics this was not the case, and these instructors were obliged to enlarge their emoluments through their class fees in order to supplement their fixed salaries. The significance of these financial arrangements for the subsequent development of the university, and indeed the creation of the public sphere in Edinburgh, should not be underestimated. In so far as the professors who were not responsible for the basic arts courses could only augment their incomes by boosting class sizes, there was considerable financial pressure on them to open up their courses to students who had no intention of matriculating or graduating. With the establishment later in the century of new chairs in medicine and the natural sciences with salaries which were sometimes deliberately kept low, there were thus institutional imperatives prompting the professors of these subjects to attract non-university men to their lectures in order to enlarge class sizes (and hence fees). The opportunity for generous remuneration, therefore, acted as a stimulus within the university for the creation of a wider audience for natural knowledge.

This structural feature of the revamped curriculum helps us to understand a pattern which is implicit in the evidence regarding the teaching of the natural sciences at Edinburgh during the eighteenth century, namely that the newer professors were (on the whole) more conspicuous in their efforts to recruit students. It is regrettably difficult to establish the extent to which successive professors of Natural Philosophy attempted to reach a broader public in the period. [...]

An equally important forum for the dissemination of natural knowledge within the university during the first half of the century was the mathematics class. When the Edinburgh mathematics chair was revived in 1674, the Town Council imported from St Andrews one of Scotland's most accomplished mathematicians of the period, James Gregory I, who died after only one full session of teaching. Following a brief hiatus, the Council then chose his nephew David Gregory, whose appointment marked the consolidation of a close working relationship between the town fathers and the Gregory clan which continued through the eighteenth century in the fields of mathematics and medicine. As part of their lecturing duties, the mathematical Gregories touched on aspects of optics and astronomy, and this part of the curriculum allowed David Gregory to introduce his advanced students to the intricacies of Newton's theory of light and colours and the Newtonian system of the world following the publication of the *Principia Mathematica* in 1687. In 1692 David was succeeded by his brother James Gregory II, who was appointed with a reduced salary, which may well have induced him to take on extra students in order to increase his class fees. Like David, James II apparently lectured on various branches of natural philosophy as they were related to mathematics, but a more concerted effort to mount a complete course of experimental philosophy was made by Colin Maclaurin, who took over James Gregory II's teaching duties in 1725.

Describing the offerings at Edinburgh in 1741, the *Scots Magazine* noted that in addition to his three mathematics classes (which ranged from simple arithmetic to fluxions and their application to propositions in Newton's *Principia*), Maclaurin also gave '[a]t a separate hour...a college of Experimental Philosophy...and at proper hours of the night describes the constellations, and shews the planets by telescopes of various kinds'.... We do know...that Maclaurin's courses attracted large numbers; writing to Francis Hutcheson in 1728, he noted that 'I now labour hard in the Winter having had sometimes 126 scholars at a time.' [...]

With the rise of the Edinburgh medical school in the first half of the century, related subjects like chemistry received an institutional boost, and it was in these fields that we see a definite broadening of the audience for the sciences. Prior to the arrival of William Cullen in 1755, the teaching of chemistry was largely taken up with practical matters such as the preparation of medicines, and the first two professors, James Crauford and Andrew Plummer, had mixed success in the classroom. Cullen, however, wrought dramatic changes in both the contents and the pedagogical style of the course. Rather than present chemistry as a field subsidiary to the medical sciences, Cullen argued that it was a fully independent branch of natural philosophy, with its own range of phenomena, a distinctive set of theoretical concepts, and specific modes of investigation. Moreover, Cullen also endeavoured to recast chemistry as a polite pursuit, for he emphasized that it was, in the words of John Robison, 'a liberal science, [and] the study of a gentleman'. By promoting his new vision of chemistry, Cullen was able to repeat the success he had already enjoyed at Glasgow, and his Edinburgh lectures soon became highly popular among students and townsmen alike. One should also note here that a larger audience was much to Cullen's financial benefit, because the chemistry professorship did not have a salary attached, so that he was entirely reliant on class fees for his remuneration (which he supplemented through his private practice as a physician).

Cullen's protégé Joseph Black likewise attracted large numbers of students after he was elected to the chair in 1766. With his days as a creative theorist and experimenter more or less behind him, Black turned his attention almost exclusively to his teaching and, according to John Robison, he was especially concerned to make his course readily comprehensible to pupils 'from the workshop of the manufacturer'. Thus, writes Robison, Black regarded it 'as his most sacred duty' to 'engage the attention of such pupils, and to be perfectly understood by the most illiterate'. Consequently, Black mastered the art of the demonstration experiment, and from descriptions of his lectures it is clear that he became a consummate practitioner of the art. ... By limiting his lectures to topics which could be concretely illustrated in the classroom, and by perfecting the use of experimental demonstrations,

Black managed to reach a very wide, and to some extent disparate audience. His elegance and gentlemanly manners appealed to those with polite aspirations, and his combined pedagogical skill and practical knowledge made chemistry accessible to the working men in attendance. Thus Black was able to expand the public for chemical knowledge in Edinburgh during the latter part of the eighteenth century, and this process was continued by Black's successor, T. C. Hope, who attracted audiences of 500 or more to his lectures.

Another science related to medicine which gained recognition through the establishment of a chair was natural history. Although the first incumbent, Robert Ramsay, treated his position as a sinecure, his successor John Walker tried to create a secure institutional niche for his field through his teaching and his efforts to build up a natural history museum at the university. Walker was in the fortunate position of having a fairly generous salary from the Crown but, like the majority of his colleagues, he supplemented his income through his class fees, and he therefore made every effort to create a broad audience for his subject. Following the lead of Cullen and Black, Walker presented his science as a form of gentlemanly knowledge, and he also pointed to the connections between medicine and natural history in order to appeal to the large numbers of medical students in the town. [...]

Around the turn of the eighteenth century a number of small groups involving university men surfaced in Edinburgh, and from what little is known about them it seems that most probably touched on various aspects of natural history, natural philosophy, or the economic improvement of the nation. The aspirations of these clubs were to find their most ambitious expression in Sir Robert Sibbald's proposal to found a Royal Society of Scotland in 1701 and, although Sibbald's hopes were never realized, his scheme does attest to the fact that a close alliance was already emerging between those who cultivated the natural sciences and their patrons. Moreover, Sibbald's plan and the clubs which emerged in this early period show that at the dawn of the Scottish Enlightenment the culture of science occupied as prominent a place in the public life of provincial Edinburgh as it did in metropolitan centres like London or Paris.

One reason why science was so highly valued by practitioners and patrons alike throughout the eighteenth century was that it was seen as a vehicle for improvement, construed in either moral or material terms. One of the most important organizations dedicated to the economic improvement of the nation was the Honourable the Society of Improvers in the Knowledge of Agriculture in Scotland (founded 1723).... Writing in the dedication to his compilation of the Society's transactions, their Secretary Robert Maxwell urged that agricultural practice had to be subjected to the principles of rational, scientific enquiry, and thereby introduced a theme which was to be the subject of countless variations in the literature of improvement....

Although Colin Maclaurin and the Professor of Law, James Craig, were the only local academics involved in the affairs of the Improvers, the professoriate were far more active in the Edinburgh Philosophical Society (otherwise known as The Society for Improving Arts and Sciences and particularly Natural Knowledge), which met from 1737 until 1783. At its inception, the PSE brought together the rump of a community of medics led by the Edinburgh Professor of Anatomy, Alexander Monro *primus*, and a circle of astronomers, mathematicians, and virtuosi centred on Colin Maclaurin. Out of these two groups emerged a society which was broad in intellectual scope and relatively inclusive socially. As one might expect, medical topics figured large in the transactions of the PSE throughout its history because of the presence of successive professors in the Edinburgh medical school, as well as physicians and surgeons based in the town and elsewhere. But in the early years of the society members also heard a number of papers on a wide range of subjects including astronomy, meteorology, history, Scottish antiquities, and agricultural improvement; later, chemistry became increasingly topical, as did commerce, industry, banking, and the construction of canals and turnpikes. Apart from medicine, therefore, the one area which engaged the membership throughout the lifespan of the PSE was the economic improvement of the nation and, with the demise of the Improvers after the '45, the society sometimes served as the main institutional forum in Edinburgh for efforts to improve farming and industrial practices. Moreover, those active in the PSE were also alert to the moral uses of natural knowledge, for they too stressed how the study of nature revealed the wisdom and benevolence of the Creator.

The broad scope of the PSE's proceedings reflects the fact that its membership drew on a wide cross-section of the community. At the outset, the society counted a significant number of aristocrats among its members, including its first President, the Earl of Morton, who was a serious man of science who later became President of the Royal Society of London. However, by the 1760s, the society had become less aristocratic in character; professional men were now more prominent, and new members were elected who came from industrial and commercial backgrounds. While the Society apparently made little effort to regulate the social composition of its membership, care was taken by the founders of the PSE to make sure that there was a healthy balance within the society between those who cultivated the sciences as an avocation and those whose careers depended on the pursuit of natural knowledge, for Maclaurin and his associates specified that one-third of the members were to be men who did not 'make Philosophy or Physick their particular Profession'. This provision clearly ensured that neither medicine nor natural philosophy dominated the proceedings, and it also opened the door to those with limited interest in the sciences, be they potential aristocratic patrons or gentlemen of some standing within the local community. [...]

III

In the concluding volume of his *A Tour through the Whole Island of Great Britain* published in 1726, Daniel Defoe observed that Glasgow was 'the cleanest and beautifullest, and best built city in Britain, London excepted'. But whereas the English metropolis boasted a vigorous cultural life, the same cannot be said of Glasgow. For Glasgow was, as Defoe noted, first and foremost 'a city of business'. It was also the home of an especially dour form of Presbyterianism which was not conducive to the cultivation of politeness, and it was too close to the attractions of Edinburgh to be entirely independent culturally. While there were coffee houses, dancing assemblies, and occasional musical concerts in the town, the dominant cultural institution throughout the Enlightenment was the university. Glasgow College nurtured the growth of printing in the city, for example, and without the active support of the professoriate ventures like the Foulis Press and the Foulis Academy would not have survived. The college also served as the institutional locus for many of the clubs which flourished in eighteenth-century Glasgow, although one should not ignore important groupings based in the town such as Provost Cochrane's political economy club and the Hodge Podge Club.

Moreover, the university was instrumental in bringing natural knowledge into Glasgow's public sphere. Financial need is often the cause of innovation in academe, and the effort by the college to raise money for new experimental hardware early in the century well illustrates this point. In 1710, the Glasgow regents proposed to remodel their courses of natural philosophy by incorporating 'the [experimental] Tryals necessary to Illustrat that Science', a change that reflected the shift towards the use of demonstration experiments in the classroom initiated by John Keill at Oxford in the 1690s. But the requisite scientific apparatus was expensive, and the college finances were in disarray. Consequently, the masters launched a public appeal for funds, offering subscribers preferential treatment in return for the much needed cash. ... Furthermore, the faculty made another proposal which was to have far-reaching implications for the future of public science in Glasgow, namely to give each session a publicly advertised 'Course of Experiments' open to 'all who are inclin'd to attend it, at such times as Gentlemen of Riper Years, and others tho not Ordinary Students, as well as those that are such, may have the Benefit thereof'. It is unclear when or even whether such a course was in fact offered, but following the visitation of a Royal Commission in 1726–7 provision was again made to open up the natural philosophy class to men from the town. The new regulations governing the operation of the university stipulated that 'any person... may attend the lessons of Experimental Philosophy without a gown'. With this rule in place, the university effectively became the primary local site of access to natural knowledge; itinerant lecturers did not visit Glasgow with any regularity to offer their courses, partly because in the 1720s at least the town fathers actively discouraged them.

With the introduction of the professorial system in 1727, Robert Dick *primus* became the first to hold the newly established chair of natural philosophy, and we know that individuals who were not pursuing a formal degree were allowed to attend his classes, as they did those given by his son Robert Dick *secundus*, who took over the chair in 1751. During the tenure of Dick *primus*, the university also gained the services of William Cullen, who in 1747 added chemistry to the range of subjects offered at the college and whose 'lectures [were] calculated not only for the medical students, but for the general students of the University, and for gentlemen engaged in any business connected with chemistry'. The most successful public teacher of the natural sciences at Glasgow during the eighteenth century was, however, the cantankerous John Anderson, who was popularly known as 'Jolly Jack Phosphorous'. Anderson succeeded Dick *secundus* as Professor of Natural Philosophy in 1757 and, following his predecessor's example, he used demonstration experiments and specimens from his cabinet to illustrate his lectures on mechanics, hydrostatics, pneumatics, optics, electricity, astronomy, and natural history. Although Anderson's scientific competence has sometimes been questioned, much of the criticism has arguably been misplaced, and there is little doubt that he played a major role in creating an audience for experimental philosophy in Glasgow. While serving as a professor, he increased significantly the numbers of townspeople attending his classes, and he managed to attract labourers as well as the polite and polished. [...]

The university also contributed to the popularization of science in the public sphere through the various clubs and societies in which the professors participated. Two of the earliest known coteries associated with the university were those founded by Glasgow's most distinguished mathematician of the period, Professor Robert Simson. Simson routinely met with a number of his colleagues as well as the occasional student or visitor on Fridays and then on Saturdays at the nearby village of Anderston; given the presence in both clubs of Robert Dick, the classicist James Moor (who was also something of a mathematician), and the future Edinburgh Professor of Mathematics, Matthew Stewart, it is highly unlikely that their discourse did not sometimes turn to mathematical and scientific subjects. Much later, the college's chemistry lecturer William Irvine encouraged the formation of a student chemical society in 1786 or 1787, at whose meetings various papers were read on such topics as fermentation and respiration. But the most important and long-lasting university grouping was the Glasgow Literary Society, which was founded in 1752 and which met on a more or less continuous basis into the nineteenth century. Initially the professoriate made up the bulk of the membership, but as the GLS prospered its constituency expanded to include a number of lawyers, merchants, ministers, physicians, and gentlemen from the town....

IV

When we turn to examine eighteenth-century Aberdeen, we find an enlightenment even more closely associated with academe than that of Glasgow. Lacking the highly developed institutional infrastructure of Edinburgh or the flourishing commerce of Glasgow, Aberdeen was a small regional centre in which limitations of size restricted the growth of the public sphere for much of the period. Because the town generated few of its own clubs or societies, the two colleges dominated civic culture to a much greater extent than in the other major Scottish urban centres. This being the case, it is significant that experimental science figured in the genesis of the public sphere in Aberdeen. We know from the correspondence of Principal George Middleton and George Garden that a circle of scientific virtuosi associated with King's College was active in the 1680s, but it is unclear whether they ever formally constituted themselves into a club, even though they were encouraged to do so by the Oxford Philosophical Society. A more important development was the public appeal for funds to purchase instruments launched in 1709 by the Professor of Mathematics at King's, Thomas Bower. As a member of the circle which included such early Scottish Newtonians as Archibald Pitcairne and George Cheyne, it is hardly surprising that Bower wanted to mount a course which would include experimental demonstrations, and he also moved in the direction of making natural knowledge public since he promised to make his course available to subscribers for a period of three years. Unfortunately we do not know the extent to which Bower's plans were realized, but more evidence survives concerning a subscription drive subsequently launched in the New Town.

Following the lead of King's College, Glasgow, and St Salvator's College in St Andrews, the Marischal faculty issued in the spring of 1726 a set of *Proposals for Setting on Foot a Compleat Course of Experimental Philosophy in the Marishal College of Aberdeen*, which outlined an elaborate scheme to restructure the teaching of the natural sciences at the college. The masters declared that they wanted to acquire 'Entire *Setts* of Instruments, necessary in *Astronomy, Mechanicks, Opticks, Chymistry, Hydrostaticks* and *Anatomy*', along with books and 'MODELS of the newest *Machines* in Husbandry' in order to mount an annual course of experimental philosophy which was to be open to both regular students and subscribers. Moreover, the masters made a point of assuring prospective contributors that the experiments would be explained in the clearest possible terms so that their meaning could be understood even by those unfamiliar with mathematics or technical scientific vocabulary because the Marischal men hoped to promote 'so necessary and pleasant a Part of Learning, and which in the Issue they hope will very much tend to the Improvement and Advantage of these Northern Shires'. Here too, then, the sciences effectively entered the public sphere due to financial need, but the Marischal faculty were also notably more explicit than their counterparts elsewhere in Scotland at the time in tying the accessibility of natural knowledge to the material benefit of the local community. [...]

When we turn to the clubs and societies associated with the two Aberdeen colleges during the eighteenth century, we again find that natural knowledge occupied a prominent institutional niche. The earliest known grouping of the period was a 'Philosophical Club' which met in 1736–7, and was apparently based at Marischal College. Thomas Reid's notes from this club show that while the proceedings focused primarily on moral philosophy, the members also discussed the natural theological implications of Newtonianism (perhaps with reference to the Leibniz–Clarke debate), and his cryptic jottings suggest that they explored the philosophical foundations of Newton's conceptions of space and time as well. Later, Reid helped found the Aberdeen Philosophical Society (also known as the 'Wise Club'), which flourished from 1758 until 1773. The majority of the membership of the APS was drawn from the ranks of King's and Marischal, so the society cannot be said to have greatly expanded the audience for natural philosophy within the local community, even though it did include various local ministers, one of the town's physicians, and the tutor to the son of Lord Deskford, who was the patron of a number of the members. ...

Beyond setting topics for discussion, the natural sciences gave rise to the ideology of scientism which informed the ethos of the Wise Club. While the APS covered much the same range of subjects as other general purpose societies like the Glasgow Literary Society, the members of the Wise Club carefully restricted the scope of their proceedings to what they regarded as strictly 'philosophical' matters. In the original regulations of the society the founders stipulated that:

> The Subject of the Discourses and Questions shall be Philosophical, all Grammatical Historical and Philological Discussions being conceived to be forreign to the Design of this Society. And Philosophical Matters are understood to comprehend, Every Principle of Science which may be deduced by Just and Lawfull Induction from the Phaenomena either of the human Mind or of the material World; All Observations & Experiments that may furnish Materials for such Induction; The Examination of False Schemes of Philosophy & false Methods of Philosophizing; The Subserviency of Philosophy to Arts, the Principles they borrow from it and the Means of carrying them to their Perfection.

The identity of the APS was thus closely tied to scientism, and hence its institutional character differed markedly from that of the other generalist 'literary' societies of the Scottish Enlightenment. ... What the Wise Club illustrates ... is precisely the same identification of the methods of the natural sciences with rational modes of public discourse that we have already observed in the Honourable the Society of Improvers in the Knowledge of Agriculture in Scotland. [...]

11.2 Dennis Dean, *Hutton and the history of geology**

What did Hutton believe?

Hutton's geological beliefs were for him part of a larger philosophy, or even theology, of nature that one might call Agricultural Deism. According to this outlook, a central revelation of divine benevolence to man is the continuing fertility of the earth, a life-supporting fecundity made possible only by the immediate effects of the hydrological cycle and the infinitely slower but even more important effects of the petrological cycle, in which new rock and soil are continuously being created in order to replace the old. Fundamental to Hutton's point of view was a sharp-eyed appreciation for the efficacy of erosion, through which earth's soils are constantly being removed from productivity and transported to the sea. Since an overall loss of fertility was unthinkable, Hutton necessarily postulated some kind of renewing mechanism, which he found to be uplift. This uplift, in turn, was powered by some poorly known but nonetheless ubiquitous force called subterranean fire or heat, which was also responsible for consolidating loose sediments into solid rock. His completed theory therefore affirmed a majestic but extremely slow natural revolution moving like some great pump or engine through a renovative cycle that included erosion, transportation, sedimentation, consolidation, and uplift. Despite Hutton's later concern with basalt and granite, both were primarily of use to him in helping to confirm the reality of a process in which sedimentary rocks were by far the most important.

In seeking to establish his fundamental truths, Hutton was led to consider a wide variety of geological phenomena and topics, including the evolution of landscapes; the work of rivers, glaciers, and marine currents; the efficacy of present-day forces; and the immensity of geological time. With regard to each of these areas, he proved to be extremely insightful. . . . If there was ever a time when Hutton came to doubt his own conclusions, or to despair over them, we have no record of it.

How original were Hutton's beliefs?

Quite original, in my opinion. Though some historians have sought to derive Hutton's theory from an earlier one, perhaps Hooke's or Moro's, they have not been able to do so convincingly. Hooke, for example, emphasized earthquakes, a topic Hutton barely mentioned. Moro, similarly, became very much a Vulcanist; Hutton never was. On the evidence, there is every reason to agree with [the view] that Hutton seldom read the geological theories of others, and then only to disagree with them. Hutton's

---

* Dennis Dean, *James Hutton and the History of Geology* (Ithaca, NY: Cornell University Press, 1992), pp. 264–9.

major theoretical bias came to him through the Deism and natural theology of his century. He then began with the initially disconcerting observation that precious soil was being washed away by rain and rivers and worked out the whole of his theoretical response from there: the grand paradox of loss and gain. What resemblances exist between his theory and any other are almost wholly attributable, I think, to the fact that two very talented observers were describing the same natural phenomenon.

To be sure, several of the ideas Hutton utilized had prior histories of their own. That seemingly destructive natural forces – such as volcanoes – actually serve constructive ends, for instance, was a familiar stance among the natural theologians; though Hutton *may* have derived his attitude toward volcanoes from Strabo, as is often asserted, he could just as easily have found the same idea in more contemporary sources. Hutton's originality, in this case, consisted in his applying the same kind of solution to the less immediately dramatic problem of erosion, which . . . was a matter of considerable debate throughout the seventeenth century. Similarly, the concept of subterranean fire, strictly defined, had been part of Stoic cosmology and was subsequently utilized by Christian theologians as a rationalization of Hell. Under Deist influence, however, the earth's heat came once again to be seen as a positive manifestation of divine intentions; Enlightenment thinkers did not, as a rule, believe in a punitive afterlife.

Deist rejections of biblical literalism and the miraculous also preceded Hutton's in calling attention to the efficacy of present-day geological forces and extended geological time. More broadly, Hutton by no means originated the concept later to be called uniformitarianism, which, as natural law, had been commonly assumed by Presocratic philosophers, Aristotle, and particularly the Stoics. It was then revived by Francis Bacon (who stressed the simplicity of nature) and substantiated by Isaac Newton. Four 'Rules of Reasoning' in the latter's *Principia*, book 3, emphasized that 'nature . . . is wont to be simple, and always consonant to itself'. Several Enlightenment thinkers, particularly Buffon, then applied uniformitarian assumptions to selected geological problems. Hutton's contribution in this area – a very significant one – is that he was the first philosopher to base his geological theory entirely on gradualistic naturalism, having eschewed any kind of preceding cosmological myth. He worked backward from the present earth, not forward from Genesis. With the possible exception of Desmarest, Hutton greatly outdistanced any of his predecessors in his empirical arguments on behalf of perpetually acting, uniform natural processes.

As part of the eighteenth century's pagan revival, such thinkers as Holbach, Hume, and Toulmin had adopted the classical uniformitarianism of Aristotle and Lucretius to argue, more or less openly, for the eternity of the earth. For Hutton, who claimed to believe in a finite earth (*Theory* [*of the Earth*, 1795], I: 223), uniformitarianism was primarily a methodological assumption – as it had been for Newton. Hutton's geological theorizing accepted, or at least considered, rapid continental collapse because its author remained unclear as to what held up elevated continental masses. But he certainly did not accept catastrophic uplift; any local catastrophes known to Hutton, and almost never acknowledged, were dwarfed by his ponderous regularities. He rarely claimed, moreover, to be describing anything other than the present earth. Making no far-reaching historical statements about the past, whose uniqueness in any respect he denied, Hutton argued instead that the renovative cyclical processes he had discovered would necessarily destroy any geological evidence more than a cycle or two old. However often misunderstood, his famous conclusion that we find 'no vestige of a beginning, no prospect of an end' does not refer primarily to the immensity (perhaps eternity) of geological time he believed in but to his logical assumption that the surviving record of the rocks comprises only the last few perishing leaves of a constantly self-destructing but equally regenerating natural calendar whose earlier sheets are now irrecoverably disseminated as unidentifiable constituents within later ones (see *Theory*, I: 223).

Though Hutton's theory did not derive from any of his predecessors', certain interactions between those theories and his own developing one should nonetheless be considered. In all probability, the chief influence of other theories on Hutton's was negative, in that he found either their methods or their conclusions offensive and deliberately opposed them. For example, Hutton explicitly attacked the well-established eighteenth-century tradition of biblically derived geological speculations that regarded the earth as a ruin brought about by the fall of man (a calamity in which Deists generally did not believe). That tradition included numerous editions of the theory of Thomas Burnet and would culminate, for Hutton particularly, in that of John Whitehurst; both writers are mentioned by name in his own work. Other theorists, so skeptical as to be virtual atheists, accepted much of what Hutton also believed – including natural law and geological time – but then drove their speculations to what were for him intolerable depths of unbelief. ...

Possible associations between Hutton and other members of the Scottish Enlightenment have not been adequately explored. It is altogether likely, for example, that both Joseph Black and John Clerk of Eldin conversed at length with Hutton throughout the years in which his theory took form. ... More broadly, the possibility of other, mostly philosophical, connections between Hutton and, for example, Adam Smith, Dugald Stewart, or even David Hume should not be overlooked. In all these latter cases, however, the fact remains that Hutton was specifically interested in geology whereas others, like Hume, were a good deal less so or, often, not at all.

It is of course demonstrable that Hutton utilized a variety of geological literature, most of it French, throughout his *Theory* of 1795, the chief authors being Saussure, Deluc, and Dolomieu. ... The possibility of significant indebtedness to Desmarest has occasionally been raised but never proven. Hutton's use of both Deluc and Saussure, however, remained almost entirely illustrative. Having already conceived his theory, he borrowed a selection of their data to substantiate it further (particularly after becoming bedridden, and thus no longer able to continue original fieldwork himself). Except for calling his attention to glaciers, there is no real evidence that either writer actually changed any major aspect of Hutton's already established thought.

How modern is Hutton's theory?

Hutton unquestionably made mistakes. Among these, four of the most important were his erroneous theories regarding the origin of flint and other stones; his dubious theory of consolidation; his unwillingness to attribute some vein deposits to water; and his failure to identify basalt with lava. As a result of the latter, he seriously underestimated the importance of volcanoes. Despite his inevitable knowledge of major seismic episodes in 1750 and 1755, he said almost nothing about earthquakes. Beyond these, a particularly significant shortcoming was his chronic inability to appreciate the geological utility of precisely identified fossils. More broadly, he did not accept the uniqueness of geological events, such as eruptions and earthquakes, or of geographical areas, or of past times. Traditional geological disasters had for him no theological significance and were altogether unimportant in the constitution of the globe. Nor was he interested in the peculiarities of individual districts, all of which were for him only parts of the same overall plan. The ideas of mapping, regional study, and systematic specimen collecting were equally foreign to him; thus, Britain's Geological Survey would not be Huttonian in origin. Except in its duration, moreover, our familiar geological timescale would have been meaningless to him. Hutton seems not to have realized that life on earth has a history, or at least chose to ignore the fact. His assumed perspective, one might say, was Neoclassical rather than Romantic, which is to emphasize that he was everywhere concerned with the basic, timeless, and enduring aspects of the earth – its formal essence – rather than with unimportant accidentals occasioned by time and place. This determined essentiality did not however prevent Hutton from analyzing specific landscapes brilliantly – indeed, his analytic imagination splendidly combined both Neoclassical and Romantic characteristics – but surviving records of his having done so are all too few. It was never his aim, moreover, to analyze any given landscape fully. Particulars were, in his mind, always subordinate to principles, as events were to processes. Once a valid theory of the earth had been established, he believed, correct interpretations of any given landscape would easily derive from it.

What is Hutton's rightful place within the history of geology?

One of the strengths of Hutton's theory was that it remained for the most part at a high level of generalization. ... Unlike previous geological theories, most of them only ingenious nonsense – to be savored as romance, then found out and discarded – Hutton's has proved to be a worthwhile foundation on which to erect a greatly elaborated but nonetheless kindred science.

Among Hutton's major contributions to the science of geology are his insistence on the basic orderliness of terrestrial processes; the centrality of heat; the duration of time; the significance of denudation, consolidation, and uplift; the efficacy of rivers; the intimate association between the earth's crust and subsurface forces; the nature and significance of plutonic and metamorphic rocks; and the ongoing nature of petrological creation. Other geological theorists have given us one or two good points, but none competes with Hutton in the number of sound principles established.

There is, then, a direct connection between Hutton's geology and our own, with the most significant work of many major figures coming in between. Clearly, we have never discarded Hutton; instead, we have built our modern science on the sturdy foundations he laid down. No prior figure in the history of geology can lay claim to such an unbroken connection with the present. Despite the merits of other plausible nominees, Hutton stands alone in the large number of profound insights he synthesized and transmitted to later workers. On the evidence, then, he is far more the precursor of our own outlook than any other theorist one can name – and therefore, as so many of his beneficiaries have proclaimed, legitimately the founder of modern geology.

11.3 J. V. Golinski, *Cullen on the usefulness of chemistry**

However attractive his teaching, Cullen's lecturing skills, although necessary for the success of his career, were not sufficient. An ability to secure and mobilize patronage was as important as didactic talent for one who sought advancement in the Edinburgh professoriate. Henry Home (Lord Kames) and Archibald Campbell, the third Duke of Argyll, were Cullen's most important patrons, both in the early 1750s when he began to try for an Edinburgh chair, and later in securing his successive promotions. His relations with these men were an essential component of Cullen's social and institutional context, and they also formed part of the total 'audience' for his work. In the terms of the three-fold typology of audience relations introduced earlier, one could say that Cullen was linked with those who attended his lectures partly through the institutionalized relations dictated by university practices and rituals, and partly through the market relations produced by the (potentially international) competition among suppliers of medical and chemical education. With his patrons, he was linked by relations which were partly institutionalized and partly personal. In return for

* J. V. Golinski, 'Utility and audience in 18th century chemistry: case studies of William Cullen and Joseph Priestley', *British Journal for the History of Science*, 21 (1988), pp. 6–8, 11–12.

their efforts on his behalf, Cullen took an interest in their philosophical work, served as consultant on their agricultural and manufacturing projects, and satisfied them that his teaching was dedicated to the improvement of the chemical arts.

The Duke of Argyll's support for Cullen dates from the latter's Glasgow days, when Argyll's patronage was said to have helped secure Cullen's initial appointment as Professor of Medicine. In return the following year, Cullen worked for Argyll on a project for the purification of sea-salt for fisheries. Subsequently, Argyll's influence on the Town Council was believed to have been crucial in securing the Edinburgh chair, against the opposition of many members of the medical faculty who were antagonized by Cullen's renowned anti-Boerhaavianism. Argyll and Drummond were known to be anxious to hire teachers of proven ability, and both were aware of Cullen's valuable reputation in that regard.

Lord Kames also supported Cullen in making the crucial move to Edinburgh. Their relationship went back to 1748, when they began a regular correspondence on agricultural improvement, and even proposed a joint book on the subject. In this correspondence, Cullen undertook to comment on philosophical matters, for example reporting the results of experiments that Kames had proposed, such as one on the effectiveness of nitre as a fertilizer. In return, Kames used his influence on Cullen's behalf. He advised his client to improve his chances of securing the Edinburgh job, for example by moving there from Glasgow and giving a private chemistry course, or by contributing to the Edinburgh Philosophical Society, about which he reported in 1752, 'I am turned extremely keen, now that I have got in good measure the management of it'. Kames also engineered a reward for Cullen's work on bleaching from the Board of Trustees for the Encouragement of the Fisheries, Arts and Manufactures, of which he was a leading member; and promised to 'bustle' for a similar premium for his work on salt purification.

Cullen's lobbying for premiums from bodies such as the Board of Trustees, and his reliance on aristocratic patronage to help secure university promotion, appears typical of Scottish academics at this time. The Scottish Enlightenment occurred in a society thoroughly permeated by relations of patronage, where the lowland nobility and gentry retained a paternalistic role in all aspects of national cultural and economic life. According to one influential analysis, that of Nicholas Phillipson, the interest taken by the Scottish aristocracy in encouraging economic and social development was a response by a traditional governing oligarchy to its loss of parliamentary power after the Union with England in 1707. This oligarchy fostered improvements in the Scottish universities, supported the efforts of bodies such as the Board of Trustees, and patronized the new clubs and societies which grew up, particularly in Edinburgh, to encourage 'improvement' in all areas of social, cultural and economic life. Their partners in these enterprises were the members of a newly emergent intellectual élite, comprising academics, clergymen, doctors, lawyers and other 'professionals' (in the contemporary sense of the term). This élite sought association with the nobility and gentry to reap the benefits of the patronage that they could still provide. The clubs and societies of Enlightenment Edinburgh thus served as channels for patronage in church, judiciary and universities, with social intercourse utilizing the vocabulary of a pervasive ideology of 'improvement'.

It is clear that for Cullen, as for other members of the Edinburgh professional/intellectual élite, carefully nurtured relations with his patrons were essential to the success of his career. Roger Emerson has noted Cullen's memberships of the Philosophical Society of Edinburgh from 1749, and of the Select Society from 1757. These connections show him exploiting the available social institutions, in which the alliance between intellectuals and the Scottish nobility and gentry was forged. It is also apparent that to sustain these relationships Cullen had to make attempts to apply his chemical knowledge to the practical concerns of his patrons; he had to convince them that chemistry would have useful applications to their areas of interest. Historians have recognized the importance of his patrons to Cullen's academic progress, and have uncovered aspects of the language he derived from the cultural values of his time to convince them of the utility of his science.

The cultural values to which reference has been made in this connection are those of the Scottish Enlightenment in general, and of Enlightenment Edinburgh in particular. In the discussions, for example, of John Christie and Arthur Donovan, Cullen's work is placed within a cultural milieu in which applied chemistry could be presented in the terms of a prevailing ideology of 'improvement', and hence could serve to attract and retain the favour of influential aristocratic patrons. Aspects of the contents of Cullen's chemical work can indeed be directly connected with the economic interests of his patrons. His projects on agriculture, on bleaching for the linen industry, and on salt purification for fisheries, closely matched their concurrent concerns. But Cullen's researches failed, at least in the cases of salt purification and alkali manufacture, to produce concrete results. And this fact should alert us to the cultural aspects of the negotiations between patron and client. In a situation in which economic needs and applicable knowledge failed to mesh perfectly, our attention is directed towards the language in which the prospect of applications was presented to potential patrons, the system of meanings which was exploited to convince them that useful results *would* be forthcoming. Certain aspects of this system of meanings are revealed by the 'improvement' analysis. John Christie has pointed out how the language of Scottish national identity provided the terms in which many economic needs were identified and portrayed. Projects which could be presented as likely to fulfil priority national needs would be regarded most favourably by potential patrons. Thus, in 1752, Kames encouraged Cullen's efforts to improve bleaching techniques, describing their improvement as 'a great object for this country'. [...]

We have seen how Cullen's attempts to find applications for his chemical researches, in areas such as agriculture, bleaching and salt-purification, were required to satisfy that part of his audience composed of personal patrons. We have also seen how the structure of audience relations involved here, in part personal, in part institutionalized, demanded that applied chemistry be presented in carefully-chosen terms. Publicly, Cullen's discourse carved out a place for chemistry in an ideological climate of 'improvement'. In private dealings with his patrons, it laid claim on his behalf to material reward for his 'Chemical Labours'.

To some extent, Cullen described the utility of chemistry in his lectures as a way of impressing his patrons. He wrote to Kames in his Glasgow days in 1749, that his lecture course would give full treatment to agricultural matters, 'to open young gentlemen's views upon the subject'. But his language in these lectures clearly also needs to be connected with the institutionalized nature of his relations with those who attended them. While he admitted to his students that 'there are many who would smile at any attempt to treat of Agriculture or other such practical Art in a College', he was prepared to endure what ridicule there might be, not solely to please his patrons, but because he believed that chemistry could provide the 'philosophical principles' of agriculture and other arts, and that such a conception should have a prominent place in a university course on the subject.

Crucial to Cullen's presentation of the utility of chemistry in his lectures was the notion of the discipline as a suitable component of the 'liberal' education of young gentlemen. He presented chemistry in terms of the gentlemanly educational ideal, which was making strong headway in the Scottish universities at this time. As John Robison, Professor of Natural Philosophy at Edinburgh in the last decades of the eighteenth century, was to put it, Cullen had set out to make chemistry 'a liberal science, the study of a gentleman'. He perceived his audience as well mannered and well bred, complimenting them on 'the decent behaviour which is to be expected from Men of good breeding'. For such an audience, detailed instruction in the practices of the various arts would have been inappropriate. Rather, the liberal ideal proposed an education in the 'philosophical' general principles of chemistry, the knowledge of which conferred authority upon gentlemen to direct a multitude of different practical activities. While acknowledging his institutional obligation to teach chemistry to answer the needs of medical students, Cullen therefore insisted, 'that it may do so, I find it necessary to deliver the philosophical principles'. Because 'philosophical chemistry' provided the principles of many arts, a knowledge of it was equally useful to the merchant directing chemical artisans, and to the physician supervising the activities of apothecaries. As Cullen told his students:

> Genius & Inventions are not suitable & are often debauching to the drudging Artist. It is the Merch[an]t of more liberal Education & extensive Views that must excite & direct the slow toiling Industry of the Artificer.

> Now I take the Merchant to be in respect to [*sic*] the Artificer what the Physician is to the Apothecary & I imagine the Physician unless he knows somewhat of the Apothecary's Art will make but a blundering Prescriber. Neither can the Merchant direct the hands of the Tradesman without a general Taste and some Knowledge of the Arts.

Cullen's notion of a 'liberal science' of 'philosophical chemistry' was clearly, at least in part, a response to the social situation of gentlemanly merchants and physicians in mid-eighteenth-century Scotland, but the language in which the conception was developed in his lectures descended from the early seventeenth-century work of Francis Bacon. Cullen rested his claims to chemistry's utility upon its grasp of 'philosophical principles' of the properties of bodies. He justified this link between chemistry's theoretical knowledge and its utility in Baconian terms. For it was Bacon who had provided the vocabulary in which the theoretical laws or principles which were derived from induction were identified with 'canons' or rules for the practical production of effects. And thus it was in Bacon's language that claims were made in the eighteenth century for the power of theoretical knowledge to direct practical activities.

Chapter Twelve
Science on the Fringe of Europe: Eighteenth-Century Sweden

12.1 Sten Lindroth, *Linnaeus and the systematisation of botany**

[. . .] The cult of Linnaeus, the native reverence for Linnaeus and things Linnaean, has naturally stimulated efforts at research in depth. Modern Swedish research on Linnaeus started in the decades about the turn of this century, during the period of the Linnaean jubilee festivals. An inventory of the extant manuscripts was made and texts were published. In 1903 there appeared Thore Magnus Fries's monumental biography, and just fifteen years later the Swedish Linnaean Society was founded. Linnaeus's life, his environment, and aspects of his work were becoming known in the minutest detail. But still, the traditional attitude of awe and wonder was maintained. Linnaeus's saintly halo shone with greater clarity than ever. He belonged to the Swedish people, and it was not fitting to make a critical evaluation of his person and his scientific accomplishment. Since then, matters have improved. Through the research of the last decades we know that the man Linnaeus was other than the one of the Romantic legend – the sweet and sentimental elements have vanished. His inner world also contained dark tones of depression and despair, at least in his later years. Linnaeus now emerges as more complicated than had previously been suspected. This applies also to his work, primarily his biological view of nature and his religious outlook, which have been the object of increasingly penetrating studies.

Thus the revision of the current picture of Linnaeus proceeds. But there still remains much to be done; one is never really finished with a man like Linnaeus. There are great and important questions that research has scarcely touched. And perhaps the Romantic Linnaeus legend still lives on

* Sten Lindroth, 'The two faces of Linnaeus', in T. Frängsmyr (ed.), *Linnaeus: The Man and His Work*, rev. edn (Canton, MA: Watson Publishing International, 1994), pp. 2, 11–29, 31, 33–35, 37.

in certain respects, preventing an impartial interpretation of Linnaean botany and the man behind it. [...]

The rare strength of feeling with which Linnaeus experienced nature led him to religion. His point of departure was always empirical, what the eye saw ... he ends by singing the omnipotence of the Creator. This is the real theme of his work.

Much has been written about Linnaeus's religious view of nature .... As a religious thinker Linnaeus does not lend himself to a ponderous treatment; he spoke spontaneously out of the fullness of his heart; and what he preached about God's wisdom in nature was only occasionally original. Linnaeus belonged to a time when all educated men glorified the miracle of creation as the safest way to a natural knowledge of God. So-called physico-theology was the fashion: the wonderful interrelatedness of natural things proved that an almighty and benevolent God had created them. The theme was preached in a rich literature.... The enlightened theology of the century was inspired by these thoughts and became, with the Wolffian doctrines imported from Germany, something of an official university philosophy, not least at Uppsala University.

Nothing was more natural than that Linnaeus should join the pious chorus. He was able to do this with greater knowledge than any other, and in inspired moments even more beautifully. 'I saw', he says in some famous lines, 'the eternal, all-knowing, all-powerful God from the back when he advanced, and I became giddy! I tracked His footsteps over nature's fields and found in each one, even in those I could scarcely make out, an endless wisdom and power, an unsearchable perfection.' Natural science became for Linnaeus knowledge of God; research into nature – out in the field or in the herbarium – a religious service. Man's highest task, the real reason for his presence on earth, was to contemplate nature and praise its eternal Author; no science, therefore, was nobler than natural history. With the doggedness of a preacher, Linnaeus drums in his theme of the Creator's omnipotence in nature; this became something of a fixed rhetorical formula, not least as an opening to his Latin dissertations. He beheld in the minutest things, as in a vision, proof of God's majesty. At Luleå, he saw in the water innumerable newly hatched fry; they were transparent, the eyes biggest, 'so that with wonder I had to admit the whole world was full of Thy glory'. All was signs, hints, portents – and everything had a meaning, a goal decided by the Creator. Every object, says Linnaeus in the dissertation *De Curiositate Naturali*, was created 'for the sake of something else'. This formulation expresses the central thesis of the natural science of the period, its teleological basis. Everything created was linked together in an endless chain in which each creature had its mission in the service of the whole.

Inevitably, this philosophy, even in the formulations of Linnaeus, can sound commonplace to modern ears. This is particularly true of its application to man as the crown and lord of creation, whom the rest of nature served. When the woodpecker pecked for larvae in rotten tree-trunks, this was not only to feed himself; he also disintegrated the tree, 'so that it no longer offered such a useless and depressing sight'. The herons of Egypt, claimed Linnaeus, had the task of eating up the land's many reptiles so that they would not frighten people with their terrible appearance – snakes and lizards, the 'ugly, unpleasant, naked crew', always aroused in Linnaeus intense revulsion; they challenged his optimism. But even in such anthropocentric interpretations he followed the conventional line.... God had not blundered in the work of creation, but had thought of everything, planned down to the smallest detail; it was the task of the investigator to detect His intentions. The physico-theological world view set out to find a meaning in nature, and this goal was not an unfruitful soil for the growth of scientific inquiry.

...Linnaeus saw himself as a chosen interpreter of nature, one of the prophets called by God to reveal His work, for the present and for times to come. The decisive, oft-repeated passage appears in one of his autobiographies. It is magnificent and frightening. The Lord, Linnaeus declares, has allowed him to see more of the work of creation than any other mortal; He has given him the greatest insight into nature any human being has been granted, destroyed his enemies, and made him a great name 'like unto the greatest on earth' (2 Sam. 2; 7–9; 1 Chron. 17:8). Thus Linnaeus worked under the special protection of Jehovah, like the ancient prophets of Israel. It is not only his dogmatism, his naive confidence in himself as lawgiver to all botanists in the world that must be seen in the light of this biblically inspired certainty of vocation. Reasonably enough, it also confirmed him in the belief that he had penetrated more deeply into God's decrees than any other natural scientist before him, that he had mapped out the mysteries of creation. In all of this, as elsewhere, he is extraordinarily close to his great Swedish contemporary, also called by the Lord to be the interpreter of the selfsame secrets: the seer Swedenborg.

Linnaeus reached deepest into the work of creation with his pronouncement on the Law of Order in nature...the inexorable order in nature which was an outflowing of God's immutable law.... What we 'usually call nature' is identical with the laws and decrees that are, that make up creation itself. [...]

But it cannot be denied that the matter also had another side. The deeper the aging Linnaeus penetrated the inner side of nature and of man, the more desolate did they show themselves to be.... [T]he tone darkens. His most basic declaration of faith was given in the hymn...*Imperium Naturae*, nature's empire, which opens the later editions of *Systema Naturae*. Its subject is Nature in nature, God, the upholder of everything, the Being of being and the Cause of causes, Who steers the changes of earthly life and the revolutions of the planets. He is everything; He is known under different names. 'If you wish to call Him Fate, you are not mistaken, for everything hangs by His finger; if you wish to call Him Nature, neither are you mistaken, for everything has arisen from

Him.' ... This passage is ... a mosaic of quotations from many sources.... We can be certain that it hardly pleased the orthodox theologians of Uppsala. Linnaeus had already gotten into trouble with them; he was conscious of the fact that he was treading dangerous ground – the theologians, he told his students, 'are complaining that we are confusing God with nature'. When we look at the sentences in the *Systema Naturae*, it is easy to understand them. Those obscure maxims on God as Being, Fate, and Nature were not irreproachable. They smacked of pagan thought, of Aristotle and the philosophy of fate of the ancient Stoics, which had earlier made an ineffaceable impression on Linnaeus; Seneca the Roman was one of his lifelong authorities. But one thinks also of Spinoza, persecuted and defamed, of whom Linnaeus had no doubt heard in Holland; for him, spirit and matter, God and nature, were one and the same, the only indivisible being beyond man's desires and suffering.

Linnaeus had traveled a long way. Behind the nuptials of the flowers and the jubilant song of the birds he glimpsed the outlines of a harsh, unfathomable countenance – the sovereign, divine power in its pure, almost abstract majesty.

The eternal order in nature

There are a few writings in which Linnaeus as the biological thinker applies the doctrine of eternal order in nature. This is particularly true of two or three Latin dissertations; the most important of them is entitled *Politia Naturae* and appeared in 1760. ...

... According to the God-centered world view of this time, nature was filled with meaning; each creature played the role ordained by the Creator, as part of the whole. Harmony and balance were the central concepts in this view of nature.... The universe was conceived of as packed with created beings, 'full of forms' ..., interrelated in hierarchical order from the lowest, lifeless matter through plants, animals, and man, to angels and spiritual substances.

During the eighteenth century this vision of the chain of creation, which ultimately goes back to Aristotle, became the common property of the educated. Linnaeus had it constantly in mind: 'The closer we get to know the creatures around us, the clearer is the understanding we obtain of the chain of nature, and its harmony and system, according to which all things appear to have been created.' In his *Systema Naturae*, Linnaeus wished to present the Creator's work 'in an orderly chain'. There were no sharp boundaries; nature made no leaps; the mineral kingdom merges almost imperceptibly into the plant kingdom; between plants and animals there are intermediate forms. The systematist had to know all this if he was to classify the multiplicity of creatures and reconstruct the plan of creation.

But the chain of nature ... could also be given another meaning. It did not just extend vertically like a cosmic ladder on which every created being had its fixed place. Nature was also at every moment alive and operative; its thronging creatures were dependent on each other in a network of reciprocal actions .... No creature could live in isolation; the universe was so constituted that the one served the other as food and maintenance. In this connection, evidently, Linnaeus could certainly argue in the current hierarchical spirit. The lower existed for the higher; the mineral kingdom gave nourishment to the plants, and the plants to insects, birds, and cattle; on them, in turn, lived predatory animals and human beings. But afterward everything returned to its origin in reversed order. The predatory animal was devoured by the bird of prey, the bird of prey by the worm, the worm by herbs, herbs by earth – 'Yes, man, who turns all to his own need often becomes food for predatory animals, and fish, for worms, and the earth.' The biological chain was, in fact, endless, an eternal circulation of life and death, of eating and being eaten.... [I]t was one of the basic laws of nature and filled him with holy awe. 'Thus does everything go round' – natural history was the doctrine of the metamorphoses of dust.

But this by no means exhausted the subject. In his dissertation on the order of nature, *Politia Naturae*, Linnaeus goes deeper in his investigations. The food-chain, its circulation, had a hidden significance. It was the instrument with which created beings were kept within their prescribed limits. There was a measure for each species of plant and animal not to be exceeded, a balance that was identical with order itself;...a sort of moral order presided over by divine justice ... was active in human life. When swarms of insects devoured plants, this was done primarily not to fill their stomachs but to prevent a catastrophic increase of certain species of plants. Linnaeus made a rough numerical estimate. If one tobacco plant in one year produced 40,320 seeds that were allowed to grow without hindrance, the earth would soon be full of tobacco plants; but this was against the world order of the Creator. In the same way, insects had their enemies in the food chain, parasites and birds of prey that kept them in check; the insect-eating birds were devoured by the birds of prey. Every form of life had its supervisor; the intricate interplay between the eaters and the eaten had as its goal the task of keeping the number of individuals in every species constant. Order in nature, what Linnaeus liked to call balance..., had to be maintained. How easily this delicate balance could be disturbed – it needed no more than the removal of one link in the ecological chain for it to be broken to pieces. If all the small birds were wiped out, our parks would be devastated by grasshoppers and other insects.

None of this was perceptible through superficial observation. The untrained eye saw only a jumble of eating, hunting, and chewing, as if everything occurred in the greatest confusion. Violent competition reigned in a nature full to the bursting point, a chaotic 'war of all against all' .... Nature, says Linnaeus in *Politia Naturae*, was a 'butcher's block'; the spectacle might well appear repulsive and meaningless. Only a person who had pierced the veil realized that the merciless struggle for existence was a prerequisite for the functioning of creation. Behind the seeming disorder one was met by a well-thought-out strategy that expressed divine Wisdom.

Linnaeus's thoughts on balance in nature were not altogether original. As a defense of divine world order it formed a natural part of the religious ideology of the time. [...]

...[W]hen Linnaeus stood full of wonder before the wise arrangements of the Creator, he was able to penetrate into the hidden life of nature. Most rewarding of all are... his Latin dissertations – the 'academic entertainments',...as he himself called them. There we encounter his teaching on flowers and their dispersal, on climate and plant geography, and on the ecology of the individual life forms. He writes about the position of plant leaves in sleep – a discovery that filled him with rapture – and follows the seasonal changes of the Swedish flora; he notes the host plants of insects; and in a paper on *Taenia* he investigates the nature and habits of the remarkable tapeworm. In all of this, Linnaeus is a thoroughgoing empiricist; he stands in the midst of living nature and takes its pulse. He was a pioneer by virtue of his trained insight into the demands made by plants on their environment. He dealt with this theme time and again, also in the *Philosophia Botanica*, that great summation of his biological thinking. Each species of plant had closely defined requirements for climate and type of soil, and this determined the distribution of the species. ... His goal, indicated as early as the introduction to the *Flora Lapponica*, with its famous description of the geographical plant-belts in Lapland, was to find the laws that governed the behavior of vegetation on the face of the earth. These regions also served, as he wrote in a brilliant section of *Politia Naturae*, the divine world order in a more intricate sense. In its natural locality each plant was protected from competition with intruders with other environmental demands – intuitively, Linnaeus anticipates the modern biological concept of the so-called ecological niches in nature.

Linnaeus put some of his best work into his investigation of fertilization and the dissemination of seeds. Nowhere else, it seemed, had the Creator constructed so many and such superlatively ingenious arrangements; here was a real playground for physico-theologians who wished to sing the praises of the divine architect; and it was not just Linnaeus who took advantage of the opportunity. But added to this was the fact that he was personally deeply engaged. The act of fertilization, through pistils and stamens, was for him a sort of sacrament, around which, ever since his youth, he had built up his botanical system. The study of this mechanism was a precondition of his reformation of botany. It is, above all, in his treatise *Sponsalia Plantarum* and the prize-winning answer to the questions of the St. Petersburg Academy, *De Sexu Plantarum* (1760), that Linnaeus has made important contributions to the sexual biology of plants. He describes different sorts of fertilizing arrangements; nature's capacity for variation in the matter of ensuring pollination for the various species of plants seemed to know no bounds. One of his finest dissertations, *Nectaria Florum*, was devoted to the morphology of the nectar glands and their role in fertilization through insects. With astonishment Linnaeus followed the career of the ripe seed and fruit out into the world. ... [I]t was Linnaeus who gave the biology of dissemination a scientific form and fitted it into a broader context. These amazing feats of artistry ... constituted so many pieces of evidence for the Creator's inventiveness. Seeds and fruit were borne through the air on wings or down; other species ejected their seeds from capsules; there were seeds equipped with claws that fastened in animal fur; berries were eaten by birds; other fruits or seeds were carried into the earth by earthworms and moles; fruit was washed down by floods, or carried across oceans. ...

The systematist

Linnaeus was nature's prophet, too, in the great works that established his world reputation – the *Systema Naturae* in its various editions, the *Philosophia Botanica*, and *Species Plantarum*. In these works Linnaeus wished to embrace the richness of creation and organize the multifarious forms. It was as a systematist that he made his epochal contribution; in his own opinion, and that of others, he was opening a new phase in the history of science .... So far, we have followed Linnaeus as the brilliant observer and the religiously inspired interpreter of nature's secrets. It now remains to pose the question of how, against this background, his true life's work, his botanical system in the first place, fits in the total picture. Here we turn a new corner and approach Linnaeus from another angle. It seems, at last, that we find ourselves faced by something resembling a paradox. [...]

Linnaeus's way out of the dilemma [to establish a universal system of botany in the absence of complete knowledge of the immense number of plant species] was perhaps both wise and necessary. But without a doubt it implied that he abandoned that side of his nature that we have already come to know. As a systematist – and this is what he was above all – Linnaeus was forced to relinquish empiricism as his sole guide; he began to refashion nature. That he did this without any scruples requires some explanation.

There was the general condition of botany when the young Linnaeus set to work in the middle of the 1730s. It had become increasingly more difficult to maintain order among the hordes of trees and herbs. A little over a century before, the Swiss botanist Caspar Bauhin had listed six thousand species of plants in all; since then the number had increased without interruption. The national floras of Europe were becoming ever better known, and from distant lands, through increasing foreign trade and scientific expeditions, a stream of unknown plant forms poured into Europe. The botanical gardens in Holland, England, and France were filled with exotic seeds, roots, and bulbs. ... The different plant systems competed with each other; the descriptions of the species in Latin were long-winded and obscure – it was easy to lose your way in the botanical reference books. The situation pained Linnaeus, and he felt as if chaos were threatening. ... It was precisely his very sensibility, the clarity with which he experienced the individual forms in nature, that threatened to overwhelm him in a sea of impressions. And surely it is here that the paradox lies. ... [T]he matter of ordering, of systematizing, became for him of such supreme importance that other considerations had to give way – even his allegiance to nature. The only way out was to catalog and classify rapidly, to find some sort of system, a pattern, that would give the uncoordinated mass a structure. ...

Here we must pause a while to consider another basic characteristic of the man Linnaeus. He was a born systematist. Throughout his whole life he felt a compulsion of almost demonic intensity to *arrange* everything – in groups, sections, and sub-sections, like troops on parade. 'God created, Linnaeus ordered', said an admiring posterity. [...]

...Linnaeus belonged to the same spiritual climate as the great philosophic-system builders of the previous century, Descartes and others .... At all costs he had to discover an Order in nature – if he did not find it, he would have to invent it himself. And in doing so, he did nature violence, forced it to obey the laws of logic. This characteristic, this mania, if you will, never caused him to turn his back on reality; but throughout his work there exists a tension between the demands of empiricism and the demands of order, and in situations of conflict it is nearly always order that is victorious. Linnaeus became a botanical legislator, with the sexual system as the crown of his work.

In this connection it is interesting to glance at his concept of species. Plant and animal species as unchangeable entities were the very core of the Linnaean mental world. They were the guarantees of order. ... The individual species had been created complete, once and for all, by God in the beginning. The invariability of species was the condition for order in nature..., and hence for the system. Each living creature emerged from an egg or a seed and was, therefore, a copy of the first parents.... It is noteworthy that as a botanist he never concerned himself with varieties. They caused trouble, turned his beautiful order upside down; only gardeners and worthless botanists bothered their heads about them.

...Without a concept of fixed species Linnaeus was helpless. He could only think of nature as a system of unchangeable substances.

So the system, the arrangement of objects in nature according to certain prescribed norms, became for Linnaeus the most important and all-but-exclusive task of biology. He insists on this time and again. ... Natural history, says Linnaeus categorically, is the same as arranging natural objects and baptizing them. What is a botanist? Simply, a person...who can give the right names. ... For Linnaeus, nature was an immense collection of natural objects which he himself walked around as superintendent, sticking on labels. He had a forerunner in this arduous task: Adam in Paradise. In the Garden of Eden – to be understood, according to Linnaeus, as an island on the equator – the first human being looked at the Creator's work 'as if at an entirely complete collection of natural objects'. Linnaeus developed the theme in the foreword to *Fauna Svecica*. Adam sat in Paradise, carrying out the two highest functions of science: he observed the creatures and named them with the aid of special signs, almost as though he had Linnaeus's writings to hand. Adam was the first Linnaean; and Linnaeus, a second Adam. Once again he had found a religious sanction for his scientific lifework; at least it was transfigured with something of the innocent early morning light of Paradise. For him, botanical gardens were, both in spirit and in fact, Edens, paradises. He liked to call them that and walked his own Eden . . . , inspecting the characters and distinguishing marks of plants, as once long ago our first parent did.

In this way Linnaeus narrowed down the field of botany greatly. He was monumentally one-sided – everything other than nomenclature and classification was scarcely accorded the rank of science. . . . When in the dissertation *Incrementa Botanices* Linnaeus gave a short survey of the history of botany, it all had to do with systematizing and with the glorification of himself, with not a word about anatomical and experimental research. . . . The real and only goal of science was to identify objects in the three kingdoms of nature, and Linnaeus's great reform of botany consisted, finally, only in a new way of ordering and putting labels on the vast host of plants.

This was to be done according to certain rules worked out by Linnaeus himself. Heaven help the botanist who departed from them – he was a heretic, 'heterodoxus', and Linnaeus spat him out. . . .

Linnaeus's use of the word 'method', *Methodus*, is revealing. To work methodically and, therefore, scientifically meant to establish *a priori* certain rules and then apply them to the botanical material. . . . Without such a method a botanist was lost. . . . How helpless, Linnaeus continues in *Classes Plantarum*, is the pure empiricist when faced by an unknown flower. He tried to work out the family from its general appearance; he racked his memory for something similar that he might have seen before; he thumbed through his herbarium and searched night and day through books and engravings. But the person who had a method – or *the* method, for that of Linnaeus was the only one allowed – followed its regulations, beginning with the key determination of class and then working his way step by step to the individual species. It was done in a trice, while the poor empiricist would be sweating away for years before reaching his goal. The future of botany depended on one single sanctified method, which was to be followed by all. To work out its laws and guidelines . . . was regarded by Linnaeus as his completely overshadowing lifework. . . .

. . . In the *Philosophia Botanica* Linnaeus has set down his 'philosophy', the theoretical basis for his reformation of botany. . . . Even if the species are the smallest, indivisible units, the genus becomes for Linnaeus the real building block of the botanical system, and the ability to correctly gather species to form genera is considered the mark of the true botanist. Here Linnaeus is intractable – creating incorrect genera of plants is 'supreme heresy'; only genera based upon the organs of sexual reproduction can be approved. ... In the same way the genera are brought together into higher groups – orders and classes – by means of the essential parts of fertilization (stamens and pistils). From start to finish the method of true systematization was based on fertilization. ...

We are beginning to feel at home in the Linnaean world. It was not his own; he had inherited it from preceding generations. Linnaeus was, in point of fact, a scholastic. Logic and definitions of concepts were his tools; his aim, using a fixed point of departure, a dogma if you like, was to order the world of plants into classes, orders, genera, and species. He worked in the same spirit as the learned doctors of the Middle Ages; in both cases Aristotelian logic provided the indispensable basis. [...]

...Linnaeus lived and worked surrounded by his material, thousands of plant and animal forms, and in that sense he remained an empiricist. He had examined them all, in herbaria and in gardens, in display cabinets of naturalia and in illustrated deluxe volumes; it is unlikely that any other scientist has personally examined such a multitude of forms of life. It was an almost incredible achievement; how Linnaeus found the time is incomprehensible. ... [H]e followed nature as far as it was possible to do so. His botanical system was artificial only in its upper reaches; the species and genera were natural – Linnaeus repeated this insistently; there he did not violate divine creation. *Species Plantarum*, with its innumerable diagnoses of species, marked a triumph for the empirical study of reality. But species and genera were only raw materials, rather like the chaotic jottings in Linnaeus's Swedish travel books. Order was required; the system must be created; and with that Linnaeus reached the fateful dividing line where construction took over. Nature was remolded by art. [...]

Scholastic logic is...the basis of all Linnaeus's work in systematics. It could hardly be otherwise. His aim was to order hierarchically the plant and animal kingdoms in accordance with the most rigorous norms of reason. For the achievement of this task, there was no other possible instrument than that basic Aristotelian logic with which Linnaeus had been thoroughly familiar ever since his school years at Växjö. Its categories and distinctions became indispensable for him. ...

...Linnaeus went to work as an out-and-out, thoroughgoing Aristotelian, his tools being the terms of scholastic logic: 'definition', 'genus', 'differentia' (difference), and 'species'. ...

...The empirical genius Linnaeus, the all-seeing, made no great 'discoveries', because he wanted to solve problems of a completely different character, and these problems were primarily of a logical nature. As a systematist and reformer of botany he ended up, with the other half of his being singularly suited to the task, in a science of concepts where everything depended upon clarity of thought. [...]

But he was no revolutionary, opening new vistas for the science of botany. Linnaeus himself, however, was of a different opinion. He alone had transformed botany – 'reformed an entire science and created a new epoch' – and that view has been adopted, at least in his native country. But there is no reason to accept Linnaeus's word. ... With Linnaeus an era in the history of botany reaches its culmination and conclusion: it is the end of scholastic botany. ... [H]is enormous influence, based on the lucidity and vigor that flowed from his works, most probably delayed the development of modern biology, and not just in the Nordic countries. [...]

Chapter Thirteen
Science in Orthodox Europe

13.1 B. Lewis, *The Ottoman view of Western science and technology**

By Ottoman times, the importance of mastering the Frankish art of war was becoming painfully obvious. This was particularly so in the artillery and in the navy. Though gunpowder had been invented centuries earlier in China, the dubious credit for recognizing and realizing its military potential belongs to Christian Europe. The Muslim lands were at first reluctant to accept this new device. It would appear that guns were used in the defense of Aleppo when it was besieged by Tamerlane, but in general the Mamluks of Egypt and Syria rejected a weapon which they found unchivalrous and which they realized to be destructive of their social order. The Ottomans were much quicker to appreciate the value of firearms, and it was largely thanks to their use of musketry and cannon that they were able to defeat their two major Muslim rivals, the sultan of Egypt and the shah of Persia. The effective use of cannon played an important part in the conquest of Constantinople in 1453 and in other victories won by the Ottomans over both their European and their Muslim adversaries. Significantly, the majority of their gun-founders and gunners were European renegades or adventurers. While the Ottomans were well able to deploy this new weapon, they continued to rely on outsiders for the science and even the technology needed to produce it. Much the same is true of the related corps of bombardiers and sappers. The inevitable result was that, with the passage of time, the Ottoman artillery fell steadily behind that of their European rivals.

The Ottoman interest in guns and mines was paralleled by their concern to keep up with European shipbuilding and navigation. When a Venetian war galley ran aground in Turkish waters, Ottoman naval engineers examined it with great interest and wished to incorporate several features of its construction and armament in their own ships. The question was put to the

---

* B. Lewis, *The Muslim Discovery of Europe* (Norton and Co., 1982), pp. 222–37, 328–9.

chief Mufti of the capital – is it licit to copy the devices of the infidels in such matters? The reply came that in order to defeat the infidel it is permissible to imitate the weapons of the infidel.

The question raised is an important one. In the Muslim tradition, innovation is generally assumed to be bad unless it can be shown to be good. The word *bid'a*, innovation or novelty, denotes a departure from the sacred precept and practice communicated to mankind by the Prophet, his disciples, and the early Muslims. Tradition is good and enshrines God's message to mankind. Departure from tradition is therefore bad, and in time the word *bid'a*, among Muslims, came to have approximately the same connotation as heresy in Christendom.

A particularly objectionable kind of *bid'a* is that which takes the form of imitating the infidel. According to a saying ascribed to the Prophet, 'whoever imitates a people becomes one of them'. This has been taken to mean that adopting or imitating practices characteristic of the infidel amounts in itself to an act of infidelity and consequently a betrayal of Islam. This dictum and the doctrine which it expresses were frequently invoked by Muslim religious authorities to oppose and denounce anything which they saw as an imitation of Europe and, therefore, as a compromise with unbelief. It was a powerful argument in the hands of the religious conservatives, and was frequently used by them to block such westernizing innovations as technology, printing, and even European style medicine.

There was, however, one important exception to this doctrine – warfare. The jihād, the holy war against the unbelievers, was one of the basic collective obligations of the Muslim state and community. When the war is defensive, it becomes an individual obligation of every Muslim. To strengthen the arms of the Muslims and make them more effective in waging jihād against the unbelievers is, therefore, in itself of religious merit and is, indeed, an obligation. To fight against the unbeliever, it may be necessary to learn from the unbeliever, and Ottoman jurists and other writers on the subject occasionally adduce a principle which they call *al-muqābala bi' l-mithl*, opposing like with like, that is to say fighting the infidel with his own weapons and devices. ... Authority was found even in a verse in the Qur'ān [9.36] in which the believers are enjoined to 'fight the polytheists completely, as they fight you completely'. This was reinterpreted to mean that the Muslims should use all weapons, including the weapons of the infidels, in order to defeat them.

On the whole, the Ottomans were willing to follow or adapt European practice in warfare and, more particularly, in artillery and in naval matters, where religious opposition was muted. They also made use of Western technology in mining. ... For these and other purposes, the Ottomans were willing to employ European experts in sufficient numbers to form a recognized group in the palace establishment, known as the *Taife-i Efrenjiyan*, the Corps of Franks. The Ottoman sultans and their ministers were well able to see the importance of European technology and to seek out and employ Europeans to serve their needs. But there was always opposition from religious conservatives, and while this was not strong enough to prevent borrowing and some imitation, it was strong enough to prevent the emergence of a vigorous indigenous technology. [. . .]

Apart from weaponry and seafaring, there was one other useful art in which Europe was seen as able to make some contribution. This was in the science of medicine. By the fifteenth and sixteenth centuries, things had changed very radically since the days when Crusaders sought the help of Muslim or eastern Christian and Jewish physicians. By now it was Europe that was advancing, Islam that was falling back. The intimate and personal character of the services rendered by physicians gave to medical innovation an attraction lacking in the more public and impersonal branches of European science and technology. In medicine it was the individual welfare, and perhaps even the survival, of the patient that was at stake. As in other times and places, in seeking out the best doctors available self-interest was able to triumph over even the most extreme bigotry. All the same, however, it was not unresisted and the more conservative practitioners of traditional medicine fought back.

At first the penetration of European medicine into the Ottoman realms was due largely, if not entirely, to non-Muslims – mainly to Jews and occasionally to Christians. . . . By the sixteenth century, Jewish physicians, most of them of Spanish, Portuguese and Italian origin, were common in the Ottoman empire. Not only the sultans, but also many of their subjects had recourse to these physicians, whom they recognized as representing a higher level of medical knowledge. [. . .]

By the following century, Ottoman doctors had a new and painful reason for paying attention to European medical skills. This was a previously unknown ailment which came to them from the West, and to which therefore they gave the name which it still retains in most Muslim countries – Firengi, the Frankish disease. The first Turkish treatise on syphilis, part of a collection of medical writings presented to Sultan Mehmed IV in 1655, is based largely on the famous work of Girolamo Fracastro of Verona (1483–1553), and also includes some borrowings from Jean Fernel (d. 1558) on the treatment of this disease. Other parts of this work dealing with other diseases quote the names of several well-known European physicians of the sixteenth century. The book indicates some acquaintance with European medicine, and it is even possible that the author was able to read Latin or at least had someone at his disposal to render him this service. But the difference in approach is already noticeable. Though the collection was presented to the sultan in 1655, the European works cited in it all belong to the sixteenth century. The Jewish doctors who came from Europe in the sixteenth century represented the highest level of sixteenth-century European medicine. The Ottoman Jewish doctors of the seventeenth century still represented the highest level of European medicine – of the sixteenth century. The renewal of contact through the training of Ottoman Greek physicians in Italian schools from the mid-seventeenth century onwards does not seem to have brought any fundamental change in this relationship. The leisurely pace and timeless framework of Ottoman scientific writing had already given rise to a serious time lag between Western and Ottoman science. It was to become much wider.

From these occasional Ottoman references to Western science, it is clear that they did not think in terms of the progress of research, the transformation of ideas, the gradual growth of knowledge. The basic ideas of forming, testing, and, if necessary, abandoning hypotheses remained alien to a society in which knowledge was conceived as a corpus of eternal verities which could be acquired, accumulated, transmitted, interpreted, and applied but not modified or transformed. Their works on medical and other sciences consist mostly of compilations, adaptations, and interpretations of the corpus of classical Islamic learning preserved in Persian and, more especially, Arabic, sometimes supplemented by material derived from Western scientific writings but similarly treated. There is no attempt to follow new discoveries and little awareness even of the existence of such a process. The great changes in anatomy and physiology occurring at that time pass unnoticed and unknown.

According to Muslim belief there was, in the early days of Islam, a rule called *ijtihād*, the exercise of independent judgment, whereby Muslim scholars, theologians, and jurists were able to resolve problems of theology and law for which scripture and tradition provided no explicit answer. A large part of the corpus of Muslim theology and jurisprudence came into being in this way. In due course the process came to an end when all the questions had been answered; in the traditional formulation, 'the gate of *ijtihād* was closed' and henceforth no further exercise of independent judgment was required or permitted. All the answers were already there, and all that was needed was to follow and obey. One is tempted to seek a parallel in the development of Muslim science, where the exercise of independent judgment in early days produced a rich flowering of scientific activity and discovery but where, too, the gate of *ijtihād* was subsequently closed and a long period followed during which Muslim science consisted almost entirely of compilation and repetition. [...]

The system of filtration, designed to exclude those imports which might have threatened the traditional way of life, remained effective against the more dangerous penetration of ideas – of the Western conceptions of inquiry and discovery, experimentation and change which underlay both the science of the West and the technology to which it gave rise. The products of Western technology might, after due consideration, be admitted; the knowledge achieved by Western science might in certain cases be applied; but that was the limit of their acceptance.

The question arose again in an acute form in the eighteenth century, when a series of defeats in the battlefield convinced the Ottoman governing elite that the Christian enemies of the empire had somehow managed to achieve superiority in the arts of war and that changes were necessary to restore Ottoman power. Their feelings are well expressed in a memorandum written by one Janikli Ali Pasha after the crushing Ottoman defeat by the Russians in 1774. Ali Pasha addresses himself to two questions which he tells us had profoundly occupied his thoughts. Why had the empire, which was once so strong, become so weak, and what should be done to restore her former strength? The Turkish soldier, he said, was no less brave then before, the people no fewer, the territories no smaller, and the resources of the empire were still as great. Yet, where once the armies of Islam had invariably put the infidel to flight, it was now the Muslims themselves who were put to flight by the infidel.

Janikli Ali Pasha's remedy was strictly conservative – a return to the good old ways. There were others, however, who saw the problem in the military superiority of the West and the answer in military reform. An important aspect of this was the establishment of training centers in modern warfare. [...]

This first school [The Üsküdar school of engineering, established in 1734] and the corps of military engineers established at the same time, were bitterly opposed by the Janissaries, who in due course forced their abandonment. The objective of modernizing the armed forces was not, however, given up, and in 1773 a new start was made with the opening of a naval school of engineering. The teachers of this new school included a number of Europeans. ...

This time the reactionary forces were not able to procure the closing of the school, which on the contrary grew and served as a model for other schools of military engineering, medicine, and similar matters established by Sultan Selim III and his successors. ... [S]everal schools were started, for both army and naval officers, offering instruction in gunnery, fortification, navigation, and the ancillary sciences. French officers were requested to serve as instructors and a knowledge of French was made compulsory for students. A library for the use of the trainees contained some 400 European books, most of them French. They included a set of the *Grande Encyclopédie*.

13.2 J. Scott-Carver, *Russia's contribution to science in the eighteenth century**

Any contemplated discussion of Russia's contributions to the European scientific community during the eighteenth century would have seemed

---

* J. Scott-Carver, 'A reconsideration of 18th century Russia's contributions to European science', *Canadian–American Slavic Studies*, 14 (1980), pp. 389–405.

fatuous until recently, largely because of the enduring image of Russian culture promoted by contemporary travelers, who either denied the existence of any scientific activity there, or claimed that it exerted no significant impact on Russian society. Recent Soviet scholarship has begun to modify both aspects of the image; but it is still generally believed that the scientific enterprise remained an essentially foreign activity barely able to sustain itself in the limited environs of St. Petersburg, much less Moscow. It is the thesis of this paper that by the end of the eighteenth century Russia had developed a cadre of competent amateur as well as professional scientists whose research activity was fully appreciated by the entire European scientific community. But to focus exclusively on their research contributions would be misleading and anachronistic; for, as will be demonstrated in the first section of the present paper, eighteenth-century Russian scientists participated in a variety of educational, administrative and popularizing activities which they considered at least as important as research. While these activities were essential to the domestication of a modernizing scientific culture in Russia, they also directed attention away from research matters and hence limited the possible contribution of Russian scientists to the European scientific tradition.

Although it was by no means the exclusive center of either research or dissemination, the St. Petersburg Academy of Sciences' role in both was crucial. When the Academy opened in 1727, all eleven members were foreigners; but an important shift in membership occurred during the century. The first Russian member was elected in 1733 when V. E. Adodurov (1709–80) entered as an adjunct in mathematics. In 1742, two more Russians were elected to adjunct status: G. N. Teplov (1725–79) in botany, and M. V. Lomonosov (1711–65) in physics. Following the latter's election, the Russian contingent regularly increased until 1800, when eight of the sixteen Academicians were Russian while three others were second-generation residents whose fathers had belonged to the Academy.

To an extent, then, the Academy was Russianized, but too much emphasis can be placed on the issue. Neither Russian nor foreign members displayed an overwhelming commitment to the process of Russianization. Rather, they were primarily concerned with the development of natural science, which they realized best flourished when ideas, methods, information, and personnel freely crossed political boundaries. By obstructing these exchanges, Russianization would have challenged the international character of the scientific enterprise. But on the other hand, if the well-being of the scientific tradition required inclusion of foreigners, it also mandated the election of Russian scientists, for science flourished by welcoming into its domain the most talented persons available, and there were well-educated Russian scientists capable of making valuable contributions to natural science.

The eighteenth century's cosmopolitanism further encouraged the inclusion of foreign scholars. Like their West European contemporaries, Russia's future scientists were daily reminded of the century's internationalism:

they became fluently polylingual in their early years; they received their secondary education in the midst of an international body of scholars; and they traveled abroad for their advanced education. ... Given their early cosmopolitan experience, it would have been unlikely and perhaps unthinkable for them to substitute the notion of the nation state for the ideal of the 'family of man' and to pursue a policy of complete Russianization. [...]

Ironically, patriotism further encouraged the admission of foreign scientists to the Academy; for rather than promote separatist, exclusionary attitudes, eighteenth-century patriotism stressed improvement in the nation's material and formal cultures. The national dissemination of a modernizing scientific culture was regarded as essential to both objectives, as natural science was assumed to be among the more effective means to improve the national economy, and to eliminate ignorance and superstition. [...]

Indirectly, then, patriotic insistence on the dissemination of a scientific outlook necessitated the inclusion of foreign scholars in the Academy, and at the same time patriotic sentiment mandated development of a single institution to serve as a showcase of scholarly excellence. If that implied recruitment of foreign scientists, so much the better, because their international connections helped bolster the image of Russia as a progressive, enlightened nation free from chauvinistic bias. And if that concern with excellence implied inclusion of Russian scientists, that too worked to Russia's advantage, because their presence demonstrated Russia's successful efforts to enlighten its society.

Thus the dynamics of the scientific enterprise, the century's cosmopolitanism, and the imperatives of patriotism fostered the inclusion of foreign scholars. More immediately practical concerns also moderated any efforts to Russianize. From the start, the Academy had been considered more than a research institute or display-case demonstrating Russia's cultural parity with other European states. It was intended to be useful to the state, and the state regularly turned to it for advice on subjects like the design and placement of lightning rods, ship corrosion, tobacco culture, commerce, bridge construction, and the use of coal. Significantly, however, the state's conception of utility was broader than that, and this more generalized view led to the regular displacement of many from careers in natural science.

The state's insatiable demand for educated administrators continually decimated the ranks of potential scientists. Of the first three Russian members of the Academy, for example, only Lomonosov remained a scientist. After three years in the Academy, Adodurov was reassigned to serve as language tutor to Catherine the Great, as official in the Orenburg government, as Curator of Moscow University, and as President of the Manufacturing College. Similarly, Teplov was reassigned to a number of administrative posts including work in Little Russia and in the Commerce College. [...]

The state's need to develop educational facilities also ran counter to any aspirations to Russianize the Academy, as indicated by the careers of M. I. Afonin (1739–1810) and F. G. Politkovskii (1753–1809). ... Afonin was sent to the University of Uppsala where he studied natural history, chemistry, metallurgy, physics, mathematics, astronomy and a smattering of humanistic subjects. Having completed advanced training under Linnaeus, he returned to Russia but rather than receive election to the Academy, he was sent to his alma mater, the University of Moscow. There he taught natural history, chemistry, and a series of more practical courses including assaying and agriculture. ... After studying at the Gymnasium and University of Moscow, Politkovskii was sent to the University of Leyden for medical training. Having completed his degree, he spent two years studying physics and chemistry in Paris where he gained firsthand knowledge of the chemistry reforms Lavoisier was then propounding. With Academician V. M. Severgin (1765–1826), then, he came to play an instrumental role in the rapid popularization of the new chemistry in Russia. In addition, as professor of natural history he instructed Moscow University's medical students in practical chemistry, gave public lectures in natural history to the city's interested residents, and served as curator of the University's rapidly expanding natural history museum.

Thus, because of its practical concerns the state repeatedly assigned promising young scientists to posts which certainly promoted the dissemination of a modernizing scientific culture, but which also precluded them not just from active participation in the Academy but also from involvement in research activities. [...]

Through its international composition the Academy came into regular contact with the entire scientific community, and it quickly achieved a prominent place within that community. But if professional scientists esteemed Russian science, the popular European press did not. If they mentioned the issue at all, travelers, for example, disparaged Russian scientific activity and claimed that Russia was somehow fundamentally inhospitable to science and its application. Thus Chappe d'Auteroche alleged in an account published in 1768 that shortly after its inception:

> the Academy lost its repute, and the Arts sensibly decreased, as the great men first invited into Russia, either died, or left the country. The sovereigns still continued to supply their subjects with able masters, and to encourage and protect men of abilities; but notwithstanding these advantages, not one Russian has appeared in the course of more than sixty years, whose name deserves to be recorded in the history of the Arts and Sciences.[1] [...]

Authored by polemicists, the travel literature was generally written to support a set of Enlightenment values and to malign a set of traditional ones. Russia's image suffered accordingly, for her autocratic political system,

---

1 Chappe d'Auteroche, *A Journey into Siberia* (London: T. Jefferys, 1770), p. 320 (the French original appeared in 1768).

her serfdom, her aggressive foreign policy, and her highly ritualized religiosity made her a target for proponents of more modern values. ... Even those without firsthand experience promoted an image of Russia as a culturally backward wasteland devoid of interest in modern science. The *Encyclopedia Britannica* boldly stated in 1797, for example, that 'science has made but a very small progress among them; and the reputation of the imperial academy of St. Petersburg has been hitherto supported by the exertions of foreigners'.[2] [...]

William Tooke, a man of vast experience in Russian affairs, was one of the few to recognize Russia's scientific establishment. Writing in 1798, he lavished praise on the scientific travel accounts of Academicians P. S. Pallas, J. Güldenstädt, and others who had explored southern and eastern Russia between 1768 and 1774. 'The journals of these celebrated scholars even furnish such a great quantity of materials, entirely new... that it would require whole years, and the labour of several literary men, only to put these materials in order, and properly to class them. ... '[3] Notably, Tooke cited accounts of foreign rather than Russian scientists, and while the intrinsic merit of those accounts surely rendered them attractive, the popularity of fictional as well as actual travel literature enhanced their appeal. For the idle and disinterested, various descriptions of exotic customs, strange beliefs, and unusual behaviors satisfied curiosity and stimulated the imagination. For those who were either impressed or alarmed by Russia's emerging status in Europe, the accounts of her political processes, social cohesiveness, and demographic as well as natural resources must have seemed useful information. Still, the purely scientific element of the travelogues contributed to their popularity as descriptions of the geographic distribution of plants and animals, accounts of climatic variation from region to region, and discussions of topography helped answer some of the questions of primary concern to eighteenth-century naturalists.

For the same reasons, the travel journals of Russian scientists achieved notoriety in Western Europe. The journal by Academician Krasheninnikov (1711–55), *Description of the Land of Kamchatka* (1755), was considered especially important by the scientific community and was translated into English, French, German and Dutch between 1764 and 1770. The expeditionary accounts of Academicians I. I. Lepekhin (1740–1802) and V. F. Zuev (1754–94) were also published in West European editions, as were the works of several amateur scientists. ...

The information contained in these accounts represented an important part of the Russian contribution to natural science, but papers published in

---

[2] 'Russia', *Encyclopedia Britannica*, 3rd edn, 18 vols. (Edinburgh: A. Bell and C. Macfarquhar, 1797), XVI, 573.

[3] W. Tooke, *The Life of Catherine II, Empress of All the Russias*, 2 vols. (Philadelphia: William Fry, 1802), I, 316.

the Academy's *Mémoires* represented a more substantial contribution. ... In 1733 Daniel Bernoulli wrote his one-time colleague, Academician Leonard Euler, that 'I cannot tell you often enough how eagerly people everywhere ask for the St. Petersburg *Mémoires*. ... '[4] And England's leading authority on quadrupeds, Thomas Pennant, implicitly confirmed the contention half a century later when requesting copies of thirty-five volumes of the *Mémoires* for his personal library.

The contributions of scientists from western Europe such as Euler and Pallas certainly contributed to their importance, but they were not exclusively responsible. The twenty studies in botany and zoology by N. Ia. Ozeretskovskii (1750–1827), the forty-seven papers in mathematics, physics and astronomy by S. Ia. Rumovskii (1734–1812), and the twenty-four articles in geology, mineralogy and chemistry by Severgin contributed to the important place the *Mémoires* came to occupy in professional circles. Indeed, as suggested by the response of *Journal encyclopédique* to an article by Rumovskii, the professional scientific community regularly expressed confidence in Russian scientists: 'The Russian Academicians are on a par with others in the most sublime knowledge. M. Rumovskii has given the solution and supplied the proof to a problem concerning the *maxima* and *minima*. In truth, the problem relates to M. Euler's celebrated doctrine of isoperimeters which is in no way exhausted. Yet, M. Rumovskii's solution has combined very great elegance with the honor of having surmounted the most perplexing difficulties.'[5] Importantly, Rumovskii was cited as specific but not unique, and his example served to support a generalized assessment made about Russian scientists. In line with that assessment, Ozeretskovskii was elected to four West European scholarly societies, Severgin to eight, and the prolific amateur Count G. K. Razumovskii (1759–1837) to seven. Membership was, of course, contingent on factors other than scientific achievement. Diplomatic considerations and personal friendships undoubtedly played an important role, but membership nevertheless signalled a degree of professional recognition as did references to Russian scientific publications in the writings, for example, of Linnaeus and Buffon. [...]

Linnaeus was remarkably familiar with Russia's natural resources, but because he frequently failed to acknowledge his sources it is virtually impossible to determine precisely the extent to which Russian Academicians contributed to his comprehensive survey of natural history. Undoubtedly, he relied on his student, J. P. Falk of the Russian Medical College, and on G. W. Steller and E. Laxmann, both of whom he recommended for membership in the St. Petersburg Academy of Sciences. The industrialist G. Demidov, who sent his sons to study with Linnaeus, also contributed

---

4 Cited in A. Lipski, 'The Foundations of the Russian Academy of Science', *Isis*, 44 (1953), p. 349.

5 *Journal encyclopédique*, 7 (1764), pp. 16–17.

through the frequent exchange of plant and animal specimens, and Linnaeus must have received additional information from Russian students including Afonin and A. M. Karamyshev (?–1791) whose doctoral dissertation was the first systematic treatment of Siberian plant life. The various studies by Academicians Pallas and both S. G. and J. F. Gmelin also served as important sources of zoological, botanical and geographical information. Last, the numerous zoological studies of at least one Russian academician, Lepekhin, enriched Linnaeus' *General System of Nature*.

Buffon was similarly familiar with Russia's natural resources, but he also frequently failed to acknowledge his sources of information. Academicians Pallas, the Gmelins and G. F. Müller were most frequently cited, but the zoological descriptions by Krasheninnikov served at least occasionally as sources of information. [...]

Additional references to Russian scientific works could be cited and, perhaps, some of the unspecified sources of information could be traced; but neither resort would substantially alter the conclusion that by the end of the eighteenth century Russian amateur and professional scientists had gained entry into the scientific community and had become respected, contributing members of it. Like the vast majority of scientists, they worked within existing paradigms and hence made no memorable discoveries or innovations. Rather, they made standard contributions, including the discovery of new plant and mineral species, the refinement of astronomical data, and the collection and standardization of meteorological information. ...They were paradigmatic in yet another fashion: for like their British contemporaries who met monthly as the Lunar Society, or their French contemporaries who met at Berthollet's residence, the Russian scientists were as intent on the dissemination and utilization of natural science as they were concerned with its internal development. In that, they contributed not just to the scientific tradition but also to the domestication of a modernizing scientific culture in Russia.

13.3 Michael D. Gordin, *The early St Petersburg Academy of Sciences**

'I have to harvest big stocks, but I have no mill; and there is not enough water close by to build a water mill; but there is water enough at a distance; only I shall have no time to make a canal, for the length of my life is uncertain, and therefore I am building the mill first and have only given orders for the canal to be begun, which will the better force my successors to bring water to the completed mill.' Statements like this one by Peter the Great, which concerns his plan to import an academy of sciences into the fledgling city of St. Petersburg, have often been construed as part of a utilitarian

* Michael D. Gordin, 'The importation of being earnest: the early St. Petersburg Academy of Sciences', *Isis*, 91 (2000), pp. 1–10, 15–17, 19, 21–2, 29–31.

design to harness technology and science for the ends of the state.[1]... Existing work on the place of the Imperial Academy in Russian culture focuses almost exclusively on the academy after its opening in 1725 by Empress Catherine I – wife of Peter the Great, the tsar who had created the Russian academy but died earlier that year. Since these studies concentrate on the course of Russian academic science in the eighteenth century, matters pertaining to its genesis and inception have largely been taken for granted. Instead of exploring the social and cultural roots of the *idea* for an imperial academy, these works explore how scientists were recruited to the academy from abroad, how it fared during the rough early years, and how it became Russified after the 1750s.

This essay concerns the question of inception. The standard accounts stressing utilitarianism are clearly largely correct, but they necessarily leave elements of the academy's history unexplained. For example, if the academy was founded for purely utilitarian reasons, why did the tsars encourage the study of 'speculative' topics like planetary vortices? Why did the bureaucracy not simply continue to rely on the technical advisors regularly imported from abroad, without organizing them into an academic body? The answers to these and related questions are embedded in the other cultural advantages that were to be extracted from the academy. Peter the Great not only imported a politically useful educational institution; he also knowingly imported a particular etiquette regime of refined manners that characterized Western natural philosophy in the eighteenth century. This effort was part of a broader program to establish new social classes through educational stratification and to change the way those classes behaved through cultural reforms. Reading the academy as a part of these two wider purposes complements the standard accounts and broadens our understanding of the links between Peter's courtly life and his scientific policies. [. . .]

A number of recent studies on academic natural philosophy in the seventeenth and eighteenth centuries have argued that the organization and structure of academies varied – but not arbitrarily – depending on whether the setting was the princely court, gentlemanly society, or the state bureaucracy. The variations in academic protocols followed the etiquette patterns of each specific community of natural philosophers. As Mario Biagioli has argued, the extant etiquette regime – understood as the set of practices and protocols that undergirded mannerly behavior – served as the condition of possibility for a particular academic structure.[2] In the Russian case, however, where poorly developed native etiquette codes were juxtaposed with

1 Quoted in B. H. Sumner, *Peter the Great and the Emergence of Russia* (New York: Macmillan, 1951), pp. 208–209.

2 Mario Biagioli, 'Etiquette, interdependence, and sociability in seventeenth-century science', *Critical Inquiry*, 22 (1996), pp. 193–238.

a decidedly advanced academic structure, Biagioli's directional argument ceases to apply. Nevertheless, a great deal may be gained by inverting it. Instead of looking at a set of etiquette protocols as the condition of possibility for the Russian academy, I consider how that academy served as a condition of possibility for the dissemination of a particular set of etiquette codes. Of course, it is not the argument of this essay that the academy was *entirely* an institution devoted to a particular vision of etiquette. Rather, the academy had two faces: one that turned toward the natural philosophical community in Western Europe and one that turned toward St. Petersburg society. The first was concerned with the technical details of mathematical philosophy and is much less specific to the Russian context. I wish to examine the second face here: how the academy functioned with respect to the culture of St. Petersburg and how Peter's project can be understood as more than just the establishment of an institution for abstract natural philosophy and practical technical advice.

The St. Petersburg academy – structure, scholars, etiquette protocols – was imported from abroad, but not without modifications. There are many reasons why Peter *needed* to import these features. I am suggesting that there was also something he *wanted*: that the Russian 'public' – to be understood throughout this essay as a very narrow stratum of the social elite – be exposed to a particular form of life that he saw embodied in this ready-made academy. The plan for the St. Petersburg academy, in fact, was specifically altered from European exemplars to highlight the public aspects of academic natural philosophy. To explore why Peter the Great might have come to associate academic science with mannerly culture, we need to begin as he did: with the idea of and the model for an academy of sciences.

Peter the Great drew the plan for the St. Petersburg Academy of Sciences from a set of impressions he had received from a variety of sources, including the Royal Society of London (which he visited in 1698) and the Académie Royale des Sciences in Paris (visited in 1717). Although some historians have isolated one of these two as the dominant model for the academy, the most significant influence was in fact that of Gottfried Wilhelm Leibniz, the German natural philosopher, and his Berlin Academy of Sciences. Not only can we trace a direct connection – Leibniz actually corresponded with Peter and other major figures of the Russian court for about twenty years – but the Imperial Academy as it was finally formulated bore remarkable similarities to Leibniz's own Berlin academy and to his vision of a global system of academies. While it is important to trace the origins of the model for the St. Petersburg academy, it would be a mistake to focus exclusively on institutional precursors. Instead, after examining the relations of the St. Petersburg academy to the three dominant models (London, Paris, and Berlin), one must ask why any of these appealed to Peter in the first place. There were structural constraints on the type of academy Peter could establish in the

Russian context, but there was also, I contend, a particular type of natural philosopher that he wanted to introduce to Russia, a type exemplified by Leibniz.

Since this argument hinges on the structural constraints facing Peter, we must start with the models open to him and determine why he ended up appropriating many (but not all) of the features of the Berlin academy. Some historians have made the case that Peter the Great based his academy on the English Royal Society, which he had visited during his 1697–1698 'Great Embassy', a full twenty years before he met Leibniz. ... While it is true... that many of the technicians – navigators, shipbuilders, and the like – Peter brought to Russia came from England through links established by the Russian Jacob Bruce and the Englishman John Colson, precious few academicians did. Most hailed from the German states. The evidence from Peter's visit to England offers only shaky proof that he actually visited the Royal Society in session. What he did see was its museum; it is possible that this gave him the idea of establishing his own personal museum, the *Kunstkamera*, which eventually became attached to the Imperial Academy, but it is unlikely that the influence extended any further. As many studies of the Royal Society have shown, the role played by the free society of gentlemen in the creation and functioning of that institution was enormous. Russia did not already have an educated genteel society; rather, the Royal Society members were exactly the type of gentleman natural philosophers that Peter wanted to *introduce*. If he could have founded a Royal Society in Russia, he wouldn't have needed to build the Imperial Academy. This is not to claim that the only function of the academy was to express mannerly behavior, but to emphasize that a club type of academy required a particular pre-existing etiquette regime in order to sustain its technical work. Russia lacked such codes of manners, and thus a Royal Society-style academy could not have performed the necessary technical functions.

A far better case can be made for the relation of the Imperial Academy to the French Académie Royale des Sciences. Peter visited the French academy in 1717, after Leibniz's death. And in 1721, just before he worked in earnest on the plans for his own academy, he was elected a member of the Académie Royale 'above all ranks', a privilege never before extended to a monarch. The French academy was a more suitable model than the Royal Society for the Russian context: the type of academy that a powerful prince such as Louis XIV required would have seemed appropriate to Peter's imperial pretensions. But he visited the Académie Royale fifty years after its founding, long after Louis XIV had passed away and during a time when its relation to the monarchy was very different. Moreover, the work the members pursued had taken a theoretical and mathematical tack that Peter regarded as suspect.

The third major model, and the most likely candidate for importation to the Russian context, was the Berlin Academy of Sciences, Leibniz's model. This institution had the advantages of being at once suitable for the

Russian context and personally promoted by its creator, who was fairly adept at insinuating himself with powerful patrons. To make the claim that Leibniz impressed Peter the Great as a model of the decorum to be expected in natural philosophers, one would have to demonstrate that Leibniz was (or at least appeared to be) a well-mannered courtier and not, like Isaac Newton, handicapped by a brusque and difficult temperament. Fortunately, several historians have already done a remarkable job of showing just that. After receiving his law degree from Altdorf, Leibniz traveled to several princely courts and aristocratic circles, ending up at Hanover after a long detour in Paris, and demonstrated remarkable skill in advancing himself as a factotum – a natural philosopher for all seasons – able to answer whatever queries his patron put forth. This was partly a strategy Leibniz developed to advance himself at court, but it was also consistent with his belief that, since philosophers did not have the powers of statesmen, it was in their interest to persuade statesmen to take their advice, thus conjoining worldly power and philosophical ideals. This professional tactic was tied to a vision of the necessity of philosophers for the Baroque state. ...

Peter the Great was in many ways Leibniz's ideal patron. For years Leibniz had sought an 'enlightened' patron who was both powerful and interested in natural philosophy. ... In October 1711 he managed to meet Peter in Torgau, where the tsar's son was marrying. A correspondence began in 1712 after the two met again in Carlsbad, and Peter appointed Leibniz a counselor and issued him an annual salary until Leibniz's death in November 1716.

Leibniz had been trying for some time to fulfill his ideal of a global network of scientific academies that would pursue coordinated research, a vision conceived at least partly in reaction to the disunified learned societies that surrounded him. ... [He] saw the Russian case as a potential instantiation of his monadic ideal of state and academy. Besides having one of the European monarchs most amenable to his purposes, Russia stretched from the Arctic to the southern steppe and across Asia. Such a broad climatic differential offered a basis for wide-ranging scientific findings, a promise eventually realized in the Imperial Academy's early geographical expeditions. Furthermore, it connected Asia and Europe and thus could provide a stable land link for findings to travel from East to West, transmitting knowledge and material from Leibniz's Jesuit connections in the Far East to European academies. Finally, Russia was a land untouched by philosophy – or so Leibniz felt – and so it was a virtual blank slate for the best of Western natural philosophy, which it would be able to assimilate rapidly. ...

If Peter was Leibniz's ideal patron, Leibniz was just what Peter thought a natural philosopher should be. He was courteous, learned, and especially interested in developing the untapped potential of Russia. Or that, at least, is what Peter saw. Leibniz, in a fairly typical fashion, tailored his proposals, especially at their first meeting, to what the tsar wanted to hear, emphasizing, for example, the reform of Russian education and internal navigation. If Peter intended to create in Russia an academic infrastructure that would produce generations of practical natural philosophers, Leibniz – the self-proclaimed 'Solon of Russia' – exemplified the kind of philosopher he hoped to turn out. An anecdote illustrates Peter's perception of Leibniz. During one of their meetings, Peter complained of a partial paralysis in one arm that hindered him in mechanical actions such as firing a pistol. Leibniz quickly threw together a simple wooden device that restored the motion in the tsar's arm. His response was practical, courtly, and philosophical, all at once. . . .

The argument that Peter and Leibniz embodied each other's ideal has limits. The two men wanted different results from their relationship: Leibniz eagerly sought patronage from Peter and the fulfillment of his goals for a network of academies; Peter desired practical advice from Leibniz on navigation, education, and the structure of a scientific academy. Neither individual thought of their interaction in terms of courtly roles. This was not a transaction *about* etiquette, but a transaction *conducted through* the language of etiquette. Each party negotiated by the courtly protocols he was accustomed to, but this does not mean that they went through these motions for their own sake. Etiquette was not an end in itself; it was a means for achieving specific goals in the court cultures of early modern Europe. . . .

. . . Before Russia could make its own Leibnizes, some tinkering with both the structure and the functions of the academy had to be done. This tinkering is visible in the outline of Peter's final academy project, written up [in a document known as the Project] just before he died.

[. . .] Beyond those pertaining to personnel issues, the most formal statement in the Project is that the academy was to serve as a model (*obrazets*) to the rest of the country. . . .

The Project envisioned two ways in which the academy could fulfill this self-perpetuating role. The first was through educational reform, a broad restructuring of the entire system along more 'Western' lines, with the academy at its pinnacle. The second involved integrating the academy into the Petrine manners reforms, deliberately *showing* Russians what it meant to belong to polite society. . . .

[. . .] Education was a way of creating new social stratifications; making the occupants of those strata behave properly was another matter. The Imperial Academy of Sciences was supposed to accomplish both tasks by showing the newly created elites proper manners. It was necessary, then, to acquaint the public with the academy. . . . It was also necessary, however, to show the public the *manner* in which science was conducted, not just the knowledge it produced. The most important attempt to do this was the series of public scientific assemblies that were to be held three times a year.

As the Project stipulated: 'Thus even for foreigners there will be a great game, since three annual public assemblies (*asamblei*) will be set up and a conversation by one member of the academy will be made from his science, and in this praise of the protector-defender [the tsar] will be introduced.' ... The tsar's invoked presence both bestowed legitimacy on the academy and further attracted the nobility, who were expected to attend such events. Although members of all classes (and, significantly, both genders) were permitted to attend, nobles were 'to freely go before those of lower rank at public assemblies where the court is gathered'. This statement shows that such events were considered more important for the nobility than for lower classes. While praise of the ruler was not unusual even for Western academicians, the explicit order that praise be offered was. ... These [public] meetings were not mere hypothetical conjectures on Peter's part; they actually went into practice after his death. A common topic of the academy members' lectures was heliocentric theory, which was often discussed in this particular public forum, although rarely anywhere else.

Just as education was an explicit mirror of other Petrine reforms, so did these public assemblies reflect the project of Peter's etiquette reforms. Etiquette protocols serve a central function in any courtly context. ... A ruler bent on changing the structure of court life must introduce a corresponding change in his court's etiquette protocols; etiquette serves as both a principle of ordering the court and a means of controlling the network of nobles. For Peter the Great, who was actively trying to adopt not only Western technical advances but also Western cultural procedures, it was vital that the change in hierarchy and the change in etiquette be effected at the same time. [...]

During this period in Russia, it was the state's role to provide the lead in establishing elite cultural patterns. All nobles had to serve the tsar in some capacity – administrative, military, or courtly – and thus major cultural transformations were generally transmitted through the service network of the nobility. Although Peter the Great conducted a great many political and military reforms during his active reign, his cultural reforms are generally perceived as the most dramatic and cataclysmic for the old Russian way of life. From the image of the West in Russian *belles-lettres* of this period to the major influx of Westerners – from footmen and courtiers to shipbuilders and scientists – the emphasis was on how to make Russians look and act like 'Westerners'. The first step was, as it so often is, to change the appearance of things. Of all Peter's cultural reforms, the most discussed and criticized were those pertaining to dress and personal appearance. Shortly after his return from the Great Embassy, Peter began to cut the traditional beards of the older nobility and to proscribe caftans in favor of Western clothes, perceived as immodest by older nobles. Even the appearance of houses was regulated, down to the type of plaster used on the roof, as was the kind of coffin one could be buried in. These cultural changes – which sometimes went so far as the brutal stripping and forced shaving of nobles in the middle of the street…– of course had their political side. Public humiliation and state intrusion were intended to curb the power of the old ruling families. But the aspect of 'culture building' should not be ignored. Even in the Naval Academy – which trained officers culled from the nobility – dancing was taught to improve 'posture', in the hope of creating a noble who was both an officer and a gentleman. […]

Changing outside appearances was only part of the reforms, however. Peter was very concerned that Russians start acting the role he had carved out for them, not just look the part. He undertook a series of popular etiquette reforms designed to make the elite conform to his ideal of Western society. Two particular means deserve to be singled out for closer examination: etiquette handbooks and assemblies. Both of these functioned under a principle common to Peter's other court reforms, the use of the tsar (or some other exemplar) as a model or *obrazets* to be copied. The first Russian etiquette book, the 1717 *Honest Mirror of Youth; or, A Testimony to Social Intercourse Collected from Various Authors*, contained general and specific Western etiquette protocols. Among other things, it instructed readers not to eat with their mouths open or spit while talking to ladies. Public violation of any of these codes could result in corporal punishment.

Assemblies were taken even more seriously. … Attendance at these events was not optional. The nobles who had to come (and even the host, for that matter) were sometimes informed on very short notice. One of the most important features of the assemblies was their part in advancing the position of women in Russian society. Whereas Muscovite society had traditionally excluded women from public life, Peter turned the assemblies into salon-like gatherings where women played the crucial role of facilitating conversation and dancing. … Recall now the assemblies that were ordained for the Academy of Sciences: just as the conventional assemblies were schools for the social graces, the scientific assemblies were schools for intellectual conduct. In both contexts, Peter's presence had to be announced and praised – or else.

There is a connection between the civilizing process that Peter was engineering at court and in society and the role he perceived for the Academy of Sciences. While in some circumstances – for instance, in matters of personal conduct – the only proper *obrazets* was the tsar himself, the cultural *obrazets* for civilized, nonviolent interactions among individuals was to be the academy, whose members were culled from the most civilized group that Peter had ever interacted with: the Republic of Letters. …

Implicit behind this ideal view of the academy was the fact that some groups would suffer as a result of the reforms. The academy was widely perceived as an attack on the Church and the old noble culture. The reasons why the academy was seen as anti-Church are fairly clear; but unless it is understood as a purveyor of new courtly protocols, we might be hard pressed to conceive how it threatened the nobility. The academy reflected a more general trend in the restructuring of court dynamics and the establishment of secure links between the court and the evolving autocratic state. It was an indicator of general cultural change at the same time as it brought that change about. The third loser in this process was the lower ranks. As the elites were becoming more enlightened, the lower classes became comparatively more ignorant. The immense technical development of Russia required a great deal of unskilled labor, which the lower classes furnished. Providing a setting in which good manners can flourish requires work; that labor came from the bottom of society.

But nevertheless the project succeeded, if only gradually. During most of Peter's reign the status of nobles and their property was tenuous, as Russia mutated from a patrimonial state to an autocratic regime with an elaborate state apparatus. The nobility's status stabilized only to the degree that the etiquette protocols imported from the West took hold. The nobility legitimated the reforms by participating in them, and they were in turn legitimated by their participation. For the same reasons, after the 1747 charter it was no longer necessary to have the academy – the link to the West – connected to the university – the link to the elites. The protocols had been internalized to such a degree that the two institutions could be uncoupled. The academy was, so to speak, initially a hybrid creation, with roots in courtly protocols, and not purely an appendage to the bureaucracy. The fact that the academy later lost its close connection to the court is evidence of Peter's success in establishing firm distinctions between the court and the state.

[...]

Situating the Imperial Academy of Sciences in the Petrine reforms requires looking beyond the utilitarian benefits Peter hoped to extract from its establishment. Instead of viewing the academy as simply part of general reforms intended to tap the technical potential of Russia, situating it at the crossroads of the education and manners projects provides a much-needed perspective on the academy's origins in Leibniz's courtly proposals, on its structure, and on how early disputes and publishing projects under its control were conducted. Further implications can be taken from these observations. For example, there has been a substantial debate in the secondary literature on whether Russia underwent an 'Enlightenment' and, if so, when it occurred and what its relation to the French Enlightenment was. The reevaluation of the academy proposed here provides a different angle on some of these older debates.

The standard view of the Russian Enlightenment dates back to the first biography of Peter the Great, written by Voltaire in 1764, which basically extended the vision of Peter presented by Fontenelle in his eulogy of the tsar before the Académie Royale des Sciences. Peter's reforms were a subject of intense debate in the French Enlightenment, and positions ranged from the almost uncritically positive (Voltaire) to the damningly negative

(Rousseau). Voltaire's picture of Peter, related in his two-volume *Histoire de Russie*, was that of a cataclysmic transformer who came upon Russia in the slumber of the Middle Ages and brought it forth to civilization through nothing but the force of an indomitable will. Voltaire's Peter had no master plan, no set of agendas he wished to accomplish beyond taking advantage of the resources available to him to solve contingently pressing problems. Peter coerced both his environment and his people to accomplish these ends, but he had no rational program.

This view of Peter, which has been repeated to the present day, is compelling, and many facts support it. It does, however, specifically deny Peter the status of an 'Enlightenment ruler' – at least by the usual definition. Such rulers, it would seem, have a rational framework that they try to impose on their countries, utilizing the principles of 'Enlightenment' to achieve 'progress'. The view of Peter constructed so adeptly by Voltaire has led historians who speak of a Russian Enlightenment to observe that epoch in the reign of Catherine the Great (1763–1796). Given the claims I have advanced here, however, it appears that such a conception misreads both Peter and the Enlightenment. Peter had more of a master plan than many have given him credit for, as the integration of the Academy of Sciences into the educational and manners reforms indicates. And it was by no means characteristic of Enlightenment rulers to try to apply 'Enlightenment ideas' – usually the armchair recommendations of French *philosophes* – to practical situations. Rather, Peter and his associates were exemplary Enlightenment figures in their selective appropriation and application of elements of Western European thought to advance certain ends. This view of 'Enlightenment' as opportunistic appropriation provides an alternative understanding of Peter the Great as an Enlightenment ruler that avoids the pat classifications of Soviet Marxists. ...

This view of Peter as a specific kind of Enlightenment ruler is accurate, but only partially so. He was without doubt a man interested in gaining practical advantages for his country: that much cannot be disputed. But, as noted earlier, there has been significant debate as to whether Peter's reforms were part of a master scheme or were merely *ad hoc* responses to pressing needs. It is difficult to find a concrete master plan in his reforms – the network of activities is far too heterogeneous. But this does not mean that there were not elements within a set of chaotic reforms that formed part of a 'minor plan', a plan with something beyond pragmatism as its goal. The Academy of Sciences was one such 'minor plan'. It involved the integration of several levels of reform in order to give the newly formed Russian elite a code of conduct and an etiquette regime in keeping with the Western and Central European pattern of nationhood being adopted by the ruling classes. At moments like this within his tumultuous reign, one sees the nature of Peter the Great as an Enlightenment transformer on a par with Leibniz, waiting to grind his cultural crops with his newly built mill.

Chapter Fourteen Establishing Science in Eighteenth-Century Europe

14.1 C. Meinel, *Chemistry's rise in status**

1 The 'Chemical Revolution'

During the Age of Enlightenment, chemistry broke with its humble role as an auxiliary adjunct science to medicine, to attain the position of a well-respected and even fashionable science, represented on the faculties of nearly all universities. The dynamics involved in this process have always attracted the attention of historians of science. . . . For not only did the entire social structure of the scientific community alter itself, but also the position of chemistry within the hierarchy of both the sciences and society was changed. Chemists began to see themselves as representatives of an independent academic discipline, experiencing the necessary acknowledgement from outside their field and forming regional or vernacular communities, enabling them to develop the common forms of scientific communication still in use today. Chemistry can lay claim to the fact that it produced the first specialized journal in a scientific field, published in 1778, and the first international scientists' gathering ever held, in 1786. [. . .]

2 The status of the field

The origins of chemistry as an academic discipline lie in the seventeenth century. The Paracelsians had introduced the therapeutic application of chemical medicines, thereby founding the field of chemiatrics. . . . Two

* C. Meinel, '". . . to make chemistry more applicable and generally beneficial": the transition in scientific perspective in eighteenth century chemistry', *Angewandte Chemie International Edition, English*, 23 (1984), pp. 339–47.

generations later this discipline was already to be found at numerous universities, functioning as an auxiliary adjunct to medicine.

Initially, of course, chemistry had no easy position. The representatives of academic science greatly mistrusted the air of alchemical obscurity surrounding it. More important, with chemistry a whole new type of science had penetrated the halls of the traditional educational institution, the University. For chemistry's real place was not at the rostrum, but in the laboratory, where, although no research and experimentation in the modern-day sense was conducted, the work was performed manually with a practical and purposeful intent. [...]

This situation was not to change very quickly, for the low status of chemistry within the academic hierarchy was institutionally and structurally defined. Professorships exclusively devoted to one particular subject were unknown; instead, each faculty maintained its own specific hierarchy, where advancement was largely based on seniority.... Within the faculties of medicine, this almost always meant that the medical professor of lowest rank lectured on chemistry as well as on pharmacy, botany, or anatomy. As soon as the opportunity arose, this professor then naturally advanced to the next highest position....

The chemists of the Enlightenment, therefore, did everything in their power to weaken the prejudice that their discipline was simply a practical craft, a kind of 'ars mechanica', without scientific character. They most certainly had recognized that their own dilemma was not so much a result of inner disciplinary deficiencies as due to the institutional conditions peculiar to their field of study. In this way a special literary tradition of doctrinal publications developed, the goal of which was to solicit support for chemistry as a distinct science in its own right and to make a broader public conscious of its true value.

3 *Pure and applied chemistry*

Under the circumstances described here, [new] doctrinal guidelines were formed for the further development of the discipline enabling chemistry to overcome its formalistic division into a theoretical science, on the one hand, and a practical handicraft, on the other...; this was the distinction between 'pure' and 'applied' chemistry, which exists to this day. ... This represented more than a change in terminology, however, and, more accurately, it originated in the desire to formulate a new scientific concept of chemistry in order to endow its level of knowledge and its capabilities with a new purpose. ... In the case of pure chemistry, this goal was to delve into the causes and apparent laws governing natural phenomena related to matter; applied chemistry was to exploit these insights for the benefit of mankind by solving concrete everyday problems. The fact that both pure and applied chemistry were to closely ally theoretical reason with experimental practice was seen as an obvious prerequisite.

Fortunately, it is possible to pinpoint the circumstances which led to this new concept of chemistry more precisely. In 1749, the first chair of chemistry was set up at the University of Uppsala. Up to this time, lectures in chemistry had not existed in Swedish universities, despite the fact that this country, rich in mineral resources, exhibited a pragmatically oriented mineral-chemical and metallurgical tradition. ... The government commission responsible for the institution of the new chair in chemistry had ordained the integration of the chair into the philosophical faculty, where students of the mining profession, as well as of economics and administrative science, obtained their general education. ... The position was offered to Johan Gottschalk Wallerius (1709–1785), who had maintained a private chemical and metallurgical laboratory at the University of Uppsala from 1738. He was well-known, even outside Sweden, as the author of a handbook on mineralogy. In July 1750, Wallerius accepted his new position as the first professor of chemistry in Sweden. [...]

The decisive factor which led to the acceptance and propagation of 'pure and applied' chemistry was the general chemistry textbook by Wallerius ('Chemia Physica', Stockholm, 1759), which he first published in Swedish and then personally translated into Latin, the language of academic Europe (Stockholm, 1760). Shortly thereafter, translations into German also appeared. This work played an important role in the history of chemistry textbooks in that it gave birth to a whole new generation of textbooks, written in the vernacular, the express purpose of which was to represent chemistry as an integral part of the physical natural sciences without reducing it to the level of medical or pharmaceutical interests. Its title, 'Chemia Physica', was intended to express just this general scientific concern. In this textbook Wallerius took the opportunity to develop his scientific concept of chemistry even further, for now he had realized that by consistently classifying the field in accordance with its objective and effect two things would occur: not only would the old separation of theory and practice lose weight, but so would the tiresome dispute over the definition of chemistry as a science or as an art. Instead, he regarded all subdivisions of applied chemistry as independent self-contained sciences which comprised both theoretical and practical aspects. [...]

4 A concept of science in the Enlightenment

In principle, the idea had been around for some time and it was Wallerius who merely succeeded in pinpointing and naming this new orientation of chemistry, which had reached a turning point around the middle of the eighteenth century. For almost simultaneously, it was possible to find Mikhail Lomonosov (1711–1765) making very related deliberations in Petersburg, in addition to the circle of authors of the French Encyclopédie. All of these are related by their common attempt to overcome the earlier contemplative ideal of science with its overrating of the theoretical, in order to replace it with a new bourgeois concept of science, which in turn contained the idea of progress and of actively molding the world. This new evaluation of chemistry and the higher status of its range of application was thus an inherent part of the purpose of the Enlightenment in putting rationality and the individual pursuit of happiness into social practice. The reason that this process began earlier and more noticeably in chemistry than in the other natural sciences lies in the fact that applied chemistry had already achieved undisputed successes; and due to its level of development, further useful discoveries could be expected.

Chemistry's understanding of its own role and its status was influenced more and more by utilitarian thought in the eighteenth century. Hardly an author wrote a new work without expressly informing the public of its far-reaching and direct benefits. . . . It would be premature to dismiss this only as rhetoric. Of course, it was also a question of safeguarding social and institutional support for their discipline, as well as obtaining new professorships, finances for a laboratory, or funds for an expensive experimental lecture. At the same time, it expressed chemistry's new understanding of its own function. The most respected representatives of the discipline made utilitarianism their own deep concern and stepped forth with treatises on the properties and preparation of foods and industrial goods, on questions of heating and lighting, dyeing and stain removal.

Of course, the content of such popular treatises promising to be of directly beneficial use did not always fulfil what the flowery titles and prefaces foretold. Just the same, they must have had a considerable effect. By means of their thematic variety and high circulation, they reached an extraordinarily wide public from gentleman farmers and manufacturers to ministerial officials. Thus the contribution this kind of scientific literature made to popularizing chemistry and to the public's acceptance of its rational and utilitarian image cannot be overvalued, even if the historiography of chemistry is all too partial to the development of theories and cognitive progress, disapproving of this kind of commercial literature.

5 *Chemistry and the economy*

In the German tradition, chemistry's preoccupation with domestic and political economy as well as its leading role in stressing its economic benefit is especially marked. The reason for this is to be found in the discipline's very close ties to cameralism, the German variant of mercantilism, related to the administrative needs of small territories. The goal of the cameralists was the welfare of the state and the 'best for all', to whose end mercantile thought and the national quest for power should strive. Their programme of economic reform envisioned increasing the working population as well as the government's intake of revenue by limiting imports and deliberately encouraging trade. By means of accessibility to and better exploitation of domestic raw materials and mineral resources, it was believed feasible to economically strengthen the country, if possible to the point of self-sufficiency. The initial cameralistic encouragement of industry, primarily due to government initiatives in the production of textiles, glass, and ceramics, achieved evident successes. As a result, new problems arose in the production and raw materials sectors, presenting a challenge to chemistry's professional competence.

Johann Joachim Becher (1635–1682), a very versatile man, known to chemists as the father of the phlogiston theory, was also one of the founders of cameralism and at a time when his chemical and alchemical works were almost forgotten, his contributions to cameralistics were still the accepted standard. In 1676 Becher founded his 'Kunst- und Werkhaus' in Vienna, with the support of emperor Leopold I, who was greatly interested in alchemy. It was a technical teaching and research institute meant to enliven Austrian manufacturing methods. Obviously, a chemical laboratory, glassworks, and a metallurgical laboratory were included in its endowment, for encouragement of trade and technological development were accepted as the central mandate of modern government. Around the year 1700, Georg Ernst Stahl (1660–1734), the great theorist and respected teacher of a whole generation of German chemists, had expanded on Becher's hypothesis of 'fatty earth' to form the phlogiston theory and was thus able to account for a great number of chemical reactions. He had also adopted Becher's doctrine of economically oriented chemical practice. [...]

Last but not least, the University of Halle was also founded in this period (1694), to develop a short time later into the leading Prussian university of its time. That same year, Stahl, who came from Jena, was made Professor secundarius on the faculty of medicine in the ambitious, newly founded university. He distinguished himself there with treatises on chemical metallurgy, assaying, dyeing techniques and the extraction of saltpeter. His chemical-technical masterpiece 'Zymotechnia Fundamentalis' (Halle, 1697) laid the foundation for the chemical interpretation of such processes as the brewing of beer, alcholic fermentation, and the preparation of vinegar. In the year of Stahl's death, 1734, it was recommended by the translator of the German edition as a manual, with the aid of which a clever statesman could save millions in imports by improving domestic production. [...]

Later, it was none other than the Stahlians who carried on this impulse, freeing chemistry from its subordinate role as an auxiliary adjunct to medicine and attempting to integrate it into the economic-cameralistic doctrine of the modern state. The treasury's vital areas of interest made the relationship between economy, production, and science a most important objective of enlightened absolutist thought: mines and foundries, saltworks and porcelain manufactures, mint and glassworks, and finally the production of saltpeter for military purposes presented tasks enough for chemistry. Even the early cameralistics (*Kameralwissenschaft*), which originated from the domestic and rural economy, eventually recognized how much it could profit from chemically founded knowledge of substances and processes.

Johann Gottlieb Eckhardt's 'Vollständige Experimental-Ökonomie' (Jena, 1754) was one of the first works from the tradition of husbandry to apply scientific principles to farming, thus beginning a new epoch in agricultural science which was then to result in a still rather primitive type of agricultural chemistry. [...]

The theoreticians and reformers of cameralistics in the eighteenth century consciously viewed knowledge of chemical science as a basic prerequisite of their discipline, demanding the establishment of independent teaching positions in chemistry in order to instruct government administration and department of finance employees. In Göttingen, then center of cameralistics in Germany, Johann Heinrich Gottlob von Justi (1717–1771), its most important systematist, held lectures on economy, polity and administration, but also on chemistry and mineralogy. In his 'Staatswirthschaft', published in 1755, the classic textbook in this field, he demanded that an independent economics faculty be established, in which chemistry, natural history, mechanics, and politics should be instructed on a very practical level with emphasis placed on their social benefits. [...]

6 *The emerging academic discipline*

Thus, chemistry had conquered university territory, where it neither had to assert itself against the overpowering tradition of the medical faculties nor to defend itself against the charge of being only a handicraft. Instead, it could test its knowledge and capacities as a science in accordance with the interests of the respective governments. 'To make the gifts of nature easier to use for human benefit', was the way Johann Georg Menn (1730–1781) put chemistry's real objective. 'Is the real purpose of chemistry not clearly aligned with the most important intention of the State as a whole to work for the common good? Is it not also in agreement with the judgement of those who want nothing more than to see that those sciences prosper which are advantageous to each and every part of the state up to and including the tradesmen?'[1] [...]

The persuasive power and success of the new doctrine can be explained by the fact that the capacities of eighteenth century chemistry with regard to knowledge of substances, theoretical interpretation, quality of apparatus and the technical ability to realize its objectives was a great deal better adapted to the new challenges posed by the field of metallurgy, glass and ceramics manufacture, textile processing and dyeing; better, that is, than the much more complex questions in the medical and pharmaceutical traditions, as they had been posed by Boerhaave's corpuscularian physiology and by the analysis and pharmacology of herbal substances, or by the chemo-cosmological doctrine of Paracelsism. Chemical theory was able to

---

1 J. G. Menn, *Rede von der Nothwendigkeit der Chemie* (Köln: Universitätsdrukkerei, 1777).

and was also permitted to prove itself, as this practical aspect now met with social approval and enjoyed official support. [...]

Primarily institutional reasons therefore determined that the earliest efforts of the discipline, as 'chemia physica' and as a part of general natural science, to break away from the medical faculty and become part of the philosophical faculty seldom met with lasting success. For, along with this transference, the right to examine medical candidates was almost always forfeited, resulting in the loss of a large number of the potential and paying students at lectures. For example, Johann Friedrich Gmelin (1748–1804) accepted a full professorship in chemistry on the philosophical faculty in Göttingen in 1775, which was associated with extraordinary membership of the medical faculty. Three years later, however, he was forced to give up this position to accept a professorship on the medical faculty, despite the fact that his main scientific interests still lay in applied and industrial chemistry. [...]

In surveying the development of chemistry in the eighteenth century, it can be noted that neither a mere improvement in external conditions nor a revolution in Kuhn's sense had enkindled the far-reaching process of transformation enabling the field to liberate itself. In chemistry, the developing scientific community centered its attention not so much on a new and binding doctrine or theory as on a scientific orientation complex composed of rational argumentation and social action, which gave the discipline's level of knowledge and capacities a new direction; this provided chemistry with a more up-to-date form of institutionalization strategy and made a re-evaluation of its cognitive content possible. This shift in orientation equally influenced its objects of study and methods used, its formulation of objectives and self-understanding, its social role, and its institutional integration. At its turning point stood the concept of a 'chemia pura et applicata', which set the stage for the autonomous development of the discipline about to take place. [...]

When Herman Boerhaave (1668–1738) accepted the chair of chemistry at the University of Leyden in 1718, he complained anxiously that chemistry was regarded as 'uncouth, repulsive and tiring, excluded from the community of the learned, unknown or mistrusted by scholars; it was supposed to stink of fire, smoke, ash and rubbish, having hardly anything appealing to offer'.[2] Three generations later Johann Friedrich Gmelin (1748–1804), professor of chemistry at Göttingen, a supporter of the phlogiston theory and one of the first historians of his field, could already maintain, full of pride, that chemistry was 'the idol before which all peoples, all classes, princes and subjects, clergy and secular, learned and unlearned, of high and low rank, kneel down; the favorite science of the great, the practice of which promised mountains of gold, quick reinstatement of ruined finances and of

---

2 H. Boerhaave, *Sermo academicus de Chemia suos errores expurgante* (Leyden: P. v. d. Aa, 1718), p. 2.

ruined health, . . . the retreat of the wise searching for light and enlightenment; the most important aid to the researcher of nature, imparting knowledge to him where other sciences fail; the key to many of the secrets of nature; the chosen guiding light in the labyrinth of countless industries which nurture, bless and enrich people and states; the rational basis for the existence of foundries, of many factories, arts and handicrafts . . . .'[3]

Hardly eighty years had passed between these two assertions. During these years that far-reaching process of new orientation took place in chemistry, which first created the conditions necessary for its subsequent development as an autonomous academic discipline and for that change of paradigms, which we call the Chemical Revolution.

14.2 A. M. Ospovat, *The origins of Werner's geological theory**

I certainly do not contend that regional influences were of no importance to Werner's theory. But I do not believe that the geology of Saxony inspired the theory, or that Werner's observations in that region played any great part in the development of the theory. Of course, there was interaction between the theory and the observations – the observations supported the theory, and the theory influenced the observations – but I believe that Werner, like other scientists, had formulated the basic tenets of his theory before he made extensive observations. [. . .]

A question of time

His first important work, *Von den Aeusserlichen Kennzeichen der Fossilien*, gives no indication of belief in or even interest in any geological theory at all. . . . This, of course, does not mean that he had never given the matter any thought or even that he did not already incline toward one theory or another. It may mean simply that he did not think that geological theory had any place in a description of the external characters of minerals. His own statements, however, imply that by the time he returned to Freiberg in 1775 he was already inclined toward neptunism, or at least opposed to vulcanism. In a note to the section of the *Kurze Klassifikation* which deals with basalt, he wrote:

> When I returned to Freiberg in 1775 I found the system of the vulcanists, and in it among other things the volcanic origin of basalt, generally accepted. The novelty and interesting features of this theory along with

---

3 J. F. Gmelin, *Geschichte der Chemie seit dem Wiederaufleben der Wissenschaften bis an das Ende des 18. Jahrhunderts*, vol. 1 (Göttingen: Rosenbusch, 1797), p. 2.

* A. M. Ospovat, 'The importance of regional geology in the geological theories of Abraham Gottlob Werner: a contrary opinion', *Annals of Science*, 37 (1980), pp. 433–40.

> the superior art of persuasion of its defenders and, to a certain extent, the persuasiveness or glamor of the matter itself soon procured for it an unusual number of adherents. If from the very beginning they seemed paradoxical to me, I had at first too much respect for the reputations of most of the mineralogists who adhered to the theory to at once declare myself against it.[1]

This passage, it seems to me, clearly implies that at the time, Werner was already opposed to the theories of vulcanists, including their theory of the origin of basalt. [. . .]

Some questions on regional influences in general

As a preliminary to a discussion of Werner's reasons for becoming a neptunist, a few remarks on the whole idea of the influence of regional geology on Werner and his contemporaries seem in order. It has often been stated that errors in Werner's theory were the result of his lack of travel and the resultant limitation in his field of observation. If this were so, a logical question would be 'If Werner had traveled in volcanic countries, would [he] have been a vulcanist?' It seems to me, however, that it is more pertinent to ask: 'If Werner had been a vulcanist, would he have seen what he saw [in Saxony]? Would he not have detected some sign of volcanic activity after all?' Perhaps not. J. F. W. von Charpentier, whom Werner described as 'one of the most ardent vulcanists', agreed with Werner at least to the extent of saying that the basalts of Saxony did not appear to be of volcanic origin. Charpentier was familiar, as was Werner, with the work of von Born, Ferber, Sir Joseph Banks, Solander, Desmarest and Sir William Hamilton. He was particularly impressed with the writings and illustrations of Hamilton on the volcanoes of Italy, but he refused to accept Hamilton's assessment of the volcanic origin of the basalts of Italy until he could see them for himself, remarking that despite Hamilton's exact illustrations, he did not dare to make any comparison between the basalts of Saxony and those described by Hamilton.

Nevertheless, Charpentier remained essentially a vulcanist, as did J. C. W. Voigt, a fact which raises another question. If regional geology plays such an important role in the formation of universal theory, why did Voigt and Charpentier disagree so thoroughly with Werner? All three men lived and worked in the same area, Charpentier was Werner's colleague both at the Bergakademie Freiberg and in the Saxon mining service. Voigt

---

1 A. G. Werner, 'Kurze Klassifikation und Beschreibung der verschiedenen Gebirgsarten', *Abhandlungen der Böhmischen Gesellschaft der Wissenschaften*, 2 (1786), pp. 272–97 (p. 294).

was Werner's student. Yet their views on the theory of the earth were very different. It has been said that Werner's tenacious adherence to neptunism was due, at least in part, to the fact that he had never seen mountains of any significant relief. [...]

Just as Charpentier and Voigt, who lived and worked among the sedimentary rocks of Saxony and adjacent regions, remained vulcanists, many geologists who had observed extensive volcanic formations remained neptunists. Steno, who might almost be called the father of neptunism, had seen Vesuvius and studied the surrounding area; but he still concluded that volcanoes had played very little part in the formation of the earth's crust. In Werner's own time, there were Guettard, Dolomieu and Desmarest, all of whom had visited the Auvergne and other volcanic regions but all of whom remained essentially neptunistic in their over-all geology. Desmarest, who discovered the Auvergne, so to speak, who did more than any other man of his time to spread the idea that volcanic activity had been much more extensive than had previously been believed, and who came to the conclusion that all basalt is of igneous origin, nevertheless continued to regard water as the most important agent in the formation of the earth's crust and the role of heat as negligible. His theory of the cause of volcanoes was similar to Werner's.

A question of reasons

In previous writings, I have shown in some detail that Werner's motive was not religious. ... Werner's background was not traditionally religious... and, most important, ... nowhere in his published works or in his lectures did he refer to the biblical story of creation or try in any way to make his system conform to that account.

My own explanation of why Werner became a neptunist rather than a vulcanist is, quite simply, that, being familiar with the geological literature available in his day and the opinions of the foremost geologists of his time, he chose the theory which seemed to him most reasonable and best supported by the evidence presented by geologists who had lived and worked before him. When one considers the extent of volcanic activity known in Werner's day and the tools of observation and experimentation available at that time, the choice seems logical. [...]

[To] most of the inhabitants of the civilized world, the destructive and renovative actions of water and the phenomenon of sedimentation were easily observable, while the actions of volcanoes and the forces of internal heat were not. The universal ocean itself seemed not so implausible. How else could one explain the presence of marine fossils far inland and high in mountain ranges, in areas where no volcanic activity was discernible?

The greatest general objection to the idea of the universal ocean in Werner's day, and before his time, was that it was impossible to explain how, within the earth's history, enough water to have stood above the tops of even the highest mountains could have shrunk into the boundaries of our present oceans. Werner himself never tried very hard to formulate an explanation. He felt that his observations and those of others amply supported the previous existence of the ocean, and he always contended that the inability to explain a phenomenon should not lead to rejection of the evidence that it had existed. However, he did remark that a part of the difficulty lay in the failure of most observers to put the magnitude, or perhaps one could say more accurately the minuteness, of the unevenness of the earth's surface into proper perspective. He pointed out that, in proportion to the size of the globe, the unevennesses of the surface are small indeed. ... And when one adds to this fact the consideration of the almost incomprehensible length of time which we now believe to be the age of the earth, the gradual diminution of the universal ocean to its present bounds does not seem impossible after all.

If we assume, then, that Werner's theory had its origins in the study and evaluation of the theories of earlier geologists, we may also ask whether Werner was particularly influenced by any one of his predecessors. As far as I have been able to determine, there is no hard evidence on the subject. I have found nothing in either his published works, his manuscripts, or the notes of his students in which Werner acknowledged any particular debt to any particular geologist. The similarities between Werner's theory and Lehmann's have often been noted. [...]

It has not been so well noted, however, that the theories of both Lehmann and Werner bear marked similarities to those of Nicolaus Steno. ... An edition of the *Prodromus* was brought out in 1763. If there had been no interest in Steno's work, why would a new edition have been published? ...I do know that Werner owned copies of both the first edition of 1669 and the edition of 1763, and Steno's name is included in a list of prominent geologists which Werner compiled. [...]

Conclusion

Thus, it seems to me that regional geology influenced the origin and development of Werner's theory only in a supportive way. Already assuming the correctness of his theory, Werner found in the rocks of his native region much evidence to support it. This is not to say that the only support for his theory came from Saxony and the surrounding area. As a beloved and admired teacher and as a famous geologist and mineralogist, Werner received reports and specimens from all parts of the Western world, including the Americas. He had a large private mineral collection, and he was curator of the mineralogical collections of the Bergakademie Freiberg. Both his published and his unpublished writings contain remarks to the effect that the rocks of even the remotest corners of the earth are similar to those of Saxony. One might infer that these reports from the far corners of the earth gave rise to the theory of the universal ocean if one did not know that the theory was not original with Werner and that Werner's neptunism was well established before he became a beloved teacher or a famous geologist. As it is, it seems much more reasonable to assume that it was very easy for Werner to believe that the rocks of distant lands were similar to those of Saxony because, according to the theory of the universal ocean, they would necessarily have been.

I do not mean to say that Werner was not influenced at all by his observations. If he had not been, we would have to conclude that he had a completely a-scientific mind. In fact, he was a keen and diligent observer, and even those of his students who came ultimately to reject his theory often expressed their gratitude to him for having educated them so well in the art of observation. [. . .]

The geologist, unable to rely upon experiment, confronted with immense periods of time, and dealing with phenomena most of which took place long before the beginning of written history, must rely more upon deduction than upon observation.

14.3 Henry Lowood, *Scientific forestry management in Germany**

In the second half of the 18th century, few occupational groups rivaled government officials in their attention to numbers. Government officials employed in the duchies, kingdoms, and free cities of German-speaking Central Europe pored over the data on population, imports, and taxes that a growing fiscal apparatus produced in unprecedented volume. Those concerned with the prosperity of the prince and his subjects, from low-level tax assessors to ministers of state, developed an attachment to the quantitative spirit proportionate to the expansion of the state's economic agencies.

Reasons of state and forces of social change brought on the bureaucratization of the state financial apparatus in the 18th century. Rather than dulling their initiative, this bureaucratization created new opportunities for officials, professors, and instructors. In the Age of Enlightenment, the improvement of fiscal administration and resource management was seen as requiring a *science* of state finances, while the proliferation of economic facts and figures raised issues of numeracy and appropriate training for office-holders charged with applying the principles of this new science, which became known as the 'cameral sciences' in Germany. The term derived from the *Kammer* (chamber) in which the prince's advisors traditionally deliberated. The subject matter ranged from economics, finance, and *Polizei* to mining, agriculture, and trade.

---

* Henry Lowood, 'The calculating forester: quantification, cameral science, and the emergence of scientific forestry management in Germany', in T. Frängsmyr *et al.* (eds), *The Quantifying Spirit in the 18th Century* (Berkeley: University of California Press, 1990), pp. 315–32.

First introduced in Prussia at the universities of Halle and Frankfurt an der Oder in 1727, the *Kameral-* or *Staatswissenschaften* were firmly established in the university curriculum throughout Germany by the last third of the century. The call for professional training in cameral science and its gradual emancipation from the faculties of law led to the creation of new professorial chairs and schools for teaching a body of theory and techniques needed for the administration of the state and its domains. ... *Kameralwissenschaft*... subject[ed] a variety of economic, administrative, and social practices to rational or 'scientific' scrutiny.

Forest management was one aspect of state administration thus scrutinized, in order to fit 'scattered pieces of knowledge... into systems' and to transform 'all sorts of activities previously left to habit... into a science'. The glue that held these new systems together was economic rationalization. The forest displayed the size of the task of managing the resources from which the prince of the late 18th century ultimately derived his wealth. Discharging the task forged new links between administration and science. The result was quantification and rationalization as applied to both the description of nature and the regulation of economic practice. [...]

...The origins of rational forest management in the quantitative 'forest mathematics' of the last half of the 18th century constitute the subject of this chapter. It will demonstrate that the first advocates of forestry science quantified in spirit in order to bring profits in practice; in the process, they established a tradition of quantitative resource management.

Better management

As a substantial portion of the prince's domain, forests constituted one of the largest sectors of the state economy in central Europe. Other forested lands in Germany belonged to the cities and the landed nobility and provided indispensable products for the local and regional economies under their control. ... Before the age of coal, which would not begin in many parts of Germany until the middle of the nineteenth century, wood was king.

After the acute and widespread devastation and neglect that resulted from the Seven Years' War (1756–63), the state fixed its gaze on economic recovery. The specter of shortages of wood fuel caught the attention of a small group of conscientious foresters and enlightened bureaucrats, who saw evidence that the deterioration of the woodlands, reported here and there since the Middle Ages, had dramatically accelerated. In the Palatinate, for example, a survey of the forests carried out between 1767 and 1776 spoke of 'woods in places so ruined that... hardly a single bird can fly from tree to tree'. The state of Germany's forests reached its nadir just when rulers like Frederick the Great sought to encourage population growth and force the expansion of industry and trade, measures bound to increase the pressure of demand for wood and other forest products. The fear of impending crisis in the supply of wood lodged in the minds of government officials throughout the remainder of the century, and was periodically intensified by reports of rapidly rising prices.

Officials vigorously pursued economy in the use of wood. But redesigning fireplaces, door-frames, and spoons offered help only on a limited scale; to expand that scale would be a tedious undertaking. Better understanding of the nature of combustion and material properties of wood offered some hope for greater efficiency in wood burning, and scattered experimental reports on these matters of forest physics appeared before 1800. The alternative of expanding the wood *supply* promised larger gains. Here a bold innovation might succeed in increasing the amount of firewood and lumber available to an entire town, city, or region. Almost in proportion to the potential payoff, however, the complex problem of proper forest management exceeded the meager qualifications of the vast majority of foresters. As a rule their primary appointments as caretakers, game wardens, and masters of the hunt required neither practical nor theoretical training in forestry. In Prussia, for example – even under Frederick the Great – posts in the forest administration, which carried the revealing title of *Jäger* [huntsman], served as sinecures for military retirees. In the absence of qualified personnel, how could a new approach to forest management arise?

After the middle of the century, the establishment of private forestry schools and publication of books and even journals devoted to forestry began to raise expectations for the training and competence of future foresters and forestry officials. The last year of the Seven Years' War saw the foundation of the first forestry school . . . , the appearance of the first book to use 'forestry science' in its title . . . , and the first journal devoted exclusively to forestry . . . . One of the first points to settle was the very definition of forests. Traditional privileges and the continued use of the forest for such agricultural purposes as grazing or mast (windfall nuts) had long discouraged a conceptually precise demarcation of the forest. Beginning in the 1760s, however, better-trained officials, equipped with publications for the exchange of ideas, promoted the notion that the forest could be defined precisely and studied objectively.

The first writers on forestry science were led by men trained in the cameral sciences – financial officials and chief foresters who expected economic disaster if the condition of the forests continued its downward slide. As these officers of the local prince consolidated their control over state-managed economies throughout Germany, they attended to the forests in their jurisdiction. Where bureaucratization and centralization of political authority extended the official's sphere of action, as in Prussia, forestry science flourished. The year 1757 marked the appearance of the first of many books on forestry geared specifically to cameral officials: Wilhelm Gottfried von Moser's *Principles of forest-economy*. Like other cameral officials, the head forester came to his post after considerable study. Every cameralist learned about forest administration, a subject of acknowledged importance: 'First, because they are a considerable source of revenue for the state, and second, because they constitute a vital necessity for the sustenance of its citizens, without which these lands – especially in the north – would hardly be habitable.'[1] Cameralist writers such as Georg Ludwig Hartig placed the new forestry alongside the 'state sciences', since the two 'make up a complete whole'.

The new breed of officials trained in cameral science described the living forest quantitatively before subjecting it to economic reason. They brought to the task a familiarity with mathematics. Mathematics figured prominently among the required subjects, especially in the first year or two of coursework, in the university curriculum in the cameral sciences and also in special forestry schools. Published curricula and schedules of lectures consistently featured mathematics as a *Hilfswissenschaft* [auxiliary science], both for the work of the future government official and as exercise for his mind. [...]

Writers on forestry presented problems and applications of special techniques, not elementary mathematical instruction. Their goal was to demonstrate how the forester should proceed mathematically, not to produce a new mathematics. With the exception of solutions to a few obscure problems..., mathematical virtuosity was not necessary. ... [T]he 'practiced algebraist', to whom calculating the value of a forest was a trivial exercise, would not be the least bit interested in applying his art to it. ... [M]ost foresters, unencumbered by such mathematical sophistication, were likely to faint at the slightest scent of a mathematical problem. ... But one could turn trees into thalers by replacing the time-worn 'routines' of the old *Jäger* with *Forstwissenschaft* [science of forestry], it was generally agreed.

This approach was decidedly German. Reforms under Louis XIV had resulted in *plans de forêts* for state-owned forests and promoted the concept of dividing the forest into annual cutting areas. Jean-Baptiste Colbert's ambitious plan for improving France's forests in 1669 had prompted new statutes, administrative reorganizations, and inventories throughout the 18th century. But a scientific forest management did not take root in France until it was imported from Germany in the 1820s. English authorities, ignoring such expressions of concern as John Evelyn's *Sylva* (1664), did not even inventory the remaining forests until the founding of the Board of Agriculture in 1793. ... In Switzerland and Austria, government officials exerted control over a lesser proportion of the forests than did their counterparts in Prussia and Saxony. Moreover, the physiocratic doctrine fashionable in late 18th-century Vienna and Bern offered a rationale for avoiding the problem by selling off woodland and converting it to farmland.

---

1 J. H. G. von Justi, 'Von der Aufmerksamkeit eines Cameralisten auf die Waldungen und den Holzanbau', in *Gesammelte politische und Finanzschriften*, vol. 1 (Copenhagen and Leipzig: Rothen, 1761), pp. 439–64, on 439–40.

Doing the work

In central Germany, particularly in Hesse and Saxony, a few foresters had applied the same enthusiasm to managing the forest as to directing the hunt. These conscientious *holzgerechte Jäger* [huntsmen who understood wood] of the midcentury set annual cuttings according to easy rules based on areal divisions of the forest. After demarcating and measuring the acreage covered by the woods under their supervision, foresters estimated the number of years that the dominant types of trees should be allowed to grow between clearings or cuttings. They then partitioned the forest into a number of divisions equal to the number of years in this growth cycle, from which they proposed to derive equal annual yields, assuming that equal areas yield equal amounts of wood for harvest each year. This straightforward method worked reasonably well for relatively short growth periods typical of coppice farming and the periodic clearing of underwood. . . .

These methods may have sufficed for a minimally trained huntsman, but not for the fiscal or forest official imbued with *Wissenschaft* [science]. The crude assumptions underlying the traditional areal division of the forest proved wholly unsatisfactory for the cash crop of forestry – the long-lived high timber, or *Hochwald*; the older the trees, the greater the variation in the timber produced by each of the divisions of the forest. Furthermore, the irregular topography and uneven distribution of German woodlands confounded ocular estimation of area without the aid of instruments. Only in the 1780s did Johann Peter Kling, chief administrator of forests in the Palatinate and Bavaria under Elector Karl Theodor, systematize forest mensuration and cartography into instructions for making forest maps of unprecedented detail.

. . . The most meticulous forest management under these methods, while an improvement over neglect, fell short of the high principles of *Kameralwissenschaft*.

After midcentury, an approach to forest economy based on the mass or volume of wood gradually displaced area-based systems. The first prominent advocate of wood-mass as the quantitative basis for sound forestry emerged from the *holzgerechte Jäger*. Johann Gottlieb Beckmann, a forest inspector in Saxony, gave the forest priority over the hunt; his knowledge of forestry derived from experience, not education. Beckmann's deep concern for preserving the wood supply led him to construct a system of forest economy that rested on a practical technique for measuring the quantity of standing wood in the forest. Beckmann instructed his team of assistants, whom he supplied with birch nails of various colors, to walk side by side through the forest at intervals of a few yards. Each member of the formation fixed his gaze to the same side and noted every tree he passed. He made a quick estimate of the size category in which the tree fell and marked it with a nail of the appropriate color. At the end of the day, unused nails were counted and subtracted from the original supply to indicate the number of trees in each category. The forester and his assistants knew from experience the approximate yield of wood from trees in each size category; with multipliers thus assigned, the number of nails used could be converted through a simple calculation into the quantity of standing wood in the forest. ...

Within a few years, a group of mathematically adept foresters followed along the trail cleared by Beckmann. ... [T]he *Forstgeometer*, a surveyor hired to demarcate the borders of the forest, prepare maps, and carry out other prescribed tasks for a set fee ..., along with the army of marching assistants, gathered the data. *Forsttaxation*, or forestry assessment – a mix of calculation, analysis, and planning – fell to the chief forester and his superiors. Forest mathematicians like Oettelt and Vierenklee were moved by a new confidence in the power of mathematics to solve problems associated with the conversion of the forest into an equivalent quantity of wood mass. Assessment, the scientific component in *Forstwissenschaft*, required general principles and techniques based on them. Without them the unrelated numbers and observations reported by foresters and surveyors would overwhelm planners and administrators. Forestry science supplied the necessary organizing principle: 'evaluation, or the ascertaining of the mass of wood, which is to be found for a given place at a given time'.[2] Identifying wood mass as the crucial variable of forestry set the stage for quantitative forest management.

Counts to calculation

Theoretical computations of tree volume began to appear in the 1760s. In the first definitive work of scientific tree measurement (*Holzmesskunde*), Carl Christoph Oettelt's *Practial proof that mathematics performs indispensable services for forestry*, the problem of estimating the quantity of wood on a tree without felling it figured prominently. Oettelt was an experienced surveyor and had held the title of 'Forest-Geometer' in the civil service of Saxony-Gotha before taking over the forest department in Ilmenau, where he would later serve under Goethe. ... Oettelt invoked geometry: 'A tree is the same as a cone with a circular base.' With the appropriate formula for the volume of a cone, calculating the volume and mass of trees was not so troublesome.

Oettelt's treatment of wood mass as a mathematical quantity was a radical departure. ... As abstract, mathematics-based forestry gained sway during the 1780s and 1790s, compilations of tables based on controlled measurements replaced the older crude techniques described by Oettelt. [...]

In the German tradition, the mathematician's forest was populated not by the creations of undisciplined nature, but by the *Normalbaum* [standard

---

2 Johann Leonhard Späth, *Anleitung, die Mathematik und physikalische Chemie auf das Forstwesen und forstliche Camerale nützlich anzuwenden* (Nuremberg, 1797), p. 195.

tree]. Forest scientists planted, grew, and harvested this construct of tables, geometry, and measurements in their treatises and on it based their calculations of inventory, growth, and yield. Writers and instructors gave foresters in the field the tools for reckoning the dimensions of the standard tree. Most treatises contained instructions for averaging measurements made on a test plot, but foresters were happier to use the *Normalbaum*. Tables of numbers representing measurements and calculations, or *Erfahrungstabellen* [experience tables], provided data organized by classes of trees under specified conditions. A small number of variables governed the forester's choice of one or another of these tables. For example, the wood mass of the typical sixty-year-old pine on good soil was given as a function of its height and circumference. These tables, which appeared in every complete manual of rational forestry practice, generally did not bother with regional variation . . . .

By the end of the 18th century, German writers on forest management had worked out steps for determining, predicting, and controlling wood mass. Heinrich Cotta presented the clearest and most widely read exposition of these steps in his *Systematic instruction for the assessment of woods*, published in 1804. . . .

If the standing forest is capital and its yield is interest, the forester can complete the chain of conversions from wood to numbers to units of currency: an estimate for the worth of the forest can thus be used to predict income, calculate taxes, assess the worth of the forest, or determine damage to it resulting from a natural disaster. [. . .]

The *Forstwissenschaftler*, and particularly Cotta, championed use of 'experience tables'. Their use reinforced the notion of a forest filled with standard trees. The forester was to instruct his assistants in the use of these tables so that a mental picture of a tree encountered in a forest corresponded to an entry in the tables. With sufficient repetition, a good forester could make an instant association from the mental picture triggered by the tree to the value of the wood mass contained in the table. The next step was to generalize: every tree of the same height has the same mass (or volume). The standard forester was trained to find the standard tree. [. . .]

By 1800, the forest assessor trained in the cameral sciences specialized in theoretical principles, mathematical preliminaries, and the cumulation and analysis of data, a far cry from Beckmann with his colored nails and squad of assistants. An array of numbers stood for the quantity of wood in the forest. The forester or cameralist trained in forestry science felt no need to step off every acre with the exactness given to the test plot, the geometrical abstraction, or exact measurements of the volume of cordwood. Instead, he could sample and generalize. The work of the assessment and management of the forest thus required only standard trees and *Erfahrungstabellen*. As Cotta argued, the crucial quantities of his science were 'determined mathematically' from the 'premises' of forestry science, not through '*direct real measurement*'.[3] The scientific forester had abandoned Beckmann's empiricism in favor of 'sure mathematical deductions, experiments and experiences in the given and understood units of measure'. Under the banner of *Wissenschaft*, the new breed of qualified forester breathed the quantitative spirit into administrative practice.

3 Heinrich Cotta, *Systematische Anleitung zur Taxation der Waldungen*, 2 vols (Berlin, 1804), pp. 5–7, 117.

Chapter Fifteen
The Chemical Revolution

15.1 C. Perrin, *Lavoisier's chemical revolution**

Lavoisier's ambitious project crystallized into a clearer pattern during the winter of 1772–1773 and was committed to paper in the research memorandum that opened a new laboratory notebook on 20 February 1773. In the interval he had begun (in customary fashion) to collect everything that had been published on the elastic fluid coaxed from bodies by fermentation, distillation, and various reactions. He referred specifically to the experiments of Hales, Black, MacBride, Jacquin, Priestley, de Smeth, and Heinrich J. N. Crantz. Each of these contributions Lavoisier regarded as separate portions of a great chain that still remained to be forged into a continuity – a task that he took upon himself and approached from an original point of view. 'An important point that most authors have neglected is to draw attention to the origin of that air that is found in a large number of substances. ... This way of viewing my goal has made me feel the necessity first of repeating and then multiplying those experiments that absorb air, in order that, knowing the origin of this substance, I might follow its effects in all the different combinations.' The operations he identified as those by which air might be fixed were vegetation, animal respiration, combustion, calcination, and certain chemical combinations; with these he proposed to begin his investigations. By February 1773 Lavoisier not only had demonstrated the material participation of air in combustion and solved the persistent puzzle of the weight gained by metals in calcination but had embarked upon an investigative program that he believed would bring about a revolution in physics and chemistry. His new vision went beyond the aims of his principal inspirer, Hales, and beyond the burning-glass investigations he

* C. Perrin, 'Research traditions, Lavoisier, and the chemical revolution', *Osiris*, 4 (1988), pp. 74–8.

had himself conceived the previous summer – which were but a modest extension of Hales's program. His objective was to trace the progress of air as it entered into chemical combination and was transferred from one combination to another. The means that would allow him to do this was his balance-sheet approach – he would follow the meanderings of fixed air by measuring weight and volume changes. Finally, the key that opened up this entire new corridor of research was his discovery of the quantifiable absorption of fixed air in common operations of combustion and calcination.

As Lavoisier progressed in his reflections during this period, he experienced growing doubts about phlogiston until he arrived, by the late winter of 1773, at the verge of total rejection of the concept. Two factors shaped his dismissal of phlogiston at that time. In the first place, experiments with the large burning lens had persuaded him that calxes of metals like iron or mercury could be reduced *without addition of phlogiston* (i.e., without charcoal); but the necessity of phlogiston for metallic reductions had been a linchpin of Stahl's theory. In addition, Lavoisier's own guiding theory on the nature of air (elaborated in an essay he composed in April) led him to infer that air – which must release its combined matter of fire upon entering chemical combination – was the probable source of fire manifested in combustion. What role, then, was left for phlogiston? Lavoisier admitted, in the first draft of a paper on calcination that he planned to read at the Academy's Easter public meeting: 'I have even come to the point of doubting whether what Stahl calls phlogiston exists, at least in the sense he gives to the word. It seems to me that in all cases one can substitute for it the name of matter of fire, of light and of heat.' But in the revised version that he did read at Easter, he excised or moderated comments on Stahl's theory, protesting that 'my experiments are not yet complete enough to dare enter into the lists with this celebrated chemist'. Still he ventured his opinion that deeper study of the phenomena of fixed air would lead 'to a period of almost complete revolution' – his first *public* reference to a revolution in chemistry.

Had Lavoisier somehow disappeared on the eve of his Easter address to the Academy, his work would have had little impact; for the transformation of chemistry that he glimpsed was only just beginning. His striking experimental results and his radical reflections remained private. Even from that moment in the fall of 1772, when he first committed to paper his vision of a revolution in chemistry, Lavoisier sensed that it would require time and effort to support his claims and win over his colleagues. Nevertheless, he probably underestimated the task ahead of him. If we compare the system of chemistry that he ultimately published in 1789 with his essays in the spring of 1773, there are remarkable differences. The transformation of chemistry was no mere social process of persuading the community of an intellectual *fait accompli* but involved further conceptual and epistemological departures.

The new system of chemistry was generated by dynamic interaction of Lavoisier's investigative program with competing programs pursued by his contemporaries.

To capture the nature and significance of the Chemical Revolution, two approaches have already yielded insights and hold the promise of further harvests to come. The first might be described as a genetic analysis that traces the development of the Lavoisian system from its roots in eighteenth-century traditions to its mature expression in the *Traité élémentaire de chimie*. Such an analysis has both social and intellectual dimensions, not only because Lavoisier's conceptual framework was modified through socially mediated interaction with the ideas and discoveries of his contemporaries, but because completion of the transformation required communal acceptance of the new system. The second approach is a conceptual analysis that dissects the mature Lavoisian system, comparing it with the complex of assumptions and theories that prevailed during the middle decades of the century. The scope of this paper does not allow me to develop either analysis here in any detail; I can only indicate some lines that each might follow.

A genetic approach reminds us, in the first place, that the framework of the *Traité* did not spring fully developed from Lavoisier's head, but was laboriously articulated over many years. In a sense, the remainder of his career can be regarded as the fulfillment of the program sketched in the February memorandum. Over the next twenty years, until his life came to a premature and tragic end, Lavoisier persevered in his examination of each of the themes identified there. Although a particular line of inquiry might be temporarily stalled or set aside, he repeatedly returned to his central interests in calcination, combustion, respiration, and vegetation – as well as to the earlier theme of heat and the formation of vapors. During the mid 1770s he was preoccupied with establishing his claims on calcination and determining the principle in the atmosphere that metals absorb when they are calcined. His interest in respiration, inspired by Joseph Priestley's work, remained a subsidiary theme, as did his curiosity about fermentation. In the late 1770s his emphasis shifted to combustion and the related formation of acids. During the same period he engaged the collaboration of Pierre-Simon Laplace in an experimental study of vaporization and of Jean-Baptiste-Michel Bucquet in a systematic repetition of standard chemical preparations (in order to resurrect the neglected part played by air). The early 1780s were devoted to elaboration of the caloric theory of heat and publication of his joint researches with Bucquet. Then, in 1783, the synthesis of water changed the thrust of Lavoisier's investigations, for knowledge of the composition of water (and of carbonic acid) opened new perspectives on fermentation, plant nutrition, and organic analysis in general. It was toward those themes – and respiration – that Lavoisier's laboratory researches shifted as the decade wore on. During the same period the active campaign against phlogiston, the creation of the new nomenclature, and the elaboration of his views for the *Traité* consumed much of Lavoisier's time and energy.

Viewed in this light Lavoisier's researches possessed continuity and internal momentum. Under closer inspection, however, the transformation of chemistry was no mere elaboration of the framework conceived in 1772–1773. Lavoisier's own further experiments and reflections, as well as the findings and criticisms communicated by his peers, worked back on the assumptions and concepts that guided him. In the process his conceptual framework was transformed and his guiding assumptions were changed or given explicit articulation. As examples, consider the fate of the key concepts 'matter of fire' and 'element'. In the 1760s Lavoisier favored the concept of an 'igneous fluid' that permeated interplanetary space and served as a medium for propagation of light but was also capable of becoming fixed in chemical combination. (In the latter form he tended to identify it with the chemists' phlogiston.) In the spring of 1773 he spoke somewhat ambiguously of 'the matter of fire, of light, and of heat', distinguishing it from phlogiston, whose existence he had come to doubt. By the early 1780s – when his manipulative techniques and understanding of heat had been further honed by collaboration with Laplace and familiarity with the British work – he began to treat heat as a matter apart, variously denoted as 'principe échauffant', 'thermogène', and finally 'calorique'. Ultimately, in the *Traité*, the igneous fluid was replaced by *two* imponderable fluids, caloric and a provisional matter of light; Lavoisier no longer spoke of a matter of fire per se. Similarly, Lavoisier began with the conventional notion of the four elements. But in the early 1780s, when he sketched an outline for an exposition of his views, he drew up a tentative list of simple substances that would replace the four elements. He was influenced in part by the growing practice in mineralogy of defining composition of substances by their immediate analysis into isolable components (rather than their hypothetical elements), in part by Condillac's philosophy, which encouraged an analytic and empirical approach to language and the objects that words represent. But the intrinsic motivation to replace the four elements came from Lavoisier's own experience of the disintegration of the traditional Stahlian hierarchy of composition. He had personally shown that metals and combustibles were relatively simpler than the calxes and acids of which they were allegedly composed. Elemental air had given way to a host of aeriform fluids or gases whose composition was then a moot point. Chemists and mineralogists spoke of four earths, rather than one. Fire was separating into distinct matters of heat and of light. Only the element water remained intact (though not for long). In place of that collapsing system Lavoisier substituted a provisional list of simple substances, pragmatically defined as those isolable components of bodies that had resisted all attempts at further analysis.

A genetic approach offers the possibility of tracing the descent of the diverse concepts and theories that make up the system of the *Traité*. A conceptual approach compares the anatomy of the articulated system of chemistry in the *Traité* with its mid-century predecessors, such as the chemistry elaborated by Venel for the *Encyclopedia*. That approach peels back successive layers of the Lavoisian system to reveal its components and to identify breaks with tradition. On the surface, it is evident that phlogiston is gone; effects previously attributed to it are now explained by some combination of Lavoisier's caloric and oxygen – the two concepts that dominate the theoretical discussion in Part I of the *Traité*. This is the most palpable aspect of the shift and the one that dominated public controversy. However, the implications of the new system run deeper than the replacement of phlogiston by oxygen (or caloric, or both). On a second level, Lavoisier succeeded in his quest to establish the chemical role of 'air'. Indeed, one might say that the joint efforts of Lavoisier and the pneumaticists added a new aerial dimension to chemistry, so that the atmosphere came to be regarded as a reservoir of key participants in natural and laboratory processes. The states of matter – particularly the gaseous state – constitute a major theme of the theoretical part of the *Traité*. On a third level, the inclusion of air overturned traditional views of chemical composition; in the end, the four elements gave way to the set of operationally simple substances. Bodies previously regarded as simple were now seen as compounded, and vice versa. The new chemical hierarchy was expressed in the new chemical language, displayed in tables of binary combination in Part II of the *Traité*. On a fourth level, the Lavoisian system had an inadvertent but major consequence. Although Lavoisier continued to speak of chemical principles (calling his oxygen the principle of acidity and his caloric the principle of heat), his attack on phlogiston undermined the very foundations of that system. Almost imperceptibly, the Stahlian analysis of properties to infer the presence of generic principles was discredited and replaced by a view of chemical reaction as a combinatorial process of simple substances, whose properties bore no necessary relation to those of the compound. Finally, on yet another level, the episode had methodological consequences. Lavoisier's success in transforming chemistry owed a great deal to his adaptation of physical methods to chemical problems. Part III of the *Traité* displayed (in addition to the traditional operations of chemistry) the instruments and quantitative techniques that had become the hallmark of Lavoisier's approach and the key to his success.

The impact overall was dramatic: phlogiston banished, the four elements abandoned, the doctrine of principles undermined, and the chemist's reliance upon qualitative methods repudiated. In this light, introduction of the new chemistry appears as a discontinuity or break with tradition – a transformation of major proportions. It was accompanied by a revolution in chemical pedagogy, as text-books and lecture notes were rewritten to present the new concepts, language, and methods.

15.2 H. G. Schneider, *Nationalism in eighteenth-century German chemistry**

The national opinion of German chemists of the Eighteenth Century is distinguished by the fact that it was based not on a few opinions voiced more or less haphazardly, but rather on a model example of organized journalistic power. This power lay first and foremost in the hands of one man, the Controller of Mines and Professor of Chemistry at Helmstädt, Lorenz Crell (1744–1816). Crell used the founding of influential professional journals for the purpose of promoting and furthering national identity and uniformity on chemical matters, and to transform chemical research into a patriotic concern. Through his chemical journals, he not only made a decisive contribution to the formation of a scientific community of German-speaking chemists, but simultaneously steered this nascent scientific community towards the nationalistic creed and patriotic undertone which later influenced the fate of German chemistry to a degree unparalleled in any other country where similar motives may have applied.

Lorenz Crell defined the task of German chemistry as not only a patriotic duty, but one superelevated to a religious level. The basis for this creed was the idea that Germany was the 'Fatherland of Chemistry'. This claim, which today has rather delicate overtones, seemed a matter of course to Crell. He wrote: 'Nature itself seems to have intended us to be chemists, by embuing us with national characteristics different from those of other nations.' Crell linked this claim to spiritual leadership in the field of chemistry with the exhortation that German chemists should show themselves to be worthy of their duty. 'If our nation is to remain the acknowledged teacher of other nations, it must continue along the laborious path on which it has won fame and advantage.' Such were the manifesto-like words with which Crell began his career as editor of the 'Chemisches Journal'.

It would be misleading and unhistorical to doubt the integrity of Lorenz Crell's character, to misjudge his pure, idealistic intentions or to interpret them in the light of later events. Crell was without doubt a deeply honest, reasonable and unbiased scholar, highly educated, fluent in several languages and respected throughout Europe. Perhaps it is precisely this which lies at the root of his seductive power and enormous influence. As a chemist, Crell is not remembered for any particular achievements. That is neither to his discredit, nor does it distinguish him; rather he is one of the overwhelming majority of professors of whom exactly the same is true. On the other hand, as a journalistic pioneer, as the motivator and organiser of a nationalistic attitude among German

---

* H. G. Schneider, 'Fatherland of chemistry: early nationalistic currents in late 18th century German chemistry', *Ambix*, 36 (1989), pp. 14–19.

chemists, Crell is one of the foremost figures in late eighteenth-century scientific history.

...Between 1783 and 1785, Lavoisier's minor works from the seventies, which heralded a radical upheaval, had appeared in German, published by the Professor of Chemistry at Greifswald University, Christian Ehrenfried Weigel (1748–1831). Moreover, in 1783, Henry Cavendish (1731–1810) in Great Britain had carried out the experiment that led to the discovery of the compound nature of water. Lavoisier had immediately incorporated this discovery into his new chemical theory as a key principle, so that the basic structure of a new chemistry, no longer dependent on phlogiston, was becoming visible.

How did Lorenz Crell and other leading German chemists react to this situation, where epoch-making innovations seemed to threaten the predominance of phlogiston, which was German in origin? Their reaction was to take begrudging note and to reject the new theory out of hand. This attitude eventually led to a high degree of self-isolation. The flow of information from abroad was broken off, due to the pressure of conformity and an instinctive resistance; the climate of debate inside Germany became more and more heated. Merely to wish to learn what it was that Lavoisier was saying was seen as an offence against the patriotic spirit of German chemistry.

This process is easy to discern from contemporary German literature. In 1785, Christian Ehrenfried Weigel discontinued his translation of Lavoisier's works. This was the very year in which Lavoisier began his general attack on the phlogiston theory in the Paris Academy of Sciences, attempting to find a hearing amongst the tumultuous uproar in the learned audience. In the same year too, opinion in France, Holland and Great Britain began to swing in favour of the new antiphlogistic chemistry. It was in this very year, 1785, that Professor Christian Ehrenfried Weigel stopped bringing Lavoisier's writings to the attention of the German public, and finished his reporting of the innovations in France. Not until 1792, seven years later, was his work of publishing a German edition of Lavoisier's writings taken up again by a younger colleague, Professor Heinrich Friedrich Link (1767–1851). [...]

Meanwhile, a climate of fear prevailed among more sensitive minds. Johann Andreas Scherer (1755–1844) complained in 1790 that 'no one in Germany had the courage to shatter this old theory'. He noted that 'this would perhaps have been a daring undertaking'. The Swedish scholar Johann Carl Wilcke (1732–1796) wrote to Lorenz Crell from Stockholm: 'I hardly dare touch on the great question of whether phlogiston exists.' Johann Friedrich Gmelin (1748–1804) later wrote with hindsight that 'many an honest but timid researcher ... had been stopped in his tracks' or at least 'deterred from giving his true opinion'.

A glance at the literature is enough to show the delay in acceptance of Lavoisier's theory in Germany. The English edition of the major antiphlogistic work on the new chemical nomenclature, published in 1787, appeared as early as 1788, but not in German until 1793. Lavoisier's major work, the *Traité Elémentaire de Chimie* (Paris 1789) was published in English immediately in 1790, but not until 1792 in German. The English edition was followed by several more editions in 1793, 1799 and 1801, but a second German edition did not appear until 1803. However, the most important defence of phlogiston, Kirwan's *Essay on Phlogiston and the Constitution of Acids* (London 1787), was immediately translated into German by Lorenz Crell. It was published in Berlin and Stettin in 1788. There is a delay of acceptance in comparison to other countries too. In Italy, the first edition of the *Traité* appeared in 1791, with further editions in 1792 and 1796. In Holland, several comprehensive descriptions of Lavoisier's system had appeared since the late eighties, *e.g.* by Martinus van Marum in 1787, Alexander Petrus Nahuys in 1789, Pieter Nieuwland in 1791. In Germany, on the other hand, Göttling reported in 1790 that 'several of my readers have expressed the desire to read about the new chemical nomenclature'. Sigismund Friedrich Hermbstädt (1760–1833) complained in 1789:

> Why is there so much opposition to something which is based purely on fact, and why are there so many violent attempts to keep hidden the great truth which lies covered under these facts? Why, on the other hand, is there so much eagerness to uphold a theory which is so incapable of explaining physical phenomena as that of phlogiston?

It was in the nature of things that the lack of knowledge about the new theory in Germany would give rise to the most absurd misunderstandings. Yet in 1791, Christoph Girtanner (1760–1800) reported:

> The new chemical theory of Monsieur Lavoisier, and the important discoveries it has led to, are still very imperfectly understood in Germany, partly because of widespread prejudice against it among our illustrious German chemists.

Thus, the following year, Heinrich Friedrich Link could state that 'we are plagued with so many refutations of this theory; some are annoying, some ridiculous'.

Nevertheless, 1792 was the year in which the dams broke. In this year, three of the younger generation of chemists, Girtanner, Hermbstädt and Link, ensured that at least Lavoisier's works were widely available to the German public. Hermbstädt in Berlin translated ... the *Traité* ... into German, and published it in 1792. Also in 1792, Girtanner published a comprehensive guide to antiphlogistic ideas in his work *Anfangsgründe der Antiphlogistischen Chemie*, and Heinrich Friedrich Link continued the edition of Lavoisier's work stopped by Weigel in 1785. [...]

15.3 Maurice Crosland, *Styles of science in French and British chemistry**

The more famous of the revolutions referred to in the title of this paper is, of course, the political and social revolution, which is traditionally dated 1789 but which developed in succeeding years, reaching its climax in the Terror of 1793–94. There were also many other major changes, for example, in education and in scientific institutions. But it is the other revolution, the 'chemical revolution' of the 1780s associated with the name of Lavoisier that is the more immediate concern of historians of science. People at the time were ready to draw a parallel between the great changes beginning to take place in society and the fundamental changes in chemistry, even if it was only a hundred years later that it came to be described as '*the* chemical revolution'.

No-one would want to argue a cause and effect connection between the two revolutions but it might be reasonable to argue along the lines of a common context. What I hope to illustrate ... is that the worlds of chemistry and politics were not insulated from each other, and I shall be concerned mainly with two countries, France and Britain, the latter providing a classic example of reaction against the political revolution. When war between France and Britain was declared in 1793 there was an additional reason for British writers to view developments in France with scepticism if not outright hostility. [...]

Some objections from Priestley

Going back to the late 1780s, we may consider the position of someone like Priestley, who provides a great contrast in character and outlook to Lavoisier. An outspoken Christian minister of religion (albeit a heterodox one), a man of modest means and democratic principles, the contrast is striking. Priestley was also modest in his claims and his language. Some of his discoveries he presented as having stumbled on by accident. He was usually content to use the language of ordinary discourse in his science, apologising if he felt it necessary to employ a new term. Yet this was certainly not from lack of learning – he was proficient in Latin, Greek, Hebrew and several modern languages.

Priestley must have seen Lavoisier as very elitist and in more ways than one. First, as an Academician, Lavoisier was one of a very small number of chemists recognised by the French state. His position hardly bears comparison with Priestley, who had merely stepped over the very low fence surrounding the membership of the Royal Society. In the eighteenth century, no great proficiency in science was required to become F. R. S. and there

* Maurice Crosland, 'Lavoisier, the two French revolutions and the imperial despotism of oxygen', *Ambix*, 42 (1995), pp. 101, 109–15.

was no restriction on numbers. The select members of the French Academy, on the other hand, drew a line between themselves and other practitioners of science. (Being a member of the Academy had helped Lavoisier enormously in the early propagation of his theory.) Priestley was anxious to present science as being open to everyone of ordinary intelligence and modest means. These ideas are most clearly expressed in the introduction to his *History of Electricity*. Therefore, quite independently of the oxygen theory, the two chemists belonged to contrasting traditions. They viewed the natural world and society from completely different standpoints.

One feature which unified Priestley's career in religion, politics and science was his hostility to authority. Already in 1790, reacting to the growing influence of the new theory of chemistry, he advocated: 'putting an end to all undue and usurped authority in the business of religion as well as of science'. He expressed a similar idea even more forcibly when writing to the French chemists in 1796 from Pennsylvania. Asserting his customary independence almost for the last time, he said that: 'no man ought to surrender his own judgement to any mere authority, however respectable'.

Persuasion, Priestley said, was possible by two means, by brute force, or by argument. Obviously in science, as in politics, the latter was preferable. Therefore, addressing the French chemists, he said:

> As you would not, I am persuaded, have your reign to resemble that of Robespierre... we hope you had rather gain us by persuasion than silence us by power.

A similar theme was taken up by Elizabeth Fulhame, the wife of a doctor and the author of *An essay on combustion*. She said that science should be open to everyone and argued against any 'dictatorship in science'. After referring to 'M. Lavoisier and other great names' she said that she was:

> persuaded that we are not to be deterred from the investigation of truth by any authority, however great, and that every opinion must stand or fall by its own merits.

Priestley as a democrat could not fail to have been impressed by the *numbers* of people supporting the rival theory. In the early days the numbers game would obviously have favoured theories of phlogiston, but increasingly in the 1790s any count of numbers of supporters would have favoured the oxygen theory. Priestley admitted in 1800 that 'great numbers' in Britain as well as in France supported the new theory. Therefore he could no longer use numerical arguments to support phlogiston. It is ironic that, in his final defence of the phlogiston theory in 1800, he himself fell back on the argument from authority. He cited the names of the German chemists: Crell, Westrumb, Gmelin and Mayer, saying: 'No person needs to be ashamed of avowing an opinion which has the sanction of such names as these.'

Unfortunately, by this time there were very few active chemists outside the German states who could be cited as still supporting a theory of phlogiston. Phlogiston had never been killed outright but it was clearly now on its deathbed. Yet it would probably be a mistake to present Priestley simply as a defender of the phlogiston theory. He was much less concerned with theory in general than Lavoisier. As a plain Englishman he often said that he was concerned only with the 'facts'. Hence a part of his antagonism to Lavoisier was based on the Frenchman's increasing concern with *theory*. It was the resentment of the practical man towards the theoretician, a feeling which is much more general than the particular case we are discussing.

Black and his editor, Robison

A British chemist of greater seniority than Priestley was Joseph Black, and it would be interesting to know what he made not only of Lavoisier's theory but also of his propaganda. On the first point we know that he accepted the oxygen theory by 1790. Lavoisier privately assured him that he would regard his support as decisive and Black may have been taken in by such flattery. As regards Black's feelings about Lavoisier's approach, we have to fall back largely on the work of John Robison, who edited Black's lecture notes for publication. Robison wrote that Black:

> always expressed a high opinion of M. Lavoisier's genius and sound sense but was much displeased by the authoritative manner in which the junto of chemists at Paris announced everything, treating all doubt or hesitation about the justice of their opinions as of the want of common sense.

In this latter comment Robison was saying that Lavoisier and his colleagues felt that their critics were lacking in *logic*, but the main point being made here was once more resentment about the *authority* assumed by the French chemists.

John Robison was professor of natural philosophy from 1774 to 1805 in the University of Edinburgh and, therefore, a respected figure in academic circles and a former colleague of Black. In 1797 he had published an extremist tract, claiming the French Revolution as a universal conspiracy. Although mainly an attack on French Jacobinism, portrayed in caricature, it also attacked those in Britain who supported the revolution, notably Joseph Priestley. In this complex story, however, we are focusing on allegations about French chemistry.

I will not waste time quoting the absurdities of Robison's political pamphlet, although it was surprisingly well received at the time of publication. I will quote only from his correspondence and the notes which Robison added to Black's university lectures, which one might assume to be a sober academic source. Robison claimed that the new French chemistry was not something worked out and published by private individuals. Rather:

> It was propagated as a public concern; and even propagated in the way in which that nation always chose to act – by address and with authority. Everything pertaining to the system was treated in council, and all the leading experiments were documented by committees of the Academy of Sciences.

In defiance of strict chronology Robison elsewhere referred to the new nomenclature of 1787 as the work of a 'Revolutionary committee': 'The new language in chemistry was not so much intended for instructing the world as for securing the sovereignty in science to the French Junto.'

As further evidence of his conspiracy theory Robison mentioned the publication of the *Annales de chimie* 'in concert' and the introduction of a new chemical language – just as the French revolutionaries had introduced a republican calendar with its strange names for the months. Indeed by mentioning certain individuals like Guyton and Hassenfratz he had evidence of an overlap between the chemists and the legislators. Referring to the new nomenclature, Robison claimed:

> A determination to be the founder of a system and a sect of philosophers seems to have seduced M. Lavoisier and made him acquiesce in measures which may be called violent and unbecoming.

In his reference to a 'sect of philosophers' Robison was using the same language as Burke.

This edition of Black's lectures was reviewed by Henry Brougham, who applauded Robison's decision to comment on the behaviour of the French chemists. He wrote:

> We rejoice that this subject is fairly brought before the public and on whichever side the decision may finally be given, the history of science, as well as the political history of our times, is likely to be illustrated by the discussion. That the French chemists formed themselves into a junto for the propagation of their system; that like all juntos, they delivered their doctrines with an authoritative tone, highly indecorous in matters of science; and they even displayed somewhat of a spirit of persecution towards those who, from ancient habits, or from a predilection for their own theories, refused their assent to the antiphlogistic doctrines, are facts which cannot be disputed.

Brougham, however, was happy to accept the oxygen theory and thought that Robison had taken too far his support of Black and his hostility to the French. Part of Robison's attitude was based on his feeling that the French chemists had not attached sufficient importance to the work of his fellow-countryman. He was most anxious 'that Dr Black should not appear like the humble pupil of Lavoisier'. But the concept of illegitimate authority and official backing is a common theme among British critics of Lavoisier. [...]

Davy's reactions

We come finally to Humphry Davy, whose very first published paper was an immature attempt to replace Lavoisier's caloric with light; he even introduced the neologism *phosoxygen*. However, even in his maturity, in his famous paper of 1807 on the decomposition of the fixed alkalis, Davy added a note to the effect that it was still possible to defend a theory of phlogiston. In his Bakerian Lecture for 1808, he proudly referred to this speculation that inflammable substances might contain something like hydrogen or phlogiston. Davy was of course prepared to accept the main thrust of Lavoisier's chemistry but only as a reasonable 'approximation' to the truth. He expected chemical theory to change in the future and I believe it was always part of his ambition to challenge the new French chemistry. He probably came nearest to this when he showed that the fixed alkalis are the oxides (or hydroxides) of previously unknown metals. This was a further blow to that part of Lavoisier's theory which associated oxygen with acidity.

For the young Davy, however, Lavoisier was a figure of the past. The Frenchman with whom he consciously competed in the most active years of chemical research was Gay-Lussac, an exact contemporary, since both had been born in 1778. The sharpness of the competition was increased by the fact that France and Britain were at war. When Davy announced his spectacular isolation of sodium and potassium using the electric pile of the Royal Institution, French government funds were used to construct a bigger and better pile at the Ecole Polytechnique. Since Davy was once quoted as saying that, if the governments of France and Britain were at war, the men of science were not, it may be worth quoting from a draft lecture he prepared in 1810 but never actually delivered, which tells a different story. He wrote:

> The scientific glory of a country may be considered in some measure as an indication of its innate strength. The exaltation of Reason must necessarily be connected with the exaltation of the other faculties of the mind and there is one spirit of enterprise, vigour and conquest in science, arts and arms.

This passage suggests that Davy sometimes saw science in nationalistic terms and this would have been all the more understandable around 1810 when Napoleon had conquered half of Europe. There were many people in

England who had a great admiration for France's contribution to western civilisation, who nevertheless saw Napoleon as a monster whose ambition seemed to amount to little less than world domination.

It is in this context that I would like to introduce a part of a letter written by Davy in the spring of 1814 to a Swiss correspondent. In the previous winter in Paris Davy had carried out some studies of iodine on a visit to Paris almost under the nose of his rival Gay-Lussac. Both have claims to have discovered the elementary nature of iodine. It was only a few years earlier that Davy had unequivocally proclaimed the elementary nature of the gas known as 'oxymuriatic acid' and which he now proposed to call 'chlorine'. The irony here is that Gay-Lussac had reached such a conclusion earlier but had been prevailed upon to claim this as no more than a possibility.

We are now in a position to understand the letter by Davy, who referred to iodine as: 'a very useful ally in my endeavour to establish the independence of chlorine and to do away [with] the imperial despotism of oxygen'. I think that the really interesting phrase here is 'the imperial despotism of oxygen'. Everyone in Britain would have been familiar with the concept of 'imperial despotism' as applied to Napoleon. But it took a chemist to use this phrase as a metaphor and apply it to the role of oxygen in his science. Oxygen was certainly at the centre of the new chemistry and Davy was here expressing his resentment. Twenty years after Lavoisier's death, one could hardly still name him as the leading chemist, and the other French chemists were a rather heterogeneous group covering at least two generations. Better to focus on the element that Davy had wrestled with throughout his whole chemical career, whether in its elementary state or as a compound with alkalis or as a supposed constituent of chlorine. The chemistry of the time was dominated by oxygen and British chemists inevitably saw this as a French achievement, even a French imposition. When countries are at war it is only too easy to characterise any situation in nationalistic terms.

Conclusion

What does all this amount to? There may be some who will want to say that I have been discussing little more than random metaphors used by British chemists in the period 1789–1815. Although some of the quotations I have used are obvious metaphors, they are anything but random. They illustrate a common theme, a common resentment against the style as much as the content of the new French chemistry. Priestley made the accusation that Lavoisier and his colleagues were arguing from authority rather than seeking a consensus. They were using force rather than persuasion to convert people to the new chemistry. It was probably understandable that someone on the losing side in an argument should claim that he was suffering from oppression, just as it was easy for Lavoisier on the winning side to claim objectivity. The claims of the oxygen theory may not have been viewed by the opposition mainly in philosophical terms. They did not stand back as might the modern student and ask, for example, which theory provided a simpler explanation of the observed phenomena. Some, like Priestley, could not help thinking in more political terms – what *right* had Lavoisier and his colleagues to change the basis of chemistry? What *right* did Lavoisier have to ride roughshod over the traditional theory of chemistry and substitute his own ideas? Lavoisier had started with a certain authority as a member of the Royal Academy of Sciences but, by the late 1780s, he had assumed a much greater authority which drew some resentment from those who did not agree with his quantitative takeover bid for chemistry.

. . . Conversion in science, as in religion or politics, can take a number of different forms. We are all familiar with the newly converted who show a greater enthusiasm than the people whom they join; there are always some who are 'more royalist than the king'. But, if the conversion is brought about at the expense of a deep resentment, one is likely to obtain recruits who are *less* royalist than the king. The importance of this in our story is that supposed 'converts' such as Davy may continue to try to wriggle out of the new system. Davy taught the oxygen theory but nothing would have pleased him better in his research than to force a major reappraisal of the theory.

. . . There has been much discussion recently about the importance of *persuasion* in science. I agree that this is an important issue to examine but I would not want to confine the agenda to rhetoric. This is only a part of the story of the rise of the oxygen theory. There is the question of a new language which might be interpreted as thought control. More fundamental than the style of Lavoisier's *Traité* was his decision to write a textbook at all and, arguably more important than the *Traité*, were the *Annales*, not a single volume but an on-going publication soon to become monthly, which hammered home the new chemistry.

I would also want to argue that there is further evidence in the story I have told of national styles in science. Several of the British critics saw an association, however indirect, between the French government and science, which they considered morally indefensible. Living under a constitutional monarchy acting in conjunction with a parliament, British writers made much of the rhetoric of liberty, and would criticise the French people living in the 1780s under an absolute monarchy. When this political system was replaced by the Jacobins and later by the rise of Napoleon, British critics had further examples of extreme authoritarianism. It was easy for critics to claim that similar authoritarianism existed in science. Indeed the *dirigisme* of the French state clearly influenced the presentation of French science and may be contrasted with a British *laissez-faire* philosophy. Through official agencies such as the National Institute, there emerged an *official* science in France, and the French were trying to spread their ideas in other countries. There only needs to be a grain of truth in an idea for it to be believed by critics – especially in time of war. The same comment may be made about the accusation that the French chemists constituted a 'junto', an accusation that may seem very strange to us today. It informs us first about a certain image, but this image was not totally unwarranted by the facts. In the *Méthode de nomenclature chimique* of 1787 and in the *Annales de chimie* from 1789 onwards the French chemists had presented themselves as a group, united in advocating the new chemistry and the new nomenclature. We may contrast this association with the position of the British chemists who worked and published very much as individuals....

It is possible to think of science as a game, played according to generally agreed rules. Unfortunately, in the case of the oxygen theory, the French and the British played the game by different rules. The British had a keen sense of fair play and it was not long before they were crying 'foul'. It is not our business today to award points to each side but this analogy may help us to understand the difference in attitude of the French and the British chemists.

15.4 Michael Conlin, *Priestley's American period**

As one of the last prominent and practicing phlogistians, Joseph Priestley precipitated a chemical controversy in America. Though historians have generally found Priestley's defense of phlogiston theory not only to be rational but also to be meritorious in many ways, no one has carefully examined his American defense. Consequently, Priestley's greatest success in America – the conversion of antiphlogistian James Woodhouse to phlogiston theory – remains unappreciated and Priestley denied his due credit....

...I argue that isolation, or more precisely the lack of social contact, played a significant role in Priestley's resistance to antiphlogistic theory.

Of course, social contact alone did not decide the Chemical Revolution. In part, Priestley's rejection of antiphlogistic theory was based on his resistance to the different epistemological assumptions and experimental methods of the antiphlogistians. Whereas Priestley was an inductive empiricist, who believed in the gradual accretion of 'facts' from numerous experiments conducted by individuals, the antiphlogistians were theoretical rationalists, who negotiated the results of selected experiments, guided by reason and theory. Whereas Priestley emphasized the importance of an individual's passively recording the results of relatively simple experiments, the antiphlogistians divided the theoretical and empirical labor amongst themselves, employing complex and expensive apparatus to manipulate the phenomena in accordance to theory.

Social contact combined with Priestley's empiricism played an important role in his American defense of phlogiston. Priestley's geographical isolation

---

* Michael Conlin, 'Joseph Priestley's American defence of phlogiston reconsidered', *Ambix*, 43 (1996), pp. 129–33, 142.

in Northumberland, Pennsylvania and theoretical isolation as a phlogistian contributed to his 'die-hard' advocacy of phlogiston theory, by insulating him from personal contact with antiphlogistians. Priestley never met his staunchest critics, Pierre-Auguste Adet and John Maclean, excepting a political discussion with the former before the controversy. The only relationship they maintained with Priestley was of a literary nature. They did not replicate Priestley's experiments. Indeed, they did not perform any new experiments in their refutation of Priestley. Subscribing to the antiphlogistic view of science as a 'collectivist enterprise', they cited appropriate passages from antiphlogistic textbooks and journals, and laboratory work they had done before the controversy.

Furthermore, personal contact and a shared empiricist orientation were crucial for the successful replication and similar interpretation of Priestley's experiments. Priestley converted Woodhouse, the only antiphlogistian to personally witness the experiments offered against the new chemistry. An empiricist himself, Woodhouse successfully replicated some of Priestley's experiments and accepted part of Priestley's phlogiston theory. William Cruickshank was the only other antiphlogistian to repeat these experiments. Priestley's literary skill notwithstanding, articles alone were not sufficient to convert Cruickshank. Lacking personal contact with Priestley and his empiricism, Cruickshank replicated some of Priestley's results, but remained an antiphlogistian. In this case, it seems that social contact and shared methodology were necessary and that literary accounts and experimental techniques alone were not sufficient for the successful replication and similar interpretation of experiments.

Even before Priestley emigrated to the United States, he was isolated. The Birmingham riots of 1791 left him a 'broken philosopher', with few friends and no apparatus. Afterwards, Priestley moved to London where he missed the community of the Lunar Society of Birmingham and found his philosophical friends in the Royal Society 'cold and distant'. Aggravating this isolation with humiliation, he was reduced to begging his few remaining philosophical friends for scraps of apparatus to cobble together a laboratory.

In his American '*exile*', Priestley's theoretical isolation was greater than it was in England and it was exacerbated by geographical isolation. Priestley hardly knew of any remaining phlogistians in the world. Even in the United States, Priestley heard of 'nothing else' except the antiphlogistic system. Moreover, geographical isolation was a persistent theme in his writings. Priestley knew that a great distance separated him from the center of philosophical pursuit in Europe, which was widened by the 'slow and uncertain' correspondence with Britain and the lack of any correspondence with France because of the Quasi-War, an undeclared Franco-American naval war. Just as he had visited the Royal Society regularly when living in Birmingham, Priestley hoped to visit Philadelphia several times a year to exchange philosophical news. Though not London, Edinburgh, or Paris,

Philadelphia was the center of chemistry in the United States and the location of one of the first chemical societies in the world. Priestley, however, lived the 'considerable distance' of 130 miles from Philadelphia, a trip of four or five days. The 'difficulty and irksomeness of a journey to Philadelphia' in winter and the yellow-fever epidemics in the summer, combined with his declining health, limited Priestley to four visits, one each in 1796, 1797, 1801, and 1803.

Priestley remained in Northumberland, making his isolation profound. Unable to visit other chemists, Priestley invited Woodhouse and others to enjoy the 'conveniences my *shed* affords for making experiments'. He realized, however, that 'it would be unreasonable to expect many visitors in these *back woods*'. Northumberland was, Priestley admitted, 'inconveniently situated for carrying on ... experiments' because of the 'want of a ready communication with Philadelphia'. This difficulty, Priestley noted, was exacerbated by the fact that America 'furnishes but few fellow laborers, and these are so scattered that we can have but little communication with each other; and they are equally in want of communication with myself'.

Unable to perform many experiments during the time of his travels from Birmingham to Northumberland, Priestley reiterated previously-published work in *Considerations on the Doctrine of Phlogiston* (1796). Admitting that he had 'nothing materially new to advance', Priestley penned the pamphlet because he felt several of his earlier publications had 'not been duly attended to, or well understood'. Priestley addressed *Considerations* to the 'surviving Answerers of Mr. [Richard] Kirwan', among whom were Claude-Louis Berthollet, Antoine François de Fourcroy, and Louis-Bernard Guyton de Morveau. Kirwan, a noted phlogistian and a friend of Priestley, converted to the new chemistry in 1792 – a triumph for the antiphlogistians. With the execution of Antoine-Laurent Lavoisier two years earlier, Berthollet, Fourcroy, and Guyton de Morveau were the most eminent antiphlogistians remaining. Priestley responded to the antiphlogistians, but too late. In 1796, the antiphlogistians were not interested in fighting battles they believed to have been won several years earlier. In their minds, the Chemical Revolution was complete.

Priestley always maintained his readiness to adopt those parts of the new chemistry which were vindicated in his laboratory. He accepted oxygen as the imponderable principle of acidity. Moreover, Priestley conceded that oxygen gas – what he called dephlogisticated air – was usually absorbed by a calx when roasted, though he believed that phlogiston was released at the same time. In *Considerations*, Priestley limited his attack on the antiphlogistic theory to a calx of mercury and iron, the composition and the decomposition of water, and the new nomenclature. Priestley emphasized qualitative indicators and volumetric measurements over gravimetric results because his interest was demonstrating that metals were compounds, and that hydrogen and oxygen were not the constituent parts of water. Priestley objected to the new nomenclature because it presumed the antiphlogistic explanation of the composition of metals and water – treating as matters of fact the very issues which he contested. [...]

Priestley attacked the famous experiments of Fourcroy, Armand Séguin, and Nicholas Louis Vauquelin, which produced water by exploding hydrogen with oxygen. As an ardent empiricist, Priestley did not accept the results of experiments he was unable to replicate. Repeating these experiments with reactants purer than those employed by the French, Priestley obtained 'highly phlogisticated nitrous acid' and a little water. Priestley explained the water as proof of his gas theory, 'that the greatest part of the *weight* of all kinds of air is water'. Moreover, Priestley attacked the result of these experiments, dismissing it as an experimental anomaly resulting from unusual methods and complex apparatus. The antiphlogistians conceded that the production of water was successful only when the 'inflammable air was burned *in the slowest manner*'. If they conducted the experiment 'in any other manner', Priestley contended, nitrous acid would be produced. 'After making all the allowance they could', Priestley noted, the French chemists still produced a considerable quantity of phlogisticated air which they could not 'well account for'. Furthermore, Priestley complained that the antiphlogistic decomposition of water 'requires so difficult and expensive an apparatus, and so many precautions in the use of it, that the frequent repetition of the experiment cannot be expected'.

Lastly, Priestley complained that the new nomenclature was premature, for facts should be known and principles determined before names were given to things. In particular, Priestley believed hydrogen ('water-begetter') was too hastily named. If inflammable air was nothing more than a component of water, as its antiphlogistic name implies, Priestley wondered why it, along with fixed air, was produced by heating finery cinder over coal – neither reagent being a source of water in the antiphlogistic view. Anticipating antiphlogistic objections, Priestley reported that the reactants were heated sufficiently to drive any water away. 'Whether we approve of the new language or not,' Priestley lamented, 'it is now so generally adopted, that we are under the necessity of learning, though not of using it.' [...]

Being ill, Priestley made no more contributions to the controversy. He died in February, 1804 confident that his lonely defense of phlogiston had been vindicated in the laboratories of his critics. Though Cruickshank never convinced Priestley, he won Woodhouse back to the new chemistry. How long Woodhouse subscribed to Priestley's interpretation of the experiments is unclear. Woodhouse reconsidered sometime between 1805 and 1807. In his 1807 edition of Jean Antoine Chaptal's *Elements of Chemistry*, Woodhouse made a 'public acknowledgement' that he had taught 'a different doctrine' and returned to the new chemistry after reading an account of Cruickshank's experiments of April 1805. [...]

Chapter Sixteen
Conclusions

16.1 Roy Porter, *Historiography and the Scientific Revolution**

Historians write about scientific revolutions as automatically as of political, economic or social revolutions: the 'French revolution' in chemistry led by Lavoisier is almost as familiar as the political revolution which cut off his head. Indeed, the idea that science advances by revolutionary leaps has long been with us, ever since the eighteenth century in fact. For, as Bernard Cohen has shown, it was Enlightenment propagandists for science from Fontenelle and the *Encyclopédistes* to Condorcet who first began to depict the transformations in astronomy and physics wrought by Copernicus, Newton and others as revolutionary breaks with the past, creating new eras in thought.

And significantly it was through being applied in this way to epochs in *science* that the term 'revolution' itself took on its present meaning. Traditionally, when used to describe political fortunes, 'revolution' had, of course, denoted change (the fall of one prince, the rise of a rival); but it was change within an essentially cyclical system in which all dynasties and empires had their rise and fall, their waxings, wanings and eclipses, for human affairs were governed by the endless 'revolutions' of Fortune's Wheel. In the traditional metaphor, in other words, it was the orbits of the planets, so gravid with astrological influence, which had defined and governed revolutions in sublunary affairs. But, from the early eighteenth century, the old equation of revolution with cycles began to yield to a secular, directional myth of human destiny. 'Revolution' kept its massiveness – a mundane event as portentous as one in the heavens, far grander than a mere revolt; but it came to signal not endless repetition but a break with the past, a fresh start, or what the Enlightenment called an 'epoch'. Hence from the *philosophes*

* Roy Porter, *The Scientific Revolution: A Spoke in the Wheel?*, in R. Porter and M. Teich (eds), *Revolution in History* (Cambridge: Cambridge University Press, 1986), pp. 290–1, 294–303.

onwards, scientific breakthroughs actually became normative for general usage, underpinning modern views of revolution as constructive and progressive, rather than as tainted by fatalism and hubris. Condorcet's faith in human perfectibility was buoyed up on hopes of scientific revolutions yet to come, programmed into the march of mind. [...]

...Though specific episodes in science have been called revolutionary for over two centuries, the concept of The Scientific Revolution is less than two generations old, the phrase, it seems, having been minted by Koyré in 1939, and first stamped on a book title in Rupert Hall's *The Scientific Revolution* (1954). But the idea probably passed into the Anglo-American mind chiefly through Butterfield's *The Origins of Modern Science 1300–1800* (1949), which contains the most celebrated passage ever penned about the Revolution – or rather about what Butterfield often styled the 'so-called' Scientific Revolution, as if, through the hesitation, to own the term's novelty. Here is his classic statement both of the uniqueness of The Scientific Revolution and its unparalleled contribution to Western history:

> Since that revolution overturned the authority in science not only of the Middle Ages but of the ancient world – since it ended not only in the eclipse of scholastic philosophy but in the destruction of Aristotelian physics – it outshines everything since the rise of Christianity and reduces the Renaissance and Reformation to the rank of mere episodes, mere internal displacements within the system of medieval Christendom.

[...] The idea of The Scientific Revolution is not, then, part of the intellectual commons which historians have grazed time out of mind. Rather it was initially the brain-child and shibboleth of a specific cluster of scholars emerging during the 1940s, including the Russian émigré Alexandre Koyré, Butterfield, whose outline history popularized Koyré's work, Rupert Hall, who was Butterfield's pupil, and, a little later, Marie Boas [Hall]. Their scrupulous scholarship and prolific works of synthesis animated an emergent discipline, and laid down a coherent framework for future research. ... They also made no secret of their 'Whiggish' view that The Scientific Revolution was a good thing. For it marked a triumph of mind, free and fearless, underscoring the essential link between liberty of thought and intellectual advance, a lesson not to be lost on Western democracies just freeing themselves from Hitler's and Stalin's brain-washings and from the utopian Marxism that had been the opium of the thirties intelligentsia.

...This 'classical interpretation' focussed as never before on explicating the stunning conceptual displacements involved in moving from the Ptolemaic geocentric universe to Copernicus's heliocentrism, or in Kepler's courageous imaginative leap in abandoning circular planetary orbits – *de rigueur* for centuries – and embracing elliptical ones. How the bounds of the thinkable were changed by Galileo's New Science of motion, Descartes's mechanics of inertia and momentum, Boyle's corpuscular chemistry, and finally by Newton's synthesis of the laws of motion and the laws of gravity! Such triumphs – the mechanization of the world-picture, the shift from the 'closed world' to the 'infinite universe' – needed more than tireless empirical brick-laying. Copernicus's achievement did not consist in adding to Ptolemy, but in viewing old data through new spectacles. Science was revolutionized by new ways of seeing.

...For these historians, science was essentially *thought* – profound, bold, logical, abstract – and thought was ultimately philosophy. As Rupert Hall put it, science is 'above all a deep intellectual enterprise whose object it is to gain some comprehension of the cosmos in terms which are in the last resort philosophical'. This identification of science with philosophy has had deep consequences, not least in the academic world in the institutional marriage – not blissfully happy – between the history and the philosophy of science. For if science is ultimately philosophical, it is no surprise that philosophers should claim the right to analyse past science – a tendency which, given the predominant ahistorical bent of Anglo-American philosophy, has often proved disastrous for proper historical interpretation. In particular the philosophers' itch to reconstruct the rationality of great texts has characteristically played fast-and-loose with historical contexts and meanings, and has encouraged anachronistic evaluations of rationality and irrationality in the history of science. [...]

Overall, this idealist reading of The Scientific Revolution as disembodied thought sustained a heroic, even romantic, image of the scientist, typified by Newton, 'with silent face, Voyaging through strange seas of thought alone'....

Such romantic views of the agony and the ecstacy of truth-seeking were not of course new, and a noble tradition had long seen scientists as heretics, forming a dissenting academy, on the social margins. ... In the Cold War, Western scholars were glad to show how the history of science 'falsified' historical materialism's reductionist way with men and ideas.... This reading of The Scientific Revolution is thus not value-free, and should not be taken at face value. Its central assumptions – that science proceeds by heroes making 'discoveries' through 'Eureka' moments, that the great scientist himself is an autonomous agent, and that science is value-free – are historically question-begging, and play a polemical part within today's politics of knowledge.

Thus the 'classical interpretation' mystified the dynamics of theory change; but a further aspect also needs scrutiny. This is the question of just what The Scientific Revolution actually revolutionized. Clearly, it is claimed – and with good reason – individual scientific disciplines were transformed: for instance, a physics of natural and unnatural movements gave way to one of forces and inertia. Equally, scientific epistemology and methodology were radically changed. Bacon, Galileo and many others made sport with time-honoured but barren preoccupations with final causes; by contrast the accent was now on material and efficient causes, and on quantities rather than qualities. ... For this was, according to the 'classical interpretation', Europe's intellectual and spiritual coming of age when Western civilization grew out of traditional infantilizing mythologies and faced up (like a man) to the stark realities of Nature. The Scientific Revolution thus formed the great divide between the traditional or primitive *mentalité* of the 'Ancients' and the mature rationality of the 'Moderns'.... But even a superficial glance at the fabric of the new science in the seventeenth century and beyond shows that it remained permeated by precisely the kinds of human values and rhetoric – the Baconian idols – it claimed to have expunged. Indeed, much of the most stimulating scholarship from the 1970s onwards has been uncovering just how important was the continuing input into science from metaphysics, theology and human interests, long after the New Science had proclaimed its independence from these influences. And it is worth remembering that the New Science's self-image of its own 'new dispensation' was clearly taken over wholesale from Biblical eschatology. Ironically the 'classical interpretation', which casts the Scientific Revolution as a watershed in the transition from primitive thought to rationality, reproduces such myths, generates new ones about progress and modernity, and runs closely parallel to the pre-history/history teleology of scientistic Marxism.

Just because its historiography embodies myths, the notion of scientific revolution need not however be summarily dismissed. The question remains: is it helpful to picture the course of the history of science as revolutionary? Or might it not make better sense to stress its 'evolutionary' aspects, its continuities and accommodation to the wider socio-intellectual environment? These large questions matter, not least because, with the irresistible rise of specialization, scholarship becomes myopic and fragmented, and, though philosophers continue to dogmatize, historians may be in danger of defaulting on the task of assessing the overall patterns of science.

To judge whether science has had its revolutions, we need a working idea of revolution, one compatible with common historical usage in other contexts. For my purposes, I propose that a revolution in science requires the overthrow of an entrenched orthodoxy; challenge, resistance, struggle and conquest are essentials. The mere formulation of new theories doesn't constitute a revolution; neither is it a revolution if the scientific community leaps to applaud an innovation, saluting its superiority. Moreover, revolution requires not just the battering of old theories but the triumph of the new. A new order must be established, a break visible. Furthermore, revolutions presuppose both grandeur of scale and urgency of tempo. Mini-revolutions, partial revolutions, and long revolutions are terminological abuses; why dilute the word when 'change' will do stout service? Lastly, I suggest that, though it may not be indispensable that the protagonists should intend from the outset to make a revolution (revolutionaries often begin as reformers), it is vital that, at some stage, consciousness should dawn of revolution afoot. The notion of silent or unconscious revolution is next door to nonsense.

Following these guide-lines, it does seem helpful to characterize the transformations in science occurring in the seventeenth century – though not the sixteenth – as revolutionary. There is no room here, and perhaps little need, to argue this contention chapter and verse. But certain key elements are worth stressing. First, many of the protagonists clearly cast themselves as crusaders for a radically New Science, engaged in life-and-death struggles against the hidebound dogma of the schools: the very titles of Bacon's *New Atlantis*, Kepler's *New Astronomy*, and Galileo's *Two New Sciences* catch this tone of embattled innovation. Bacon and Galileo amongst others were witheringly dismissive of the dead hand and dead mind of orthodoxy to a degree that finds no parallel, for instance, in Copernicus or Vesalius. Doubtless, much of this was rhetoric; doubtless, it was largely straw schoolmen who were being slain and slain again; doubtless, seventeenth-century natural philosophers tapped the scholastic legacy more than they admitted. Yet the seventeenth century really saw intense struggle between rival natural philosophies, and the call for liberation from die-hard orthodoxy runs right through the century, culminating in the Ancients versus Moderns debate and the Battle of the Books, won in science by the Moderns (according to the Enlightenment, by a knockout).

For the standard-bearers of the New Science indeed had a struggle on their hands. Traditional doctrines had been deeply entrenched in seminaries and universities, in textbooks, curricula and in the educated mind. Not least, they were protected by those watchdogs of intellectual orthodoxy, the Christian churches, notably the papacy in such episodes as the burning of Bruno and the trial of Galileo, but also by other confessions too, as witness the conservative role of Laudian Anglicanism in early Stuart England. The Battle of the Books meant the burning of the books. In a radical gesture Paracelsus symbolically burnt the texts of Avicenna; and, as late as the 1680s, Oxford University was still making bonfires of Hobbes's writings. It would be foolish caricature to depict these as struggles between the forces of darkness and the children of light; yet the seventeenth century remains a cockpit of violent conflicts between rival natural philosophies, which often resolved themselves into struggles between Old and New, and which resulted – something much less true of the sixteenth century – in victory for the new.

Moreover, many sciences did undergo fundamental reorientations both in their conceptual foundations and their fine texture. A few examples will surely suffice. In astronomy, geostatic and geocentric systems still predominated in 1600; but by 1700 all members of the scientific elite espoused heliocentricity. In 1600, versions of the Aristotelian physics of finitude, local motion and the four elements still held the floor, in many cases, in newly refined and reinvigorated forms; by 1700, one mode or other of the mechanical philosophy had swept them away amongst leading scientists. Matter theory had come to hinge not on the traditional four elements and on qualities but on particles and short-range forces incorporating new laws of motion and principles of dynamics. The traditional divide between science celestial and terrestrial was challenged by Galileo, and bridged by Newton's universal gravitation. Methodologically, observation was set at a premium and so stimulated, and was in turn reinforced by, the development of new scientific instruments such as the telescope and microscope. This opened up both the macro- and micro-worlds, both visible and conceptual, and contributed to the general development of instrumentation which is so important a factor in modern science. Going hand in hand with this, experimentation led to a new way of doing science and a new way of promoting science's claims to 'objective truth'. Moreover, mathematical advances – pre-eminently Descartes's coordinate geometry and Newton's and Leibniz's infinitesimals – empowered science to calculate and control areas which had been impressionistic before. Such a list could be greatly extended.

These changes, it must be stressed, were not just pious hopes for a great instauration; they were substantial and permanent achievements which were built upon. Taken singly, it is true, the work of Kepler or Descartes, Galileo or Boyle, created as much chaos as it resolved. But collectively, their investigations amounted to a progression of fruitful reformulations of fundamentals until, with Newton above all, a synthesis was reached widely saluted as coherent, dazzling in scope and potential, ripe both for solving workaday problems (Kuhn's 'normal science') and for generating future investigations. Newton set the seal.

Thus, the concepts and practice of many individual sciences – kinetics, hydraulics, optics – were transformed, and new philosophies of Nature established. Confidence in science led to the extension of mechanical models to new fields, as for example in Borelli's physiology, and boosted the prestige of natural philosophy so that it could become definitive of intellectual authority – witness the enthusiasm throughout the eighteenth century for applying Newtonianism to aesthetics, social and moral philosophy, politics and psychology. For the radical intellectuals of the Enlightenment, science's successes cast metaphysics and theology in the shade. For Locke, philosophy's job should be to serve merely as science's 'under-Labourer', sweeping aside the rubbish for science's 'masterbuilders'. For Diderot and D'Alembert, Priestley and Erasmus Darwin, science was the engine of progress.

In other words, the transformations in science were revolutionary not just in techniques and concepts, but in forging an unparalleled place for science in European culture and consciousness. Above all, new conceptions of Nature and man's relation to it became dominant. The 'classical interpretation' of course acknowledged this in its claim that The Scientific Revolution rent the old veils of myth, and 'discovered' Nature as she really was: rational, regular, law-governed, mechanical. But a truer way of putting it would be to say that seventeenth-century science created and imposed its own model of Nature as a regular, mechanical order, which legitimated scientific man's intellectual and practical control of Nature. Traditional beliefs about Nature handed down via the intertwining systems of Christian theology, Humanistic philosophy and occult wisdom now seemed all too confused, difficult, anarchic, dangerous. The proliferation and ubiquity of spiritual forces, magical resonances, and providential infiltrations had confused God, man and Nature in ways that increasingly seemed scandalous and enervating to powerful intellectual currents, engaged (as Rabb has put it) in a search for stability.

Seventeenth-century scientific ideologues embarked upon their task of conceptual clarification with a will. Not without fierce controversy, new formulations of the fundamentals were hammered out. The true divide between God and Nature had to be insisted upon; typically, in the mechanical philosophy, all activity was attributed to God, but a God who was increasingly distant, and Nature was reduced to a machine, inert and passive. Similarly, man and Nature were also demarcated, most extremely in Cartesian dualism in which Nature became merely extension (matter in motion) and man alone possessed consciousness. Such programmatic segregation of the divine, the natural and the human had gigantic consequences in terms of franchising man's right, through science and technological intervention, to act upon Nature. For, once Nature was thus 'disenchanted', the New Scientists increasingly claimed man's right, as Bacon put it, to 'conquer and subdue her'. If Nature were not after all alive but just an object, it could be taken to pieces, anatomized, resolved into atoms. Passive and uniform, Nature was open to experiment, or, in Bacon's grim metaphor, to be 'tortured' into revealing its truth. The dictum that science should (in Bacon's phrase) 'penetrate from Nature's antechambers to her inner closet' became axiomatic.

Within these seventeenth-century reconceptualizations, God became more remote and Nature less sacrosanct. Man's right to progress (even to redeem himself) through the pursuit of knowledge of, and power over, Nature became central to influential visions of human destiny; and the conquest of Nature became a practical, noble and even godly goal. The material transformation of the West over the last three centuries would have been impossible without the technical capacity generated by seventeenth-century science; but it would also have been unthinkable without the sanction and encouragement given by the new visions of science and Nature formulated in Baconianism, Cartesianism and other parallel seventeenth-century philosophies. [...]

16.2 Steven Shapin, *Telling stories: the Scientific Revolution**

As our understanding of science in the seventeenth century has changed in recent years, so historians have become increasingly uneasy with the very

---

* Steven Shapin, *The Scientific Revolution* (Chicago, IL: University of Chicago Press, 1996), pp. 3–10.

idea of 'the Scientific Revolution'. Even the legitimacy of each word making up that phrase has been individually contested. Many historians are now no longer satisfied that there was any singular and discrete event, localized in time and space, that can be pointed to as 'the' Scientific Revolution. Such historians now reject even the notion that there was any single coherent cultural entity called 'science' in the seventeenth century to undergo revolutionary change. There was, rather, a diverse array of cultural practices aimed at understanding, explaining, and controlling the natural world, each with different characteristics and each experiencing different modes of change. We are now much more dubious of claims that there is anything like 'a scientific method' – a coherent, universal, and efficacious set of procedures for making scientific knowledge – and still more skeptical of stories that locate its origin in the seventeenth century, from which time it has been unproblematically passed on to us. And many historians do not now accept that the changes wrought on scientific beliefs and practices during the seventeenth century were as 'revolutionary' as has been widely portrayed. The continuity of seventeenth-century natural philosophy with its medieval past is now routinely asserted, while talk of 'delayed' eighteenth- and nineteenth-century revolutions in chemistry and biology followed hard upon historians' identification of 'the' original Scientific Revolution.

There are still other reasons for historians' present uneasiness with the category of the Scientific Revolution as it has been customarily construed. First, historians have in recent years become dissatisfied with the traditional manner of treating ideas as if they floated freely in conceptual space. Although previous accounts framed the Scientific Revolution in terms of autonomous ideas or disembodied mentalities, more recent versions have insisted on the importance of situating ideas in their wider cultural and social context. We now hear more than we used to about the relations between the scientific changes of the seventeenth century and changes in religious, political, and economic patterns. More fundamentally, some historians now wish to understand the concrete human *practices* by which ideas or concepts are made. What did people *do* when they made or confirmed an observation, proved a theorem, performed an experiment? An account of the Scientific Revolution as a history of free-floating concepts is a very different animal from a history of concept-making practices. Finally, historians have become much more interested in the 'who' of the Scientific Revolution. What kinds of people wrought such changes? Did everyone believe as they did, or only a very few? And if only a very few took part in these changes, in what sense, if at all, can we speak of the Scientific Revolution as effecting massive changes in how 'we' view the world, as the moment when modernity was made, for 'us'? The cogency of such questions makes for problems in writing as unreflectively as we used to about the Scientific Revolution. Responding to them means that we need an account of changes in early modern science appropriate for our less confident, but perhaps more intellectually curious, times.

Yet despite these legitimate doubts and uncertainties there remains a sense in which it is possible to write about the Scientific Revolution unapologetically and in good faith. There are two major considerations to bear in mind here. The first is that many key figures in the late sixteenth and seventeenth centuries vigorously expressed *their* view that they were proposing some very new and very important changes in knowledge of natural reality and in the practices by which legitimate knowledge was to be secured, assessed, and communicated. They identified *themselves* as 'moderns' set against 'ancient' modes of thought and practice. Our sense of radical change afoot comes substantially from them (and those who were the object of their attacks), and is not simply the creation of mid-twentieth-century historians. So we can say that the seventeenth century witnessed some self-conscious and large-scale attempts to change belief, and ways of securing belief, about the natural world. And a book about the Scientific Revolution can legitimately tell a story about those attempts, whether or not they succeeded, whether or not they were contested in the local culture, whether or not they were wholly coherent.

But why do we tell *these* stories instead of others? If different sorts of seventeenth-century people believed different things about the world, how do we assemble our cast of characters and associated beliefs? Some 'natural philosophers', for example, advocated rational theorizing, while others pushed a program of relatively atheoretical fact collecting and experimentation. Mathematical physics was, for example, a very different sort of practice from botany. There were importantly different versions of what it was to do astronomy and believe as an astronomer believed; the relations between the 'proper sciences' of astronomy and chemistry and the 'pseudo-sciences' of astrology and alchemy were intensely problematic; and even the category of 'nature' as the object of inquiry was understood in radically different ways by different sorts of practitioners. This point cannot be stressed too strongly. The cultural practices subsumed in the category of the Scientific Revolution – however it has been construed – are not coextensive with early modern, or seventeenth-century, science. Historians differ about which practices were 'central' to the Scientific Revolution, and participants themselves argued about which practices produced genuine knowledge and which had been fundamentally reformed.

More fundamentally for criteria of selection, it ought to be understood that 'most people' – even most educated people – in the seventeenth century did not believe what expert scientific practitioners believed, and the sense in which 'people's' thought about the world was revolutionized at that time is very limited. There should be no doubt whatever that one could write a convincing history of seventeenth-century thought about nature without even *mentioning* the Scientific Revolution as traditionally construed.

The very idea of the Scientific Revolution, therefore, is at least partly an expression of 'our' interest in our ancestors, where 'we' are late twentieth-century scientists and those for whom what they believe counts as truth about the natural world. And this interest provides the second legitimate justification for writing about the Scientific Revolution. Historians of science have now grown used to condemning 'present-oriented' history, rightly saying that it often distorts our understanding of what the past was like in its own terms. Yet there is absolutely no reason we should not want to know how we got from there to here, who the ancestors were, and what the lineage is that connects us to the past. In this sense a story about the seventeenth-century Scientific Revolution can be an account of those changes that we think led on – never directly or simply, to be sure – to certain features of the present in which, for certain purposes, we happen to be interested. To do this would be an expression of just the same sort of legitimate historical interest displayed by Darwinian evolutionists telling stories about those branches of the tree of life that led to human beings – without assuming in any way that such stories are adequate accounts of what life was like hundreds of thousands of years ago. There is nothing at all wrong about telling such stories, though one must always be careful not to claim too much scope for them. Stories about the ancestors as ancestors are not likely to be sensitive accounts of how it was in the past: the lives and thoughts of Galileo, Descartes, or Boyle were hardly typical of seventeenth-century Italians, Frenchmen, or Englishmen, and telling stories about them geared solely to their ancestral role in formulating the currently accepted law of free fall, the optics of the rainbow, or the ideal gas law is not likely to capture very much about the meaning and significance of their own careers and projects in the seventeenth century.

The past is not transformed into the 'modern world' at any single moment: we should never be surprised to find that seventeenth-century scientific practitioners often had about them as much of the ancient as the modern; their notions had to be successively transformed and redefined by generations of thinkers to become 'ours'. And finally, the people, the thoughts, and the practices we tell stories about as 'ancestors', or as the beginnings of our lineage, always reflect some present-day interest. That we tell stories about Galileo, Boyle, Descartes, and Newton reflects something about our late twentieth-century scientific beliefs and what we value about those beliefs. For different purposes we could trace aspects of the modern world back to philosophers 'vanquished' by Galileo, Boyle, Descartes, and Newton, and to views of nature and knowledge very different from those elaborated by our officially sanctioned scientific ancestors. For still other purposes we could make much of the fact that most seventeenth-century people had never heard of our scientific ancestors and probably entertained beliefs about the natural world very different from those of our chosen forebears. Indeed, the overwhelming majority of seventeenth-century people did not live in Europe, did not know that they lived in 'the seventeenth century', and were not aware that a Scientific Revolution was happening. The half of the European population that was female was in a position to participate in scientific culture scarcely at all, as was that overwhelming majority – of men and women – who were illiterate or otherwise disqualified from entering the venues of formal learning.

[...] 1. I *take for granted* that science is a historically situated and social activity and that it is to be understood in relation to the *contexts* in which it occurs. Historians have long argued whether science relates to its historical and social contexts or whether it should be treated in isolation. I shall simply write about seventeenth-century science as if it were a collectively practiced, historically embedded phenomenon, inviting readers to see whether the account is plausible, coherent, and interesting.

2. For a long time, historians' debates over the propriety of a sociological and a historically 'contextual' approach to science seemed to divide practitioners between those who drew attention to what were called 'intellectual factors' – ideas, concepts, methods, evidence – and those who stressed 'social factors' – forms of organization, political and economic influences on science, and social uses or consequences of science. That now seems to many historians, as it does to me, a rather silly demarcation, and I shall not waste readers' time here in reviewing why those disputes figured so largely in past approaches to the history of early modern science. If science is to be understood as historically situated and in its collective aspect (i.e., sociologically), then that understanding should encompass all aspects of science, its ideas and practices no less than its institutional forms and social uses. Anyone who wants to represent science sociologically cannot simply set aside the body of what the relevant practitioners *knew* and how they went about obtaining that knowledge. Rather, the task for the sociologically minded historian is to display the structure of knowledge making and knowledge holding *as social processes*.

3. A traditional construal of 'social factors' (or what is sociological about science) has focused on considerations taken to be 'external' to science proper – for example, the use of metaphors from the economy in the development of scientific knowledge or the ideological uses of science in justifying certain sorts of political arrangements. Much fine historical work has been done based on such a construal. However, the identification of what is sociological about science with what is external to science appears to me a curious and a limited way of going on. There is as much 'society' inside the scientist's laboratory, and internal to the development of scientific knowledge, as there is 'outside'. And in fact the very distinction between the social and the political, on the one hand, and 'scientific truth', on the other, is partly a cultural product of the period this book discusses. What is commonsensically thought of as science in the late twentieth century is in some measure a product of the historical episodes we want to understand here. Far from matter-of-factly treating the distinction between the social

and the scientific as a resource in telling a historical story, I mean to make it into a topic of inquiry. How and why did we come to think that such a distinction is a matter *of course*?

4. I do not consider that there is anything like an 'essence' of seventeenth-century science or indeed of seventeenth-century reforms in science. Consequently there is no single coherent story that could possibly capture all the aspects of science or its changes in which we late twentieth-century moderns might happen to be interested. I can think of no feature of early modern science that has been traditionally identified as its revolutionary essence that did not have significantly variant contemporary forms or that was not subjected to contemporary criticism by practitioners who have also been accounted revolutionary 'moderns'. Since in my view there is no essence of the Scientific Revolution, a multiplicity of stories can legitimately be told, each aiming to draw attention to some real feature of that past culture. This means that selection is a necessary feature of *any* historical story, and there can be no such thing as definitive or exhaustive history, however much space the historian takes to write about any passage of the past. What we select inevitably represents our interests, even if we aim all the while to 'tell it like it really was'. That is to say, there is inevitably something of 'us' in the stories we tell about the past. This is the historian's predicament, and it is foolish to think there is some method, however well intentioned, that can extricate us from this predicament. [...]

16.3 Margaret Jacob, *The Scientific Revolution and conceptions of identity**

Lecturing on Newtonian mechanics and dynamics around 1800, the natural philosopher John Dalton employed all the standard demonstrations in what had become by then a well-established genre of scientific education. On his tabletop he used oscillating devices, pendulums, balls made of various substances, levers, pulleys, inclined planes, cylinders of wood, lead in water, and pieces of iron on mercury to illustrate phenomena as diverse as gravitation, the '3 laws of motion of Newton', impulse or the 'great law of percussion', force and inertia, specific gravity, attraction and magnetism. There was nothing extraordinary in what Dalton was doing, first in his Quaker school then at New College in Manchester. The genre of British lecturing focused on Newtonian mechanics had begun in the second decade of the eighteenth century with the travels and publications of Francis Hauksbee, Jean Desaguliers, and Willem s'Gravesande who lectured in the Dutch Republic. Dalton was deeply indebted to their legacy. [...]

---

* Margaret Jacob, 'The truth of Newton's science and the truth of science's history: heroic science at its eighteenth-century formulation', in Margaret Osler (ed.), *Rethinking the Scientific Revolution* (Cambridge: Cambridge University Press, 2000), pp. 315, 319, 321–2, 328–32.

Historical judgments about canonicity and stature depend upon a lengthy process of shifting and weighing, of accumulation, of the new being assessed and assimilated, of values as well as creativity being defined, recognized, and then enshrined. The would-be king of Denmark got to be immortalized as *Hamlet* and the lad from Lincolnshire got to be one of the heroes of modern science as a result of a historical process. Heroes are not born, they are made. However brilliant Kepler or Newton had been, their innovative contributions would have remained relatively unknown, or esoteric, possibly even banished, without a larger transformation in the way literate Westerners conceived both nature and history. ... If we make the move to diminish, or dismiss, the depth and breadth of the intellectual transformation in the Western understanding of nature (bracketed between roughly 1540 and 1750), we want to be sure about what it is that we are giving up.

As we broaden the canon and explore the context, we need to understand that although Newton and Boyle – and who knows how many others among their contemporaries – were, in their different ways, practicing alchemists, this fact does not alter the profound character of the intellectual transformation described by the somewhat misleading, shorthand phrase, the Scientific Revolution. That intellectual transformation can be dismissed only if we believe that two elements alone define it: the heroes had to be pure, simply great 'scientists', and they alone made it happen. In effect, if the older historiography erred on the side of simplicity, then the history it sought to convey can be dismissed. The logic of the dismissal is flawed. Finally, and not least, if the history of science as a discipline abandons a central problematic, that of explaining how and why Westerners moved to mechanize the world picture, then it would lose, not enhance, one of its most important raisons d'être. In the West, at least, revolutions tend to keep their readerships. [...]

Dalton and his Newtonian predecessors added to the true science of Newton what could be presumed by his audience to be an equally true history of science. All the history one needed to know was there: Newton had been right about the vacuum and Descartes had been wrong about the vortices. The development of science was a progressive story of great discoveries punctuated by the occasional misdirected theory; such had been the swirling vortices. In the eighteenth century the history of the heroes of true science framed the presentation of natural knowledge. Their history got told with the science, almost as the filler, the enticement to spark interest among a restless, not scientifically literate, audience. Within this history lay the key to human enlightenment, to a new intellectual freedom. ...

Little wonder that the heroic narrative of socially isolated geniuses, first articulated in the eighteenth century, survived well into the twentieth. Decoupling it from the truthfulness of scientific laws has taken a generation or more of historical scholarship. It has been difficult and controversial to create a textured, nuanced, and historically informed understanding of modern science at its origins. The rise of postwar and now contemporary scholarship about science has required the dismantling of assumptions about isolation, about disinterest, about purity as the key to brilliance – in short, about the way history and human beings work – which were once taken to be as certain as the very science learned in lecture hall, pulpit, and school. The search for a richer, more textured history should not, however, lead us to turn away from some of the earliest historical associations of the new science. We cannot imagine that the revolutionary legacy of science has nothing to do with us, that in effect we have never been modern. As both modern historians with our methods, and as citizens with our expectations, we are in part science's beneficiaries. [. . .]

Nowhere in Europe did the revolution in science find greater admirers than in France during the 1790s. The task was to translate the central premises of the French Enlightenment into the life of the entire nation. . . . Just as had befitted free-born Englishmen of the seventeenth century, science now adorned the inheritance of late eighteenth-century French democrats and republicans. Not surprisingly, the French Revolution, coupled with its seventeenth-century antecedent in England, held the key to why heroic science – and the closely related concept of revolutions in science, politics, and industry – became so dominant and pervasive, particularly in Anglo-American historiography.

We all know that in the first instance Copernican science also provided one source for the notion of there being 'revolutions' in matters of state. Indeed, the language of astronomy, partly indebted to Copernicus's 'revolutions of the heavenly orbs', provided vocabulary for the profound changes in England during the 1640s and 1650s. By 1660 the terms 'revolutions and commotions' in the state had become commonplace. Under the impact of those midcentury events, both political and intellectual, revolutions in the state and in thought began to take on the modern meaning of progressive, irrevocable change, not simply a returning to a previous place or a revolt that occurs periodically with little to show for the trouble. In 1651 Robert Boyle himself used the term 'revolution' to describe the progress he expected in philosophy and divinity as a result of the civil wars: 'I do with some confidence expect a Revolution, whereby Divinity will be much a looser, & Real Philosophy flourish, perhaps beyond men's Hopes.' For Boyle and his friends the expected revolution had millenarian associations; by the 1690s a more secular understanding of time had become commonplace in England. So too had the use of the term 'revolution' to describe the events of 1688–9.

Largely because of seventeenth-century events, in the eighteenth century in English, French, and Dutch the term 'revolution' became working linguistic capital. It could function as both political and cultural currency. In 1766 Josiah Wedgwood, speaking about what would come to be known as the Industrial Revolution, wrote to a friend: 'Many of my experiments turn out to my wishes, and convince me more and more, of the extensive capability of our Manufacture for further improvement. It is at present (comparatively) in a rude, uncultivated state, and may easily be polished, and brought to much greater perfection. *Such a revolution, I believe, is at hand*, and you must assist in, [and] profit by it.' It is a myth of recent origin, perpetuated by proponents of the so-called new economic history, that early industrialists experienced the changes they themselves wrought in manufacturing as something gradual, almost imperceptible. Having available the notion of revolutions born out of profound political transformations and out of the history of science, early mechanists and industrialists like Wedgwood possessed the vocabulary to describe unprecedented transformation effected not by the sword or the air pump but by the machine.

As in most things modern, the French Revolution gave extraordinary circulation both to the word and, more important, to the concept of revolution as a sharp and irrevocable break with the past. Almost predictably by the 1790s the term 'revolution' began to be applied by the French to economic phenomena, particularly to what was happening in British industry....

...In the mind of the revolutionaries, industry, invention, genius, and taste were of a piece, and science and technology were exemplars of all that genius and invention could achieve....

...More than any other body of culture, science released the revolutionary imagination, helped to develop its fantasies, to eliminate doubt about what human beings could accomplish. After 1700 radicals and moderates on both sides of the Channel, and across the Atlantic – even conservatives interested in piety and politeness – could embrace the imaginary of revolutions whether in cotton or in regimes. The histories of science that they told were filled with exaggeration and hyperbole, with assumptions that we would now characterize as naive. But the first believers in science had also allowed for, indeed celebrated, the possibility of rapid, irreversible transformations. In societies encumbered by hierarchy, blood, and birth, they had imagined, and experienced, significant intellectual changes.

We may find it fashionable now late in the twentieth century to cast doubt on the very notion of there having once been a 'scientific revolution'. But by 1720, with the exception of Fatio, not one of Newton's close followers, not Desaguliers, or Pemberton, or Clarke, or even the millenarian Whiston, or the student of the Temple of Solomon, William Stukeley, could have understood the master's alchemy. They hung on his every word, but I think it to be the case that not one of them could have explicated his alchemical texts. Such a rapid shift may justly be imagined as revolutionary. It took historians like Dobbs, Westfall, and Karin Figala years of hard labor to penetrate a mind-set that had disappeared within one generation. By 1750 few among the literate in northern and western Europe and the American colonies would have found it remarkable that alchemy had become obscure, esoteric, and, the enlightened said, ignorant. The historiographical notion of there having been a revolution in Western thinking about nature makes for an inheritance that cannot be erased so easily. Bringing it down will entail dismantling a set of interrelated mental structures that support beliefs as basic to Western thought as the value of technological development, industry, human freedom, the rule of law, and the possibility of progress. Saying that it never happened cannot alter the gulf between Newton and his first generation of followers. We may be better served both as historians and people by a finer honing of our historiographical legacy, not by attempting its wholesale deconstruction.

Index

Printed in the United States
By Bookmasters

Give the books away?

Yes. These books were meant to given away as gifts that will instantly bond you with your prospect.

And the price? A little more than $1 each in quantities. About the cost of an audiocassette tape, but so much more impressive.

The proof is in the results. First, you'll personally love the book as it will quickly direct you to the most direct road to wealth. Second, you'll love the instant relationships this book creates with your prospects. Now you have something really important to talk about. And third, the book pre-sells network marketing so that your prospect is ready to take advantage of your business opportunity.

ISBN 1-892366-00-2